Progress in Metal Additive Manufacturing and Metallurgy

Progress in Metal Additive Manufacturing and Metallurgy

Editor

Robert Pederson

MDPI • Basel • Beijing • Wuhan • Barcelona • Belgrade • Manchester • Tokyo • Cluj • Tianjin

Editor
Robert Pederson
University West
Sweden

Editorial Office
MDPI
St. Alban-Anlage 66
4052 Basel, Switzerland

This is a reprint of articles from the Special Issue published online in the open access journal *Materials* (ISSN 1996-1944) (available at: https://www.mdpi.com/journal/materials/special_issues/met_addit_manuf_metall).

For citation purposes, cite each article independently as indicated on the article page online and as indicated below:

LastName, A.A.; LastName, B.B.; LastName, C.C. Article Title. *Journal Name* **Year**, *Article Number*, Page Range.

ISBN 978-3-03943-663-7 (Hbk)
ISBN 978-3-03943-664-4 (PDF)

Contents

About the Editor

Robert Pederson (Professor) is Professor of Engineering Materials and Head of Division of Subtractive and Additive Manufacturing at University West, Sweden. Robert has 20 years of experience within the research and development of advanced metallic materials for aerospace and space applications. He received his PhD in 2004 working on a project on titanium alloys in close collaboration with Volvo Aero Corporation (which became GKN Aerospace Engines Systems Sweden in 2012). During 2004–2016, Robert worked at Volvo/GKN as a research cluster leader, company specialist in materials technology, project leader, project manager, and responsible materials application engineer in a number of projects related to civilian aero engines as well as space rocket applications. In 2013, he became Associate Professor at Chalmers University of Technology; in 2015, he was appointed Adjunct Professor at Luleå University of Technology; and he has served as Full Professor at University West since 2016. He has supervised 5 PhD students in the completion of their doctoral degree and is currently supervising 3 PhD students. The research work conducted in his laboratory is focused on exploring and improving understanding of the thermomechanical processing–microstructure–mechanical properties relationship for titanium alloys and other metallic alloys, and the key technologies of welding and additive manufacturing processes used by the aerospace and space industry.

Preface to "Progress in Metal Additive Manufacturing and Metallurgy"

Research in the additive manufacturing (AM) of metals has witnessed a dramatic rise in global attention during the past decade. Some AM processes have evolved from conventional welding processes, while others, such as powder bed fusion processes, have been developed with the specific intent of enabling the manufacture of complex 3D geometrical objects. One key feature of all AM processes is that material is only added where it is really needed, thereby permitting near net shape manufacture utilizing starting feedstock in powder or wire form, with virtually no residual material waste if all the unmelted material can be fully recycled.

The distinct heating–cooling cycles associated with various AM processes result in different as-built microstructures and varying types of defects which are additionally governed not only by the used process parameters but also by the geometry of object(s) being built as well as by the local environmental conditions prevailing during processing. Consequently, in-process monitoring of different parameters is important to understand the process parameter–microstructure relationships during layer-on-layer manufacturing. For low-stressed, statically loaded components, the microstructure determines the average mechanical properties. However, for cyclically loaded critical parts, like aeroengine or turbine components, defects limit the lower bound of the mechanical properties and are, therefore, a major concern as they restrict the loading conditions during operation. In view of the above, post-build treatments, like hot isostatic pressing (HIP) that can minimize certain types of defects like porosity, can become relevant, depending on the material and AM process in question. Other in situ/post-build treatment/hybrid solutions have also been considered, such as inducing residual compressive stresses in built material to reduce the influence of surface topography/defects/residual stresses on properties. From an implementation standpoint, the final quality of finished parts also needs to be ascertained using appropriate non-destructive evaluation (NDE) methods, which represent an area under active development.

It is thus apparent that AM involves a complex manufacturing chain spanning a number of different expert competences that have to be coordinated appropriately to enable successful economical serial production of AM components. Since such a complex knowledge chain is challenging for any single organization to internally complete, establishing collaborative networks that stitch together complementary competences is often a key enabler.

This Special Issue intends to address the latest progress in various facets of metal AM that constitutes the entire value chain. I therefore hope the AM-related focus of this Special Issue will be of interest for a broad audience—researchers as well as companies—engaged in AM.

Robert Pederson
Editor

Review

Current Status and Perspectives on Wire and Arc Additive Manufacturing (WAAM)

Tiago A. Rodrigues *, V. Duarte, R. M. Miranda, Telmo G. Santos and J. P. Oliveira *

UNIDEMI, Departamento de Engenharia Mecânica e Industrial, Faculdade de Ciências e Tecnologia, Universidade NOVA de Lisboa, 2829-516 Caparica, Portugal; v.duarte@campus.fct.unl.pt (V.D.); rmmdm@fct.unl.pt (R.M.M.); telmo.santos@fct.unl.pt (T.G.S.)

* Correspondence: tma.rodrigues@campus.fct.unl.pt (T.A.R.); jp.oliveira@fct.unl.pt (J.P.O.)

Received: 8 March 2019; Accepted: 1 April 2019; Published: 4 April 2019

Abstract: Additive manufacturing has revolutionized the manufacturing paradigm in recent years due to the possibility of creating complex shaped three-dimensional parts which can be difficult or impossible to obtain by conventional manufacturing processes. Among the different additive manufacturing techniques, wire and arc additive manufacturing (WAAM) is suitable to produce large metallic parts owing to the high deposition rates achieved, which are significantly larger than powder-bed techniques, for example. The interest in WAAM is steadily increasing, and consequently, significant research efforts are underway. This review paper aims to provide an overview of the most significant achievements in WAAM, highlighting process developments and variants to control the microstructure, mechanical properties, and defect generation in the as-built parts; the most relevant engineering materials used; the main deposition strategies adopted to minimize residual stresses and the effect of post-processing heat treatments to improve the mechanical properties of the parts. An important aspect that still hinders this technology is certification and nondestructive testing of the parts, and this is discussed. Finally, a general perspective of future advancements is presented.

Keywords: wire and arc additive manufacturing; additive manufacturing; microstructure; mechanical properties; applications

---

1. Introduction

Additive manufacturing is nowadays one of the hot topics in the manufacturing and engineering worlds. The ability to create three-dimensional, complex, and near-net shape parts in a layer by layer deposition process is currently a major driving force for major breakthroughs. These breakthroughs are observed either in the process itself, by developing process variants with dedicated purposes to increase capabilities, but also on the materials used, since the non-equilibrium solidification conditions which occur during fusion-based additive manufacturing can lead to microstructural features not often found in conventional materials manufacturing processes.

Currently, additive manufacturing processes based on fusion are mostly focused on powder-bed systems using laser and/or electron beams as heat sources. Despite very high precision dimensional tolerances achieved with these techniques, the deposition rate is low, increasing lead times. Additionally, using powder as the feedstock materials makes the process more prone to the formation of defects such as pores, which can hamper the structural integrity of the parts, especially during dynamic solicitation conditions. Wire and arc additive manufacturing (WAAM) uses an electric arc as the heat source and a solid wire as the feedstock material. Though the precision of the as-built parts may be lower than those obtained using powder-bed systems, the depositions rates are significantly higher, allowing to manufacture large metallic structural parts in short times. WAAM is currently being embraced by both academia and industry owing to the advantages of the technique. Therefore, several research papers have been dealing with fundamental aspects associated with the process and its effects on the material's microstructure and mechanical properties. Additionally, some applications are already making use of parts built by WAAM, showing the viability of this process. This review paper intends to provide an overview of the major developments in WAAM, covering important topics, such as process variants, materials for WAAM, development of residual stresses, and post-processing heat treatment, as well as non-destructive testing. This review finalizes with current applications based on WAAM parts and a summary of areas where additional research efforts need to focus.

2. Wire and Arc Additive Manufacturing (WAAM) Process and Variants

WAAM is classified in the category of direct energy deposition according to ASTM F2792-12a [1], and is defined as the combination of an electric arc used as a heat source, and a wire employed as a feedstock material. The process is schematically represented in Figure 1. WAAM relies on the fundamental concepts of automatized welding processes, such as: gas metal arc welding (GMAW) [2], plasma arc welding (PAW) [3], and gas tungsten arc welding (GTAW) [4].

In the past years, WAAM has had different designations, such as rapid prototyping (RP), shape welding (SW), shape melting (SM), solid freeform fabrication (SFF), shape metal deposition (SMD), and even 3D welding [5].

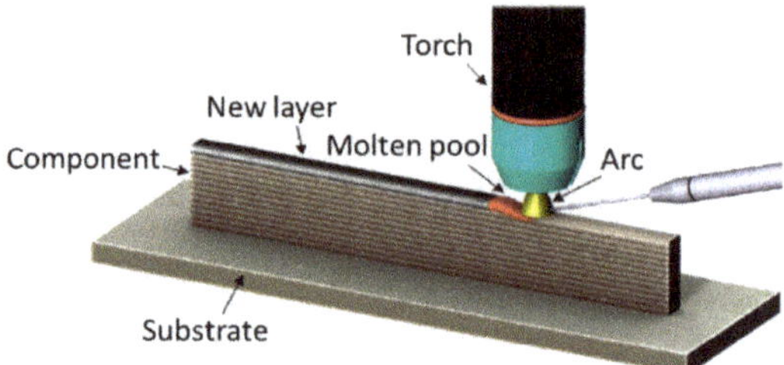

Figure 1. Schematic representation of the wire and arc additive manufacturing (WAAM) process (adapted from [6]).

From the existing arc welding processes, GMAW, also known as metal inert gas (MIG)/metal active gas (MAG), is the most used process in WAAM. GMAW is a fusion-based arc welding process where the arc is established between the tip of a consumable wire and the workpiece under the protection of an inert or active shielding gas that also protects the weld pool and adjacent material. Deposition rates range from 15 to 160 g/min using GMAW in additive manufacturing, depending on the deposited material and process parameters, making it ideal for the production of large-scale parts in short time spans [7–9].

The other two arc welding processes are GTAW and PAW. These have some similarities, since they both use a non-consumable tungsten electrode to establish an electric arc with the workpiece under an inert shielding gas without filler material. GTAW was one of the first arc welding processes that became widespread due to its high precision with almost no defects, since the electric arc is very stable [10]. However, for WAAM it needs external filler material. PAW is a high energy-density process, where the arc is forced to pass through an orifice placed between the cathode and anode that constrains the arc, resulting in an increased arc stability. Through the use of a mostly inert, plasmogenic arc, a highly localized ionized plasma forms with very high temperatures and energy. Thus, it is considered a high-density energy process with energy densities lower than those obtained by high power beams as a laser [11], but higher than other electric arc processes. The heat affected zone (HAZ) of PAW beads is narrow and thin, allowing for better control of weld bead geometry due to the increased flexibility for independently controlling the most important process parameters: current and wire feed speed [12]. By varying the plasma gas flow rate, torch orifice diameter, and current intensity, it is possible to achieve different operating modes in PAW namely microplasma, medium current, and keyhole plasma. Microplasma is characterized by requiring low welding currents, between 0.1 and 15 A. This welding technique, when applied to additive manufacturing, presents major advantages regarding the total wall width (TWW) by allowing the production of thin parts with TWW values as low as 2 mm. Though deposition rates are low, about 1.0 g/min [13], when compared to other arc based processes, the low heat input allows for good surface finish. The most commonly used operation mode in PAW is medium current, with the current typically ranging from 15 to 200 A. This operation mode has very similar characteristics to GTAW, but the plasma confinement makes the arc stiffer and less sensitive to torch stand-off variations. Wall width varies from 4 to 15 mm and the deposition rate can be as high as 30 g/min [3,7,14]. Keyhole mode is characterized by high penetration, making it unsuitable for additive manufacturing, since it fully melts the previously deposited layers compromising the wall stability and geometric accuracy.

Electric arc-based welding, in general, can be complex since several process parameters must be controlled to have a good quality of the final parts. Process parameters include: current intensity, voltage, shielding gas type and flow rate, contact-tip-to-work distance, wire feed speed, travel speed, and torch angle. Thus, for each equipment and materials involved, these have to be optimized. The right selection of parameters affects the transfer mode, which is very important to determine bead width, penetration and size, deposition rate, and surface roughness.

In the pursuit for a better and more stable process to control molten metal deposition with reduced heat input, a variant of GMAW, known as cold metal transfer (CMT), has been adapted to additive manufacturing. It is an advanced material transfer process in which an incorporated control system detects when the electrode wire tip contacts with the molten pool, and by activation of a servomotor, retracts the wire in a push and pull electromechanical process, to control droplet transfer. Variants of cold metal transfer include CMT pulse (CMT-P), CMT advanced (CMT-ADV), and CMT pulse advanced (CMT-PADV), which have been developed by Fronius [15]. When optimized, cold metal transfer is suitable for application in Ti-based alloys [16].

Another variant is tandem GMAW, in which two wires are fed into the melt pool in order to achieve high deposition rates (160 g/min) [9,17]. Nevertheless, this method requires a high amount of energy to maintain the arc, so some improvements that enhance heat dissipation are required to control the molten pool shape.

Arc welding-based technologies have been successfully used for additive manufacturing applications, especially because there is considerable knowledge accumulated on process, welding metallurgy, and mechanical performance of welded parts. However, in WAAM there are issues which still need research, such as determination of optimum torch path planning to obtain fully dense parts with minimized residual stresses, control of microstructural evolution during multiple layer deposition, and effect of temperature between layers, among others. Some research groups have developed new process variants to mitigate some of the above-mentioned issues, namely by applying mechanical deformation between layers or active heating and cooling, for example.

2.1. WAAM Process Variants

2.1.1. Cold-Work Based Techniques

High pressure inter-layer rolling is a cold-work process variant developed at Cranfield University [18]. It consists of imposing a load of up to 100 kN onto a roller traveling over the already existent deposited layers, to promote plastic deformation of the surface and thus recrystallize the grain in the following deposition. The process is schematically depicted in Figure 2.

Martina et al. [18] showed that inter-layer rolling in WAAM of Ti-6Al-4V induced prior β grain refinement, thickness reduction of the α-phase lamellae, and an overall modification of the microstructure from columnar to equiaxed. After these promising results, several studies have been conducted to extract other benefits from the high pressure inter-layer rolling, particularly to control residual stress and distortion in aluminum [19,20] and titanium [21,22] alloys. In order to strengthen the as-deposited WAAM material, Gu et al. [14,23] combined inter-layer rolling with a post-WAAM heat treatment, achieving higher mechanical properties (ultimate tensile strength and elongation) with the increase of the applied load.

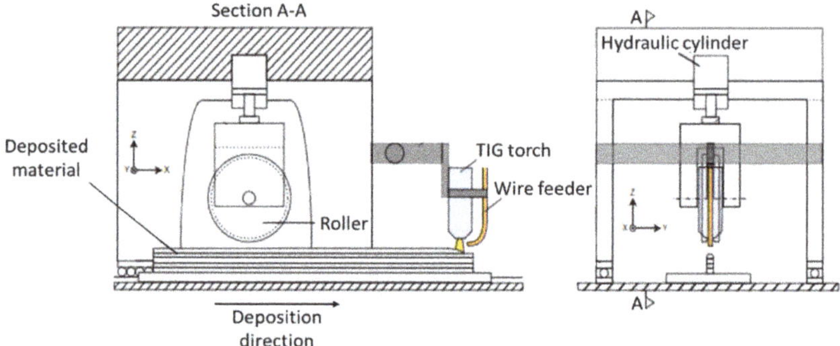

Figure 2. Schematic diagram of cold rolling WAAM process (adapted from [18]).

Porosity is a recurrent problem in WAAM, particularly in aluminum alloys, due to its low hydrogen solubility in the solid-sate having a preferential location at the layer boundaries, which causes a decrease of mechanical properties when stressed in the perpendicular direction. Inter-layer rolling decreases porosity size and quantity with increasing rolling load, which has been claimed as the reason for the ductility increase of WAAM-rolled aluminum alloys [24]. Roller design plays an important role in this methodology and its geometry must be adapted according to the produced part features (i.e., thickness) in order to achieve homogeneous grain refinement [6]. In addition, cold rolling has also been used to control the parts width, and consequently, improve the surface finish of the final part geometry [25]. An example of a cold rolling variant, side rolling, is depicted in Figure 3, in which the material is strained in both longitudinal and normal directions.

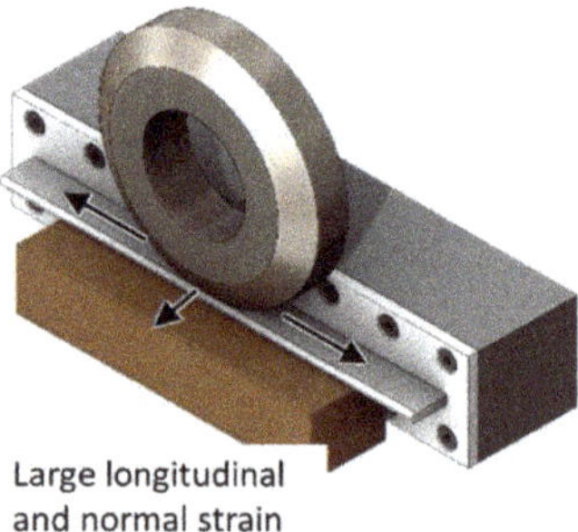

Figure 3. Schematic representation of side rolling of WAAM part to enhance surface finish (adapted from [19]).

An equally significant improvement of mechanical properties has also been made through the use of machine hammer peening [26,27] and laser shock peening [28]. Laser shock peening was performed laterally on 2319 post-machined aluminum parts and resulted in a decrease of average grain size by 22%, and in an increase of hardness [28]. This method, however, only had noticeable effects within the first millimeter of the wall and the induced plastic deformation was not enough to refine the complete wall thickness.

In summary, cold-work based techniques can significantly reduce residual stresses and distortion, improve microstructural homogeneity and mechanical properties, reduce porosity, minimize waviness, and increase final part geometry accuracy. However, most studies have been carried out on simple geometry parts and the use of this methodology may be therefore limited to specific designs. Heavy equipment is required when cold-working is to be applied, and its usage can cause an increase in lead times devaluing the main characteristic of WAAM: its high deposition rate.

2.1.2. Active Inter-Layer Cooling and/or Heating

The high heat inputs and consequent low cooling rates typically result in coarse columnar grains and anisotropy. Moreover, in WAAM, by definition, the inter-layer temperature is the temperature of the previous deposited layer just before a new one is deposited, and is of major importance on the part's final properties [29]. It determines the conditions of heat dissipation by conduction through the part, directly affecting the cooling rate and consequently, the microstructure and mechanical properties. Due to the difficulty in reducing heat accumulation, it becomes difficult to keep a low inter-layer temperature. This type of control is not suitable since a high inter-layer temperature may only be achieved with the use of double torch system. On the other hand, the time needed to cool down the part to a suitable temperature may lead to a total production time impracticable with extended idle times. Even though a high inter-layer temperature improves the wettability of the molten metal [30], when working with high temperatures the deposition becomes unstable and may even lead to the collapse of the wall. In a first approach, the inter-layer temperature can be controlled by imposing an interlayer idle time which can be optimized with the simulation of thermal behavior during part production using finite element models [29].

The utilization of process add-ons to control thermal cycles has already been tested, using compressed CO_2 gas to impose a forced cooling. This promising development in WAAM of Ti-6Al-4V presents benefits including: better surface finishing with less oxidation, refined microstructure, improved mechanical properties, and enhanced manufacturing efficiency [31].

Another alternative to control the thermal cycles in WAAM, is using thermoelectric cooling. Figure 4 depicts this approach. Heat sinks by conduction to the side walls, enabling similar heat dissipation conditions throughout the full deposition. Additionally, this technique allows the control of the bead geometry, decreasing the surface waviness by about 60% and the total fabrication time can decrease by nearly 60% as well [32], since a continuous heat dissipation condition is achieved without adjusting process parameters.

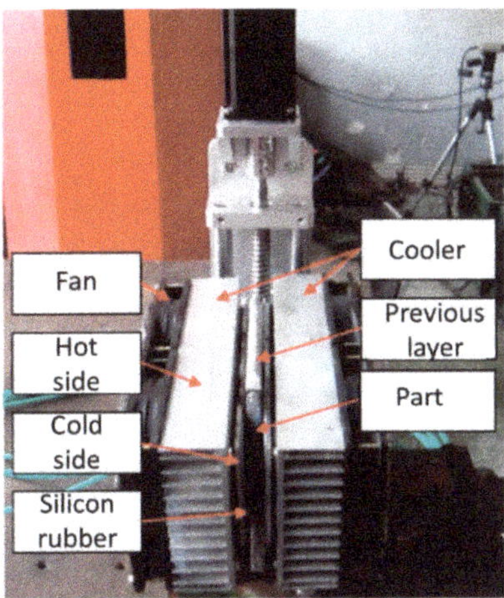

Figure 4. Representation of thermoelectric cooling setup (adapted from [32]).

While the previous technique envisaged to cool down the temperature interlayers, for some materials and applications it may be of interest to have a quasi-isotropic material with minimum residual stresses without post processing heat treatment. In this case, an innovative method was developed to mitigate residual stresses consisting of an inductor with two symmetrical coils mounted on both sides of the as-built part, as schematically depicted in Figure 5 [33]. The inductor can perform both pre-heating and post-heating, depending on its positioning relative to the arc. In general, this technique reduces the residual stresses and distortion of as-built parts. Moreover, this system has the potential to overcome the inter-layer temperature issue, opening the possibility of maintaining a constant inter-layer temperature throughout the full deposition.

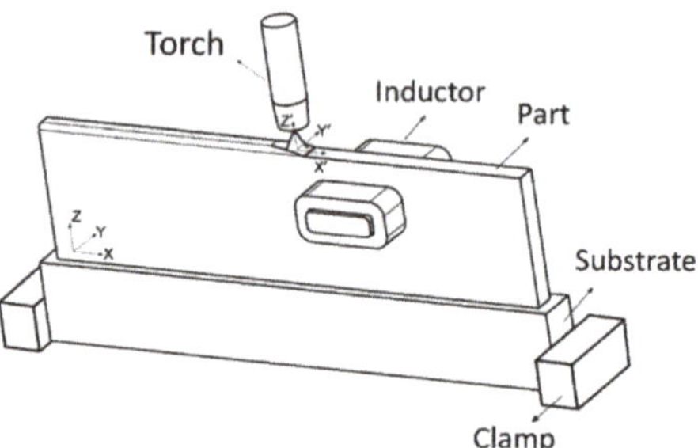

Figure 5. Schematic representation of the inductor heating system (adapted from [33]).

Another process variant named hot-wire arc additive manufacturing (HWAAM) was developed and constitutes an efficient alternative to refine the typical columnar grains in Ti-based alloys processed by WAAM [34]. This variant consists on the use of another power source, which assists in melting the filler material, and reducing the amount of arc heat input in a GTAW-like application. The secondary power source has the positive pole connected to the filler wire while the negative one is connected to the substrate. Figure 6a,b depicts a comparison between samples built with and without this process variant. Besides a visible change in the wall geometry, due to different heat inputs, the size of columnar β-grains decreased in the sample built by HWAAM, resulting in a mixture of short columnar grains and equiaxed ones. This mixture promoted elongations of 12.6% and 12.8% in the longitudinal and transversal directions, respectively, in opposition to the 23% and 9.17% obtained with conventional WAAM, thus, confirming the appropriateness of this method to produce isotropic parts.

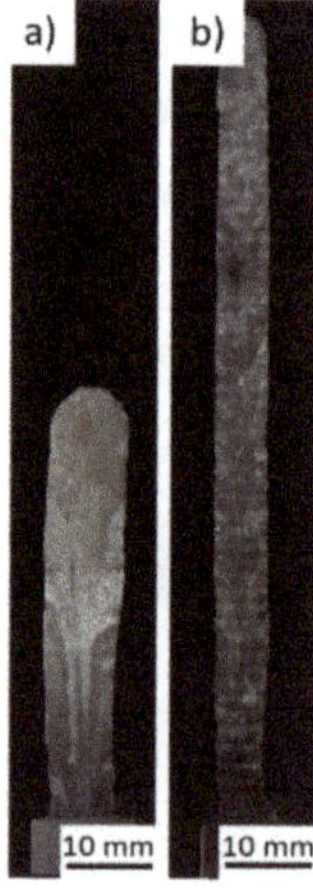

Figure 6. Macrostructure of samples produced: (**a**) with secondary heat source and (**b**) without (adapted from [34]).

Localized heating/cooling mechanisms have shown promising results to reduce the interpass temperature that can cause parts to collapse and induce a heat treatment on previously deposited layers.

Moreover, with only a constant interpass temperature it is possible to achieve more homogenous properties. However, this is a field where more research is necessary to manufacture parts with accurate pre-designed microstructures.

2.1.3. Pre-Heating of the Substrate

Pre-heating the substrate is one of the most efficient methods to mitigate residual stresses and cracking, since it reduces thermal gradients and homogenizes temperature distribution. Alberti et al. [35] compared depositions performed with and without pre-heating at a temperature of 300 °C in a PAW additive manufacturing process, and observed that pre-heating increased the wettability of each layer and enhanced the regularity of wall thickness, decreasing surface waviness.

It is known that in WAAM, the width in the first layers is significantly less thick than the remaining layers due to a rapid cooling rate, which is caused by the large area of the substrate and its initial temperature. With pre-heating of the substrate, heat conduction decreases and heat losses are minimized, resulting in smaller temperature gradients.

Figure 7 presents the temperature gradients with the increase of layer height. The benefits of pre-heating the substrate on the first layers with a heat input of 570 J/mm is visible. The maximum temperature gradient in the first layer without pre-heating is 3.82×10^5 °C/m and reduces to 3.63×10^5, 3.40×10^5, and 3.12×10^5 °C/m with a pre-heating of 200, 400 and 600 °C, respectively. In addition, the temperature gradient decreases with the increase of deposited layers. Besides reducing the temperature gradients and achieving a smoother thermal cycle on the firsts layers, other benefits include reduced thermal stresses and cracking susceptibility [36].

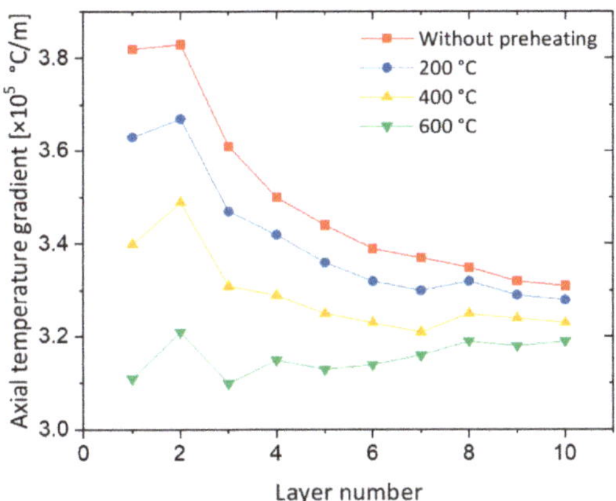

Figure 7. Temperature gradients variation from the first to the tenth layer (adapted from [36]).

2.1.4. Substrate Release Mechanisms

Usually, the substrate has a composition similar to that of the material being deposited, supporting adhesion and stability. Even though novel design approaches consider the substrate as a part of the final component [37], Haselhuhn et al. [38] tested several ways to easily remove the produced parts from the substrates, reducing material waste. The authors evaluated the energy necessary, by the Charpy impact test, to remove aluminum parts by spray-coating aluminum oxide (18.50 µm thick), boron nitride (5.95 µm thick), and titanium nitride (6.25 µm thick) on an aluminum substrate. Each one was seen to assist in substrate parts removal, but there was not any statistical difference between the adhesion strength. Additionally, the effect of deposit on the first layers without shielding gas and with reduced voltage and current, in order to decrease weld penetration, was also investigated. The authors concluded that the arc instability caused by the lack of shielding gas reduced weld bead penetration. Finally, the effect of using dissimilar materials to promote the formation of brittle intermetallics between parts and the substrate was evaluated. According to the summarized results presented in Figure 8, the authors observed that the formation of intermetallics and the non-use of shielding gas in the first layers decreased the energy required to remove the parts produced from the substrate. This can be attributed to the mechanical properties of the intermetallics and the oxides formed by the inexistence of shielding gas, which tends to require lower amounts of energy to fracture, hence facilitating detachment between the substrate and the as-build part. In future works it is necessary to assess whether these mechanisms can be used for large-sized parts.

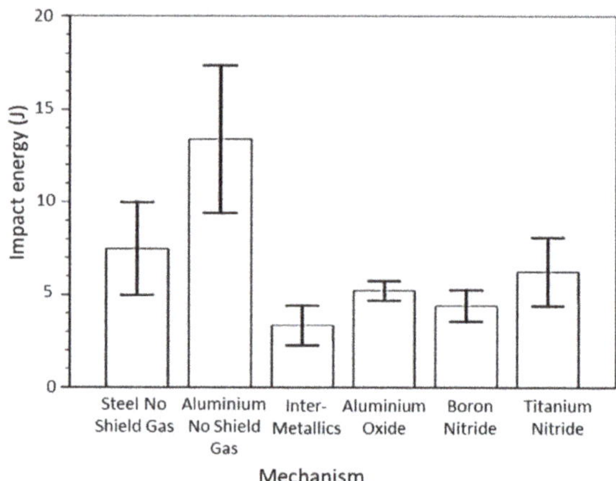

Figure 8. Impact energy test results of the samples built with different strategies and coatings (adapted from [38]).

2.1.5. Shielding Mechanism

Shielding gas is one of the most influent parameters, since it affects bead geometry, process stability, transfer mode, and bead appearance [39]. Besides its right selection, other methods have been used to improve WAAM parts quality. Xu et al. [40] studied the effect of oxides on the mechanical properties of maraging steel by varying shielding conditions. Results were obtained by making one deposition in the open atmosphere with pure argon, and the other in an argon-filled tent (chamber), in which the oxygen level was controlled below 300 ppm. Extra tent shielding substantially improved the surface waviness and deposition efficiency by 37% and 9%, respectively.

Gas shielding flow was found to be of remarkable importance regarding wall appearance. In a preliminary study [39], the effect of gas shielding flow in WAAM was analyzed, since turbulence flow caused the shielding gases to mix with the surrounding air, resulting in poor shielding conditions and increased atmosphere contamination that can lead to oxidation. A new device (Figure 9) consisting of three distinct parts was developed to achieve laminar flow of the shielding gas. The first part is a diffusion chamber that uniformly distributes the inlet gas, the second is a honeycomb wall that straightens the flow and reduces its lateral velocity, and the last part, at the end of the chamber, is a layer of metal mesh used to further improve the uniformity of the flow. Overall, this new device decreased the level of contamination up to three orders of magnitude.

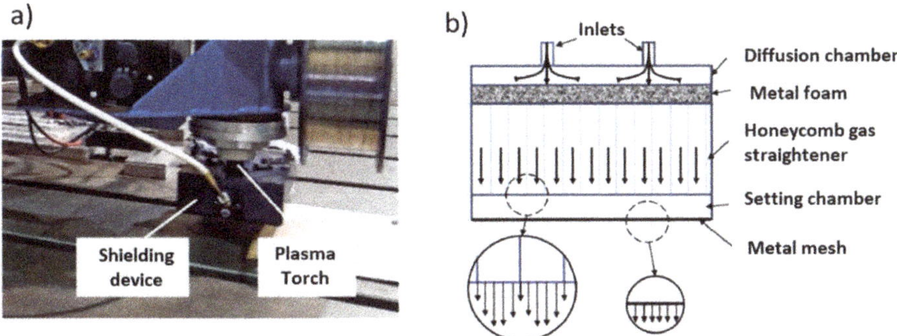

Figure 9. (**a**) Setup of the WAAM trailing shield device; (**b**) schematic representation of the shielding device (adapted from [39,41]).

As it can be observed, there are currently several impactful WAAM variants, all aiming to create parts with better mechanical properties and tolerances, and to reduce the need for post-processing techniques (machining and/or heat treatments). The abovementioned recent development shows the great activity and interest that WAAM is attracting, not only from academia, but also from the industry. Obviously, that focus cannot only be paid to the process variants in WAAM. The materials to be processed also play a critical role, and due to the vast possibilities of materials that can be used in WAAM, the same process variant may have different effects on the microstructure and/or mechanical properties of the parts. The following section is dedicated to the most recent developments regarding the different materials used in WAAM.

3. Materials

During parts fabrication, the deposited material undergoes various heating and cooling cycles that may result in different grain structures along their height. Grain structure control is of major importance since it determines the material mechanical properties. Typically, WAAM parts comprise large columnar grains, formed by epitaxial growth from the substrate aligned along the buildup direction normal to the solid/liquid interface, which has the maximum temperature gradient, thus eliminating the need for nucleation sites [42]. This type of growth results in anisotropic properties, which can be detrimental for multi-axial loading conditions. Equiaxed grains are desirable since they can reduce crack susceptibility while improving ductility, resulting in components with (near) isotropic properties. Thus, the use of add-ons that can perform in-process heat treatments, as well as post-WAAM heat treatments, is essential. Another practice to control the microstructure is the use of inoculants to refine the grain structure. A deep understanding of the mechanisms of nucleation with inoculants and grain growth, as well as the challenges for its use in additive manufacturing, is well described in [43].

Generally, any material available in the form of welding wire can be used for WAAM. The most used ones are steel, aluminum, titanium, and nickel-based alloys. Ti-based and Ni-based alloys are increasingly being studied due to their welcomed adoption by the aerospace industry. Thus, the desire to mature this process for its adoption into mass production of aerospace components comes from the ability to produce large parts with a low buy-to-fly (BTF) ratio. Other characteristics include, high specific strength, thermal and electrochemical compatibility with advanced composite materials, and the cost associated with applying subtractive methods onto these materials.

One of WAAM advantages is the possibility to process a vast range of materials, therefore the present section reviews recent results, challenges, applications, and key details of the most commonly used metallic alloys for WAAM.

3.1. Titanium-Based Alloys

Titanium-based alloys are increasingly being study in WAAM, allowing for a reduction in the high costs associated with processing these materials. Ti-based alloys have high strength, toughness, good corrosion resistance, and can tolerate extreme temperatures without significant loss of mechanical properties, making them suitable for aerospace and biomedical applications [44]. Ti alloys represent around 15% of the total weight of the Boeing 787 [45], owing to its electrochemical compatibility with carbon fiber polymer composites.

Amongst the different additive manufacturing processes, WAAM allows for a better control of the microstructure of these polymorph alloys, since these materials are highly sensitive to the thermal history. Ti-6Al-4V is the most used Ti alloy and consequently the most studied in WAAM. Typically, it is constituted by two phases, a hexagonal close-packed structure (hcp), α, and a body-centered cubic (bcc), β. The different temperatures and cooling rates result in microstructure variations through the parts height. The most common microstructure comprises fine acicular or Widmanstätten colony and basket weave lamella-α morphologies [46]. Columnar β-grains from prior layers with grain boundary-α [47] are also prominent undesired features, causing premature failure in transverse loading solicitations [46,48]. This columnar structure is difficult to avoid since in low concentrations, aluminum and vanadium have a high solubility in titanium and do not partition ahead of the solidification front, becoming irrelevant as grain refiners [49]. Although the β grains transform to fine α during cooling below the β-transus temperature, primary β grains can still have a detrimental impact on mechanical properties [50].

Wang et al. [12] manipulated process variables to refine the poor primary β grains of Ti-6Al-4V in pulsed-GTAW, concluding that the peak/base current ratio and pulse frequency had no significant effect. However, equiaxed grains were achieved with a higher wire feed speed, since more nucleation sites were provided, blocking the columnar growth.

From the previously described process variants, interpass rolling [51] showed to be a suitable process to mitigate the typical anisotropy of additive manufacturing parts including Ti-6Al-4V parts. Figure 10 depicts different electron backscatter diffraction (EBSD) orientation maps of both the α-phase and reconstructed β grain structures prior to transformation in an unrolled sample, rolled sample with 50 kN, and rolled sample with 75 kN. The β grains are represented by the color red. Reconstruction of prior β structures is often used to index this phase, due to its small scale and low volume fraction. The unrolled sample exhibited a β-phase with a strong preferential <001> crystallographic direction, and with columnar grains visible in both maps. When rolling was applied, new β orientations, associated with the deformation of α laths, were created and a strong columnar texture was mitigated. In a recent study, these new β orientations were found to arise from twinning with the residual β [52]. Another key effect observed was the refinement effect induced by the interpass rolling applied to the WAAM deposits: by increasing the applied load the grains become more refined, resulting in an improvement of the material microstructure.

The common anisotropy of additive manufacturing parts and potential presence of undesired phases can significantly reduce mechanical properties in both build up and transversal directions, urging the need to control the microstructure during fabrication. Potentially nucleant particles are being used in WAAM to refine the microstructure and enhance mechanical properties of Ti-based alloys.

By adding boron traces (up to 0.13 wt.%), Bermingham et al. [53] demonstrated the effectiveness of inoculants in eliminating the anisotropic microstructures of Ti-6Al-4V. Boron had a significant impact on β-grain morphology and TiB needles were formed. These particles were found dispersed in the microstructure, allowed to nucleate α-grains, and produced isotropic α-microstructures. The boron modified alloy exhibited an increase of 40% in the failure strain, with the average failure stress maintaining around 850 MPa.

Mereddy et al. [54] added up to 0.41 wt.% of carbon in Ti-6Al-4V parts. The β-grain density increased while the α-lath length decreased. Carbon is an effective refiner in Ti alloys with hypereutectic compositions since it nucleates TiC particles. However, for hypoeutectic compositions, refinement is a result of the segregation of carbon solute, decreasing solidification temperature, and generating constitutional supercooling and growth restriction. The mechanical properties of as-built samples with and without carbon additions, with a small amount of carbon (0.03 wt.%), medium amount (0.1 wt.%), and excessive amount (0.41 wt.%), are illustrated in Figure 11. Sample built with 0.41 wt.% of carbon formed large carbides which significantly deteriorated the mechanical properties, while the part built with a medium amount of carbon had an increased strength and ductility of 9% and 30%, respectively.

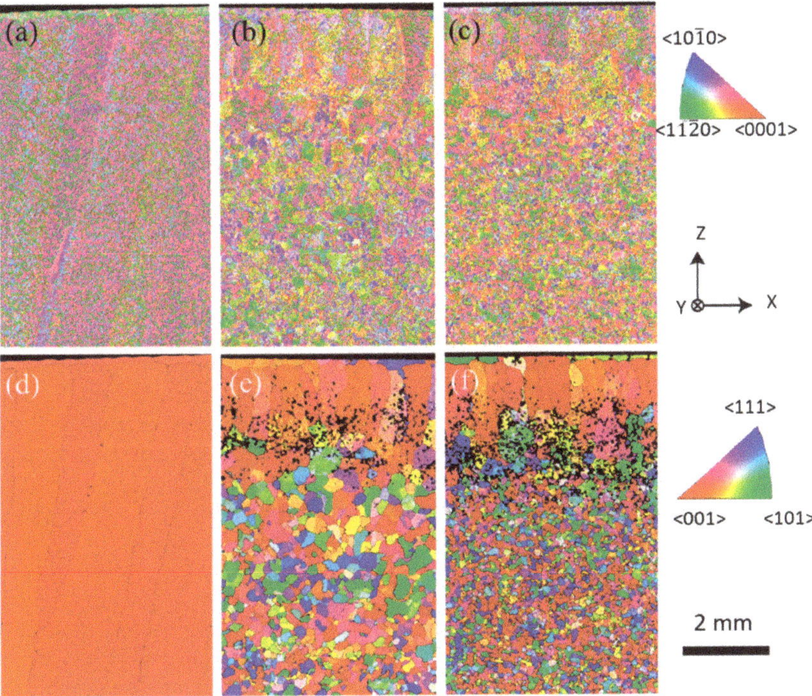

Figure 10. Electron backscatter diffraction (EBSD) maps of the α-phase (**a–c**) and reconstructed β-parent phase (**d–f**) of samples produced without rolling and with rolling loads of 50 kN and 75 kN (adapted from [51]).

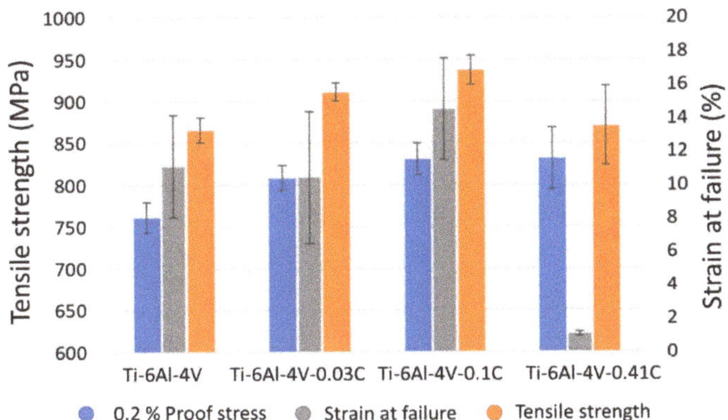

Figure 11. Effect of carbon additions on Ti-6Al-4V mechanical properties (adapted from [54]).

Similarly, the same authors added silicon to a commercially pure titanium wire, promoting refinement of the grain size, particularly on the prior β grains [55]. Silicon, however, did not fully eliminate columnar grains. Instead they became narrower with a similar length to those of silicon-free samples. Overall, silicon promoted supercooling and growth restriction, but further refinement might only be possible with additional powerful refiners.

3.2. Nickel-Based Alloys

Nickel-based alloys are a class of materials mostly used in the aerospace and nuclear industries, for instance in transition ducts and gas turbines. These alloys are characterized by high strength at elevated temperatures, low thermal expansion, and excellent corrosion resistance. Their common austenitic matrix makes them suitable to operate within a wide range of temperatures. High costs, ability to adhere to cutting edges, and the presence of abrasive carbide particles makes Ni-based alloys difficult to machine, so WAAM becomes a viable technique to eliminate the material waste and consequently the overall costs associated with processing of this alloy. Upon solidification, Ni-based alloys may exhibit the following: solidification cracking [56], liquation cracking [57], ductility-dip cracking [58], and strain-age cracking [59]. Hence, special care must be taken during WAAM of these materials.

Typically, Ni-based alloys, such as Inconel 625 and Inconel 718, have high concentrations of alloying elements that can segregate during solidification in the interdendritic spaces. Moreover, their mechanical properties are highly governed by the Laves phase, and its morphology is dependent on thermal history which consequently affects the parts final properties. Inconel is a solid-solution-strengthened nickel-based superalloy, that with the addition of substitutional alloying elements, such as Cr and Mo, provides nucleation sites and the preservation of austenite once cooled. Other phases commonly found in Inconel that are used for strengthening effects, include γ' phase [Ni_3(Al, Ti, Nb)], γ'' (Ni_3Nb, ordered bct $D0_{22}$ structure), and blocky MC carbides. Nevertheless, the mechanical properties of Inconel alloys can decrease with the formation of undesirable phases, such as the δ-phase (Ni_3Nb, orthorhombic) [60].

Inconel 625 parts manufactured by WAAM consist of vertically columnar dendrites in an austenitic phase (γ) matrix [61]. Solidification started with the L to γ reaction, and Nb and Mo precipitated in the interdendritic and grain boundaries where Laves phase will precipitate. With the increase of height during build up, the primary arm spacing varied from 13 μm, 23 μm, and 35 μm, from near the substrate, in the middle section, and in the top region, respectively. It obtained an average ultimate tensile strength of 722 ± 17 MPa and 684 ± 23 MPa in the travel and build directions, respectively. As for the elongation, 42.27 ± 2.4% and 40.13 ± 3.7%, respectively, were obtained. Fracture analysis revealed a ductile morphology with dimples. The shielding gas also played an important role on the final mechanical and geometrical properties of Inconel 625 parts. The ultimate tensile strength increased by 50 MPa when 97.5% Ar and 2.5% CO_2 as shielding gas was used [62].

Plasma arc-based WAAM was used to investigate the properties and microstructure of Inconel 718 superalloy in three conditions: as-built, with interpass rolling, and rolled with heat treatment in agreement with AMS-5662M [63]. The as-built samples exhibited the typical dendritic structure with Laves phase aligned with the build direction. The rolled samples showed refined grains near the boundaries of the produced walls with a recrystallized core in the central region. When solution plus aging heat treatments were applied, some of the Laves were dissolved and allowed to homogenize the distribution of the secondary phase particles. By X-ray diffraction, the δ phase was noticed after solution plus aging treatment of the rolled sample. However, its presence was not found in the wrought alloy. Laves phase, residual Laves phase, and δ phase precipitates of a rolled sample are depicted in Figure 12. Even tough rolling induced recrystallization, and columnar grains were still observed in some regions of the as-built samples, falling short to the wrought alloy. Consequently, the hardness was not homogenous along the wall thickness. Moreover, with the solution treatment, Nb did not completely diffuse and consequently Laves did not fully dissolve, suggesting the need for higher temperatures and/or longer periods for the heat treatment to be effective.

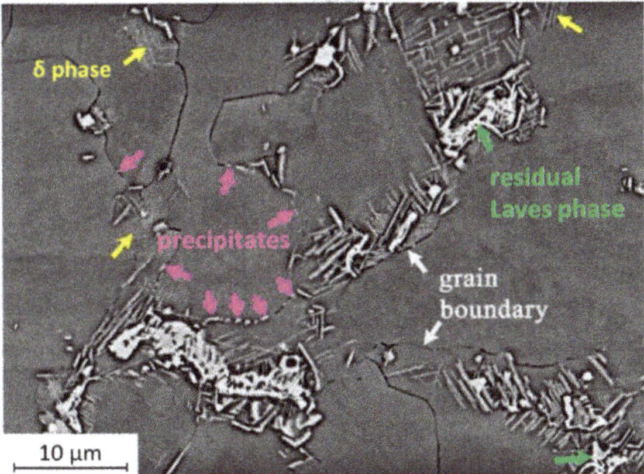

Figure 12. Precipitates found under SEM of the rolled with heat treatment sample (green: residual Laves phase, yellow: δ phase, purple: nanoscale precipitates along grain boundaries) (adapted from [63]).

Table 1 presents the mechanical properties obtained for the as-built and rolled samples with and without heat treatment. With heat treatment, the average ultimate tensile strength was increased by around 284 MPa and 232 MPa more than the as-built samples, and by 266 MPa and 284 MPa more than the as-rolled samples in the longitudinal and transversal direction, respectively. Therefore, it was concluded that a significant improvement in the mechanical properties of Inconel WAAM parts can be achieved through proper heat treatment schedules and by paying special attention to the post-processing of these materials.

Table 1. Mechanical test results of Inconel 718 parts in the longitudinal (Long.) and transversal (Trans.) directions (adapted from [63]).

Process		YS 0.2% (MPa)		UTS (MPa)		Elongation (%)		Hardness (HV)
		Long.	Trans.	Long.	Trans.	Long.	Trans.	
As-built	unrolled	525 ± 7	506 ± 2	818 ± 13	756 ± 7	33.3 ± 2.5	27.9 ± 1.3	259 ± 8
	75 kN rolled	763 ± 8	687 ± 1	1082 ± 13	1072 ± 6	26.2 ± 2.2	26.6 ± 1.3	330 ± 19
Solution treated	unrolled	790 ± 9	791 ± 14	1102 ± 78	988 ± 6	14.7 ± 1.3	12.8 ± 1.2	417 ± 16
	75 kN rolled	1057 ± 19	1035 ± 20	1348 ± 10	1356 ± 10	15.1 ± 3.3	17.4 ± 1.1	443 ± 18

Xu et al. [64], evaluated the effect of oxides and different types of wires on the final properties of Inconel 718, since this alloy is known for its oxidation-assisted crack growth mechanism at high temperatures [65]. An oxide layer of Al_2O_3 and Cr_2O_3 with 0.5 μm thickness at the top layer was noticed, thus confirming that oxides do not accumulate during parts fabrication. By comparing different wires, differences of up to 50 MPa in the ultimate tensile strength were possible due to differences in chemical composition and uncertainties in TiN particles, which acted as nucleation sites.

Improvements in superalloys might only be achieved with thermomechanical processing. Even though Nb can diffuse, and consequently eliminate, Laves phases at interdendritic areas with heat treatment, the common large columnar grains will only coarse. Therefore, it is crucial to control Nb segregation and Laves formation in situ, since it depletes the matrix of useful alloying elements, or use of cold-work based techniques [66].

Another alloy of interest from the Ni-based group is Monel. These alloys, mainly composed of nickel and copper, are characterized by their high strength and excellent corrosion resistance at high temperatures, and are used in a wide range of aerospace applications. The ability to withstand corrosive environments makes them suitable for marine applications. Monel K500 and FM 60 were tested by means of cold metal transfer. The secondary dendrite arm spacing was smaller for the Monel K500 (4–9 μm) than for the FM 60 (6–12 μm), due to a higher precipitation of TiCN in Monel K500, which delayed dendrite growth. Energy dispersive X-ray spectroscopy (EDS) results showed segregation of Cu for both alloys, due to differences in the melting point of Ni and Cu, and the difficult diffusion of Cu in Ni. Precipitation of Ti-rich particles is essential to achieve superior mechanical properties and their density can increase with the use of inoculants.

3.3. Steels

Steels are easily acquirable ferrous alloys widely used in automotive, ship, construction, and gas industries that, in combination with WAAM, can be used to manufacture parts with an overall low cost. However, some authors [67] claim that the production of these low-cost alloys by WAAM is only viable for large parts with complex geometries. Among steels, stainless steels have found applications in chemical plants and nuclear industries, where parts with high heat and corrosion resistance are required (e.g., pressure vessels). Austenitic stainless steel, such as SS 304 [68], SS 308LSi [69], and SS 316L [70,71] have been successfully used in WAAM, as well as martensitic stainless steel 420 [72].

Chen et al. [62] reported that 316 parts of stainless steel presented both austenite (γ), delta-ferrite (δ), and sigma (σ) phases with different morphologies at various positions as a result of the thermal cycles experienced during build up. After the fourth consecutive layer, the microstructure was composed of fine vermicular δ and σ phases within the γ matrix. With the increase of the parts' height, the volume of σ increased, resulting in a decrease in strength and elongation. To avoid this phase, thermal cycles should be controlled to avoid long residence times between 600 and 900 °C.

The feasibility to produce high nitrogen Cr-Mn stainless steels was shown by Zhang et al. [73], where parts were mainly composed by dendritic δ-ferrite and columnar austenite, γ, with the existence of some CrN and Cr_2N inclusion islands. The presence of oxygen in the shielding gas resulted in the formation of Mn-based oxides, which were found to be detrimental to the mechanical properties. Moreover, during post processing heat treatments the low solubility of nitrogen in δ-ferrite induced the formation of CrN and Cr_2N in the nitrogen supersaturated regions. Due to the relatively high energy surfaces between the matrix and inclusions, Cr_2N nucleated around Mn oxide inclusions, as observed in Figure 13. In conclusion, this material has shown to be deposited very effectively by WAAM with nearly isotropic characteristics due to a stable austenite matrix, while the presence of nitrogen had a work hardening effect.

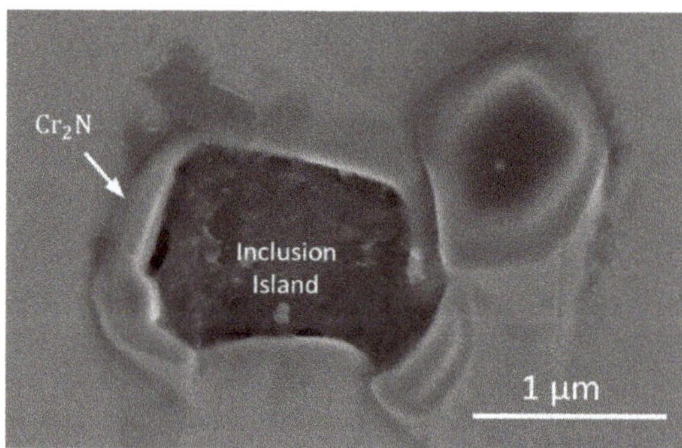

Figure 13. Cr_2N precipitations around the inclusion island after heat treated at 1100 °C for 30 min (adapted from [73]).

The feasibility of WAAM to process steel tool production was demonstrated in H13 steel [74]. The results presented in this investigation suggest that the significant differences in properties along the part's height are due to a different thermal history. The hardness varied between 300 and 360 HV and near-isotropic parts were obtained after annealing at 830 °C for 4 h. Average values of ultimate tensile strength and elongation were respectively, 1085 MPa and 10% for horizontal specimens, and 871 MPa and 7.8% for vertical ones.

Maraging steels are a class of superior steel with high mechanical strength attributed to the presence of intermetallic compounds such as Ni_3Mo, Ni_3Ti, Fe_2Mo, and Fe_7Mo_6. The aging temperature of maraging steel is relatively low (482 °C) and since this temperature is frequently exceeded during the process, overaging effects may occur. Using a plasma torch, Xu et al. [75] deposited maraging steel parts that revealed a martensitic matrix with fine residual austenite. With further aging, samples experienced an increase in the ultimate tensile strength of horizontal/vertical samples from 1118/1026 MPa to 1410/1345 MPa, while a decrease in elongation from 11.7/8.0% to 8.5/6.2% was observed. The authors reported the importance of avoiding undesired TiN inclusions in the feedstock material to obtain superior properties in the WAAM deposited parts.

Steels with a high carbon equivalent are more likely to experience cold cracking, a frequent problem due to rapid cooling, hydrogen entrapment in the heat affected zone, and residual stresses. Nevertheless, the procedures recognized to mitigate these problems in welding are intuitively used in WAAM, including pre- and post-heating, which is settled by re-heating or re-fusion of previously deposited material, which reduces the cooling rate, and subsequently the formation of brittle microstructures. Therefore, the intrinsic characteristics of WAAM are beneficial to avoid the aforementioned problems, though others may occur, such as overaging or precipitation of undesired phases as a result of the complex thermal cycles experienced during samples build up [76].

3.4. Aluminium Alloys

Welding of aluminum alloys (AA) has always been problematic, due to the formation of an aluminum oxide layer and solidification behavior. The use of WAAM in aluminum alloys is limited as porosity is of major concern. Such limitations have led to some investigations on the effect of heat treatments in WAAM Al parts. However, not every Al alloy is heat treatable. As it occurs in the welding of aluminum, during building of parts it is also preferred to use an alternate current (AC) [77] to remove the natural surface oxide film (alumina) which has a higher melting point. If not, melted remains are trapped inside the molten pool, resulting in pores and internal defects, which drastically decrease the parts mechanical properties. WAAM of Al alloys is very challenging due to the turbulent pool dynamics caused by the periodic inversing of polarity, which can result in decreases of the part accuracy. Other important properties regarding welding of the Al alloys include high thermal conductivity, high coefficient of thermal expansion, high solidification shrinkage, wide solidification temperature range, and high solubility of hydrogen [78].

Gu et al. [79] studied the influence of wire quality of final parts properties, highlighting that the pre-existence of undesired contaminants is a major driving force for hydrogen cracking. The main Al alloys used in the aerospace industry are 2xxx (Al-Cu) and 7xxx (Al-Zn) series alloys. However, they are highly susceptible to hot cracking if process parameters lead to high levels of thermal stress and solidification shrinkage. Nevertheless, Fixter et al. [80] successfully fabricated AA2024 parts without solidification cracking. The importance of these results arise from the fact that AA2024 is an unweldable alloy, and part production is enabled by the suitable selection of the Mg content in the feedstock wire.

Cold metal transfer (CMT) is widely accepted as the most reliable variant to process Al alloys. However, CMT pulse advanced (CMT-PADV), developed by Fronius, was proven to entirely eliminate gas pores, due to an oxide cleaning effect [15]. Additionally, this variant is characterized by its low heat input and by the preservation of nucleation particles. Figure 14a,b depicts the macrostructure of aluminum with conventional cold metal transfer and with CMT-PADV, respectively, where the non-existence of pores with CMT-PADV is visible.

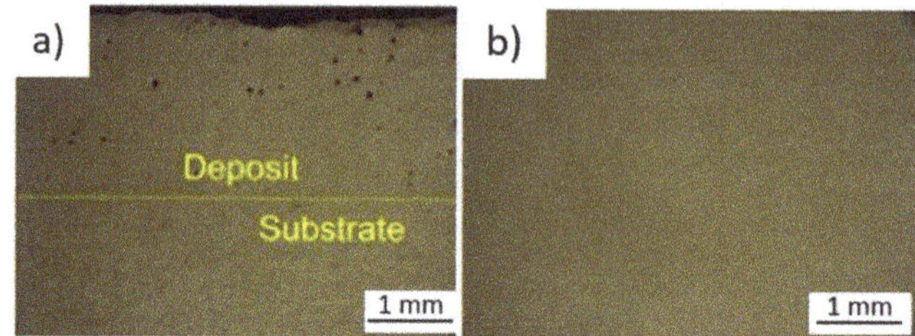

Figure 14. Porosity presence in samples manufacture with: (**a**) conventional cold metal transfer; (**b**) cold metal transfer-pulse advanced (adapted from [15]).

Zhang et al. [81] studied another similar power source mode named variable polarity cold metal transfer (VP-CMT). The typical waveform is presented in Figure 15. The pulsed arc mode results in an oscillation, that in association with alternating arc polarity changes, breaks the dendrite arms, providing heterogeneous nucleation sites. The typical columnar grains transformed into equiaxed and a grain refinement effect was observed. Despite isotopic grains confirmed by microscopy techniques, only a difference of 8% in ultimate tensile strength from horizontal and vertical samples was possible, due to the existence of interlayer pores.

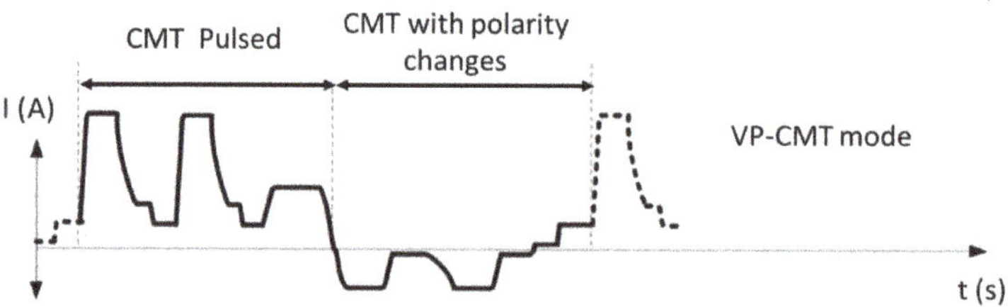

Figure 15. Current wave form of variable polarity cold metal transfer mode (adapted from [81]).

The following Al-alloys deposited by WAAM have been reported in the literature: 5A06 [82,83] Al5Si [84], AA5183 [85], Al-Mg4.5Mn [20], Al-5Mg [86], Al-6Mg [81], and Al–6.3Cu [23]. In general, WAAM has its value in the production of Al parts, but the mechanical properties obtained are not always superior to the ones achieved from machining a billet, and the existence of add-ons, such as rolling, become important. The pressure applied, besides absorbing a significant amount of atomic hydrogen, also reduces porosity. Such has been experimentally verified for a load of 45 kN, where pores were eliminated to a level below the resolution of optical microscopy [14].

The effect of inter-layer rolling with different applied loads is clearly depicted in Table 2, which compiles the results of the number of pores, mean diameter, area percentage, and mean sphericity of WAAM deposited AA2319 and AA5087 alloys. Pores were completely eliminated when a 45 kN load was applied in between two consecutive deposited layers.

Table 2. Analysis results of pores for various stated WAAM 2319 and 5087 alloys [23].

Condition	**As-Built**		**15 kN Inter-Layer Rolling**		**30 kN Inter-Layer Rolling**		**45 kN Inter-Layer Rolling**	
Alloy	2319	5087	2319	5087	2319	5087	2319	5087
Number of pores (In a total area of 120 mm^2)	614	454	192	336	5	11	Pores were completely eliminated	
Mean diameter (μm)	13.5	25.1	12.5	13	8.8	9.6		
Area percentage (%)	0.176	0.232	0.029	0.061	0.005	0.007		
Mean sphericity	0.74	0.74	0.67	0.63	0.37	0.42		

It is known that inter-layer rolling can provide nucleation sites that promote grain refinement and, depending on the inter-layer temperature, strain rate and applied load, different microstructures and properties can be achieved. The strengthening effects of inter-layer rolling for different WAAM deposited aluminum alloys are presented in Table 3. A linear improvement of the ultimate tensile strength and yield stress with an increase of rolling load in both cases, but with a consequential decrease in the elongation, is visible. Since inter-layer rolling is unable to uniformly deform beads, differences between longitudinal and transversal still subsist.

Table 3. Mechanical properties of different aluminum alloys with different build conditions in the longitudinal (Long.) and transversal (Trans.) directions.

Material	Variation	YS 0.2% (MPa)		UTS (MPa)		Elongation (%)		Ref.
		Long.	Trans.	Long.	Trans.	Long.	Trans.	
ER 2319	As-built	135	130	265	260	18.4	15.7	[23]
	rolled (15 kN)	146	140	270	265	15	14.8	
	rolled (30 kN)	185	170	290	280	13.2	11.8	
	rolled (45 kN)	250	245	322	310	8.6	7.3	
ER5087	As-built	142	-	291	-	22.4	-	[20]
	rolled (15 kN)	170	-	301	-	21.6	-	
	rolled (30 kN)	200	-	320	-	20.9	-	
	rolled (45 kN)	240	-	344	-	20.1	-	

The microstructure and mechanical properties of 5363 (Al-5Mg) Al alloys were enhanced by adding titanium powder between layers [86]. Since Al_3Ti and Al have similar face centered cubic (FCC) structures, formation of nucleation sites was improved. The addition of Ti resulted in the formation of fine equiaxed grains at the interlayer interface (Figure 16). The ultimate tensile strength and elongation respectively increased by 20.25 MPa and 3.13% in the horizontal direction and by 25.89 MPa and 6.97% in the vertical direction. This work highlights the viability to use inoculants to act as grain refiners during the production of Al WAAM parts in an attempt to obtain more isotropic and improved mechanical properties.

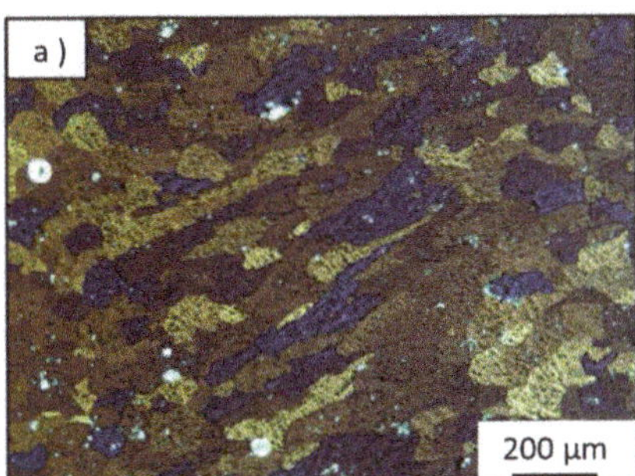

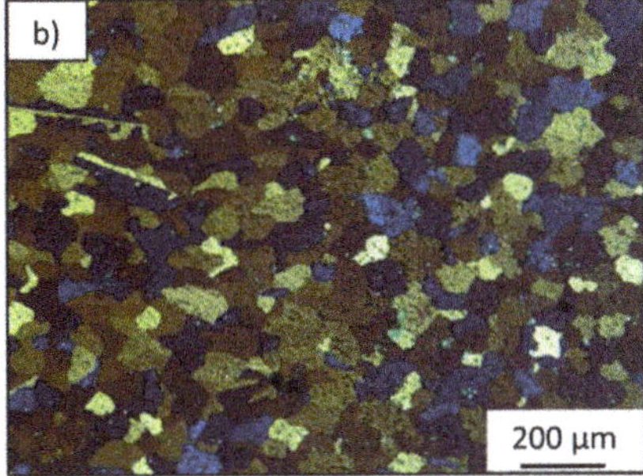

Figure 16. Microstructure of 5363 aluminum alloy: (**a**) as-built and (**b**) with Ti additions [86].

Sales et al. [87] aided the deposition of AA 5183 and AA 5356 with scandium which was responsible for the formation of Al_3Sc intermetallic particles that acted as nucleation sites. The effect of adding zirconium and scandium as grain refiners was the same as using only scandium. The ultimate tensile strength and yield stress were increased by nearly 60 MPa in both horizontal and vertical directions.

3.5. Magnesium Alloys

Magnesium alloys are increasingly being used as an alternative to aluminum to reduce the overall parts weight in the automotive and biomedical industries. Magnesium alloy advances throughout the years were hampered due to flammability risk, but with the increased interest in Mg-Al alloys, rare earth elements (zirconium, gadolinium, dysprosium, yttrium, neodymium and cerium), and other additional trace elements (Ca, Sr, Sb) were added to Mg suppressing ignition susceptibility. Magnesium is characterized by a hexagonal close-packed (hcp) structure and has few slip systems resulting in poor ductility. Owing to its structure, several defects can occur during forging or extrusion (i.e., edge cracking), therefore most magnesium products are processed via casting. Modified AZ31 and AZ61 are the most used Mg-alloys, however, only reports of the former processed by WAAM exists. As it occurs for aluminum, magnesium alloys also form an oxide refractory layer, but this can be removed easier than for aluminum. The necessity to refine magnesium structures was achieved by Guo [88] while presenting the feasibility to manufacture AZ31 WAAM parts. Such an achievement was obtained with GTAW technology by using six different pulse frequencies (1, 5, 10, 100, and 500 Hz), with the corresponding microstructures depicted, in Figure 17. Samples built with 5 and 10 Hz presented higher surface waviness, but the grain size was smaller and finer, measuring around 21 μm. The weld pool went through resonance with these frequencies, and as a result, the cooling rates decreased enhancing a finer structure. Sample built with 5 Hz exhibited an ultimate tensile strength of 258 MPa and an elongation of 25.6%, while sample built with 10 Hz exhibited an ultimate tensile strength of 263 MPa and an elongation of 23%. Overall, the samples produced indicated good plastic behavior well above the recommended value (234 MPa) stipulated by ASTM standard B91-12 [89].

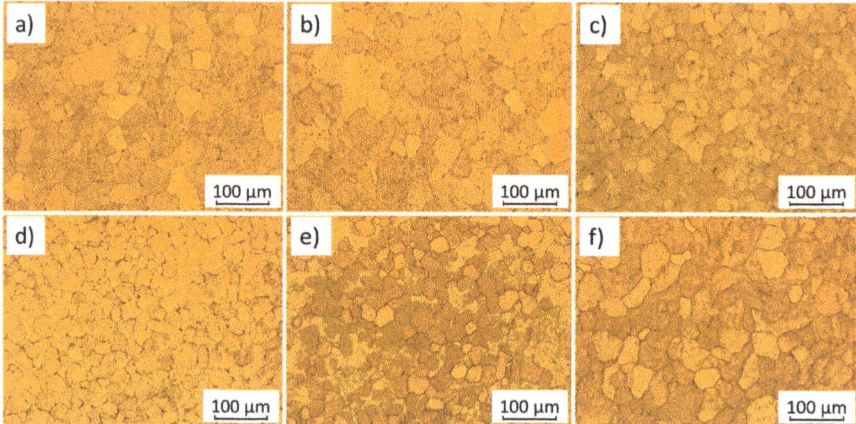

Figure 17. Microstructure of samples produced with a frequency of (**a**) 500 Hz, (**b**) 100 Hz, (**c**) 10 Hz, (**d**) 5 Hz, (**e**) 2 Hz, and (**f**) 1 Hz (adapted from [88]).

Another already processed magnesium alloy with WAAM was the AZ91D [90], in which the authors highlight the remelting of previously deposited layers as the major problem with processing this materials. Furthermore, parts produced exhibited higher corrosion resistance than a cast magnesium sample, enhanced by the formation of $Al_5Mg_{11}Zn_4$. Therefore, WAAM can be seen as a potential replacement of conventional casting processed in specific applications.

Rare elements are often added to Mg-based alloys. Since these are used for biomedical applications and are very susceptible to corrosion, ways to control and improve Mg alloys, rather than rare elements, is of major importance.

3.6. Functional Graded Materials

Besides producing bulked parts, WAAM is a suitable candidate to manufacture functional graded materials (FGM), which are an advanced class of heterogeneous materials which exhibit a controlled spatial variation of its properties (physical, mechanical, biochemical, among other) along at least one direction. Moreover, the manufacture of materials with site-specific properties is also possible [91,92]. Among the different additive manufacturing processes available today, production of FGMs is possible through binder jetting, directed energy deposition, material jetting, powder deposition, and sheet lamination. Wire and arc techniques offer unique advantages to manufacture FGM, due to their ability to vary the properties of the deposited material throughout deposition. The development of FGMs via WAAM can be obtained using the following two approaches: (i) by varying process parameters, such as wire feed speed or current; (ii) by feeding multiple wires, as schematically shown in Figure 18.

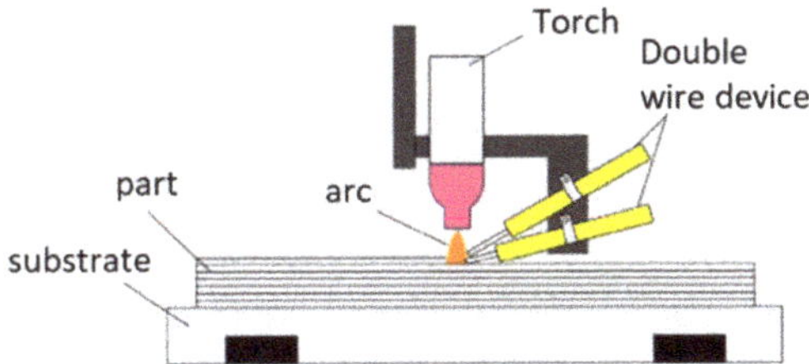

Figure 18. Schematic representation of double wire setup (adapted from [93]).

Somashekara et al. [94,95] used two separate wires with two independent power sources fitted into one torch in order to obtain flat pieces with a gradient of properties. Experimental studies allowed for a regression model to predict hardness as a function of torch speed and current of each wire with a maximum error of 6.5%.

A high purity annealed iron wire (99.5 at.%) and 1080 aluminum were combined under the electric arc of a tandem torch [96]. The content of aluminum was varied from 15 to 55 at.% every four layers by 5% increments. Such variation resulted in a compositional gradient with different intermetallics being formed in a variation of phases detected by X-ray diffraction. Near the substrate, specimens showed large columnar Fe_3Al grains. With the increase of Al content phase changed to B2 structured FeAl and large columnar grains were eliminated, and when the content reached around 50 at.% $FeAl_2$ was formed. With 36.1 at.% of Al content, an ultimate tensile strength of 315 MPa was feasible, but the mechanical properties of the specimens started to decrease rapidly for higher Al amounts. Specimens exhibited low ductility resulting in brittle transgranular lamellar fractures.

FGMs can potentially mitigate the issue of localized stress concentrations, as well as the manipulation of desirable properties and phases. However, when mixing dissimilar materials, some elements are likely to have different melting temperatures that when deposited may vaporize [97], requiring some attention. Additionally, the formation of brittle and/or undesired compounds greatly increases when fusion-based FGMs are being created. Therefore, it is critical to determine the range of process parameters that allow the production of complex shaped parts, exhibiting mechanical, chemical, or other graded property, while at the same time avoiding detrimental phases. To that sense, the use of a thermodynamic computational approaches [98] may be of great use to achieve the desired parts.

3.7. Other Materials and Dissimilar Depositions

Apart from the already referred alloys, other metals have been studied, such as copper-aluminum (Cu-Al8Ni2Fe2) [70], a dissimilar deposition of stainless steel and Ni-based alloy [99], and lastly, NiAl bronze alloys (NAB) with potential for marine applications [8,100]. Aluminum-copper (ER2319) and aluminum-magnesium (ER5087) were fed into the same electric arc, and by adjusting the wire feed speed different chemical combinations were achieved (Al-3.6Cu-2.2Mg, Al-4Cu-1.8Mg, and Al-4.4Cu-1.5Mg) [93]. The phases in Al-3.6Cu-2.2Mg were mainly α-Al and S phase and with an increase of Cu and decrease of Mg content, θ phase gradually increased.

4. Deposition Strategy

The deposition strategy is critical in additive manufacturing processes based on fusion. In this section, recent studies and methodologies regarding the building strategy for WAAM parts are presented.

Currently, there are several optimized softwares that slice 3D models and consequently create a G-code to be read by fusion deposition modeling printers. However, until now, there has not been a clarification about all WAAM constraints (e.g., residual stresses), in order to produce parts directly from CAD models. Kazanas et al. [101] proposed that the deposition of previous layers to occur in the form of pyramid to avoid the formation of humps in single inclined-walls.

Another recurrent macrostructural problem occurs when every layer has the same start and end point. The excessive heat sink at the beginning of the deposition decreases weld penetration. In contrast, at the end of the layer, low heat dissipation due to high temperatures results in layer height drop. The inconsistent height will accumulate along the deposited layers, precluding the process. To overcome this issue, the current and travel speed should be higher in the beginning and reduced gradually at the end of the deposition [102]. Another way to mitigate this problem is through the use of a zig zag approach by switching in every layer the start and end points [103]. However, this last method can result in zones that have thermal accumulation promoting higher residual stress at the walls boundaries. An easy approach might be to consider these zones as sacrificial ones.

Understanding bead geometry and its relationship with process parameters to improve quality of produced parts is of major importance during WAAM. Optimization of process parameters to obtain better surface quality is a rather a tedious work based on trial and error methods so, Geng et al. [83] unveiled two types of forming mechanisms named wetting and remelting, that are determined based on the remelting width in each deposition. Furthermore, the authors developed equations of beads cross sections based on process parameters, such as, wetting angle and radius of the spherical cap. The adequate material input and process parameters were calculated, so that the molten metal could spread vertically tangential to the spherical cap, improving surface waviness

Nevertheless, WAAM parts are not uniquely built by single walls, and once understood bead geometry behavior, its then necessary to optimize torch tool path considering build orientation selection, build sequence, design constraints and if it will be necessary to conducted post machining. Mediocre planning may result in porosity, internal defects, lack of fusion between adjacent beads and high residual stresses, a topic that will be further described in the next section. A method named tangent overlapping model was proposed to approximate the bead cross-section by means of functions (parabola, cosine and arc) [104]. An optimal distance value between adjacent beads of 0.738 times the width of one bead was achieved in order to suppress the valleys beads.

Regarding guidelines for the decomposition of CAD models, Ding et al. [105] presented a tool-path generation model that decomposed layers in polygons and each area was consequently filled. This model also automatically generated a final closed-looped tool-path by minimizing start/stops and crossovers of weld paths (Figure 19). A second approach on path planning modeling for WAAM was advanced by Ding et al. [106] through the improvement of earlier studies [107,108]. This method, named medial axis transformation, divides the geometry of a slice by producing a set of bisector segments. When more than two segments connect, those points become branch points and the paths will be generated by recursively offsetting contour-clockwise around the segment that connects two branch points as depicted in Figure 20.

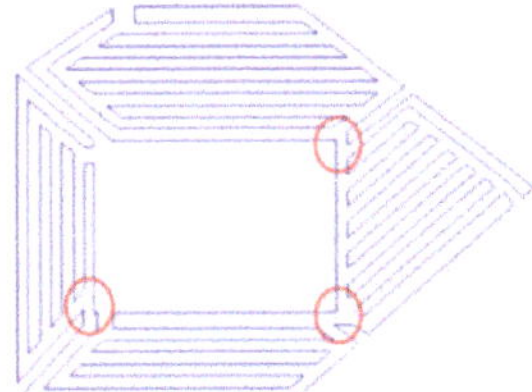

Figure 19. Convex polygonal method [105].

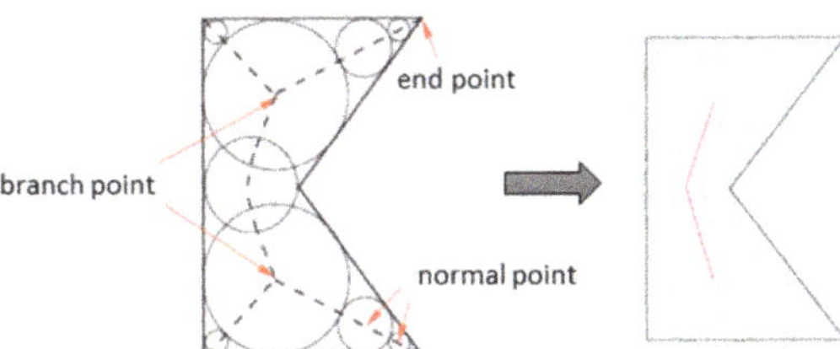

Figure 20. Schematic representation of the adaptive medial axis transformation planning method (adapted from [109]).

A path strategy regarding the manufacturing of 90° walls is presented in Figure 21 [110]. In order to assure a constant height for each layer, the strategy consisted in that after every fourth layers made with the first strategy (Figure 21a), the second deposition strategy (Figure 21b) was employed. Since one of the applications of WAAM is the fabrication of shell-part types, these successful developments are important, as they allow the production of smooth fillets of T-type structures.

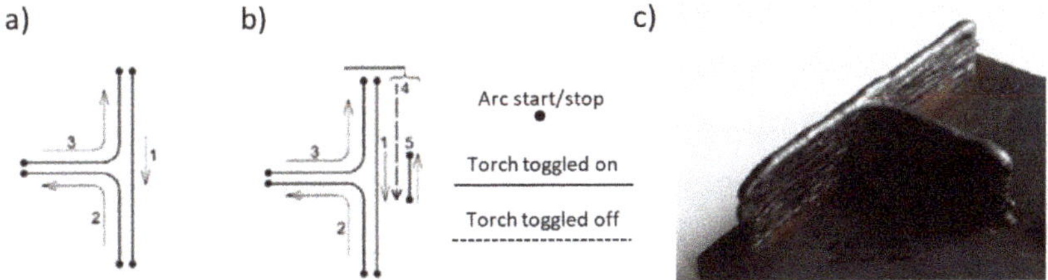

Figure 21. (a) First deposition strategy; (b) second deposition strategy; (c) final part made (adapted from [110]).

Other studies were performed to improve WAAM walls appearance, create T-type connections, and construct acute angles by adjusting the vertical distance and angle of the feed wire in a GTAW-based application. According to [82], an optimized angle of 10° with a distance of the melting wire tip to the molten pool surface of 3.8 mm guaranteed smooth layers. In a recent development, the influence of three different deposition building strategies (oscillation, parallel, and weaving) on surface waviness and porosity of maraging steels was studied [111]. Weaving deposition strategy, depicted in Figure 22, resulted in lower surface waviness and pores with reduced contact angle. In addition, Ma et al. [112] used weaving as a replacement for the multi-bead overlapping strategy to obtain better surface flatness, which required less post-machining, thus decreasing the associated production costs.

In some additive manufacturing processes, there is a need to add material to support overhang volumes, that is later removed. Some research groups [37,113] studied the possibilities and advantages, and developed models in which the substrate was mounted on a 5-axis positioning system, which allowed the build direction selection during fabrication to change, thus eliminating the need for supporting material. With such a system, the correct selection of the substrate position becomes essential based on some criteria factors: mass of substrate waste, mass of the deposited material, the number of build operations, build complexity and symmetry. A detailed analysis of the correct position of the substrate may allow the BTF ratio to be reduced.

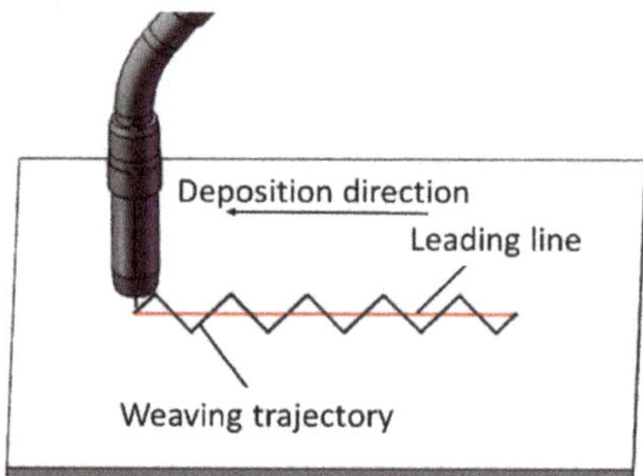

Figure 22. Weaving deposition strategy (adapted from [112]).

Since most of the software for metal deposition are limited and without public access, Yunyong et al. [114] took advantage of the free open source CuraEngine to introduce new settings to generate G-code to print metallic parts. These settings permit the torch to toggle on and off, path optimization to avoid crossovers in the same layers, the number of start/stops are minimized, and an option to pause the torch was also added for subtract cooling between layers. It is expected that these open-source software codes can be embraced by the additive manufacturing community to further expand its applications.

5. Residual Stresses

Residual stress challenges are of extreme importance within the WAAM process and are due to the complex thermal behavior and thermo-physical properties of the materials to be deposited [7]. Residual stresses are defined as stationary stresses at equilibrium in a portion of material after all external forces are removed [115]. Residual stresses are determined by their characteristic length: type I are macro-stresses varying over the dimensions of the component; type II are intergranular stresses; type III are formed at an atomic scale [116,117]. Despite the residual stresses that can be reduced by laser shock peening [28], in WAAM they can be as high as the yield strength of the material [118], negatively affecting the mechanical properties and leading to distortions and decreased tolerances. If these residual stresses exceed the local yield stress of the material, plastic deformation occurs, but if it exceeds the ultimate tensile strength then fracture is expected. These stresses are the result of the repetitive heating and cooling, which induce repetitive expansion and contraction of the material. For that reason, upon unclamping, the part balancing the internal residual stresses bends, and these can reach up to 500 MPa [22]. Considerable efforts are being made on strategies to mitigate this problem so that results can be further incorporated in path planning algorithms. Main findings include, that when building a layer, a pattern starting from the edges to the center cause less residual stresses on the substrate [119].

In others studies, several methods are employed in order to minimize the heat accumulation, and consequently residual stresses, based on the regulation of dwell time [120]; by pre-heating the substrate [36], since it reduces thermal gradients making the temperature distribution more homogeneous, it can increase the wettability of the first layers [35]; by mounting the substrate on a 5-axis system and building parts on both sides so that the residual stresses are balanced [37]. Other methods include the use of secondary heat sources to induce pre- or post-heating to obtain smoother temperature gradients [33]. Cold rolling is also used for the control of residual stresses in WAAM parts [19]. Figures 23 and 24 depict substrate distortions and longitudinal residual stresses, respectively, with three different rolling loads. As-built samples experienced stresses near the yield strength in the longitudinal direction. A vertical rolling load of 28 kN was sufficient to mitigate distortion in the aluminum parts. Rolling allowed for a decrease in longitudinal stresses that become compressive near the top layers. However, rolling induced stresses in the transversal and normal directions that did not previously exist, indicating a three-dimensional stress state, which can significantly alter the mechanical properties of the parts.

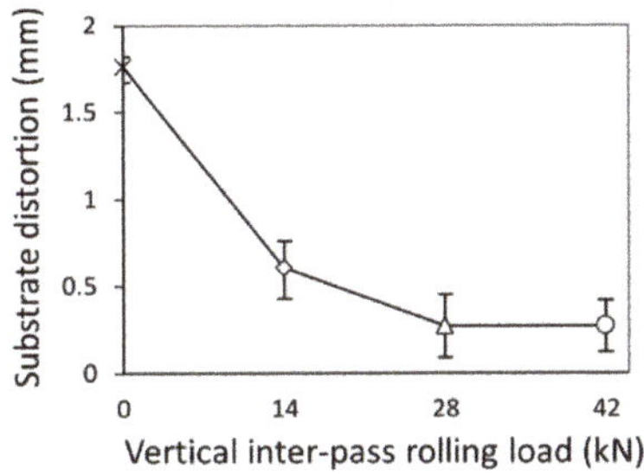

Figure 23. Substrate distortion with three different rolling loads (adapted from [19]).

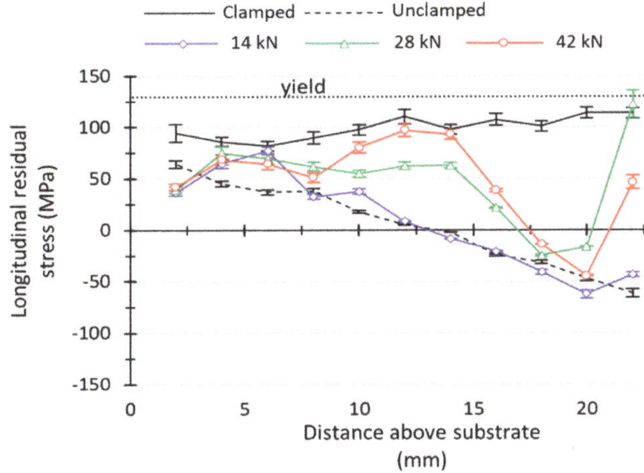

Figure 24. Longitudinal residual stresses of aluminum parts after rolling (adapted from [19]).

Ultrasonic impact testing is a cold-work treatment that induces compressive stress on the top or on the side of the walls within a very limited depth, producing grain refinement. In WAAM it was seen to be effective in reducing residual stresses by nearly four times, compared to the as-built samples [121]. The energy of the arc was not seen to be enough to remelt all the recrystallized equiaxed grains of previous layers, forming a bamboo-like macro-structure.

6. Heat Treatments

During WAAM, excessive heat accumulation can prevent the accurate control of bead geometry [122], but also might lead to differences in mechanical properties and microstructure with the increase of samples' height [123] becoming important to perform heat-treatments in order to achieve isotropic properties. These treatments might also be used to achieve the required mechanical properties, phases, relieve residual stresses, and remove internal defects.

In Section 2.1.2 in situ methods of heating and cooling WAAM parts were discussed, which can provide reliable and adequate heat treatments to achieve the required properties. However, these methods are still relatively recent and computational simulation developments are still necessary to predict phases and mechanical properties. Additionally, post-WAAM heat treatments can be used by means of hot isostatic pressing (HIP), solution treatment, annealing, aging, and other thermal treatments.

Gu et al. [14] studied the influence of post-deposition heat treatments on the porosity of 2319 and 5087 aluminum alloys, well-known for their porosity-related issues. After deposition, samples were kept for 90 min at 535 °C, followed by cold water quench. However, a combination of inter-pore coalescence (by Ostwald ripening) and hydrogen diffusion phenomena originated the growth of pores. Fully dense parts where only obtainable by combining inter-layer rolling and post-WAAM heat treatment. Other benefits from post-processing heat treatments were applied to NiAl Bronze WAAM components [8]. Parts fabrication annealing at 675 °C for six hours was applied, followed by air cooling, and resulted in the relaxation of the previously induced residual stresses in a fine homogenized microstructure composed of α phase (Cu) and intermetallics, such as NiAl and Fe_3Al, as identified by X-ray diffraction (Figure 25). These constituents were not found in the base material, and their presence resulted in a significant increase in hardness from 181 HV for the base material to 210 HV after post-WAAM heat treatments.

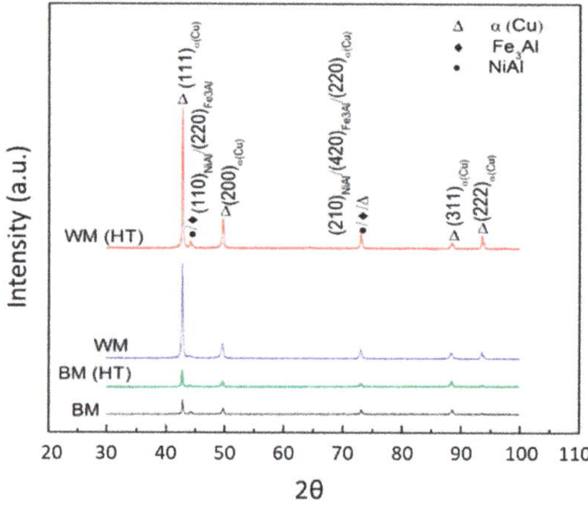

Figure 25. Phase constituents of as-cast base material (BM), as-deposited WAAM part (WM), heat-treated base material (BM HT), and heat-treated (WM HT) (adapted from [8]).

Different post-processing heat treatments were used in WAAM parts of Ti-6Al-4V in [124]: stress relief (480 °C for 2 h); hot isostatic pressing (927 °C for 2 h at 1500 bar with a heating and cooling rate of 5 °C/min); vacuum annealing (927 °C for 2 h with a heating and cooling rate of 5 °C/min); and solution treated followed by annealing (967 °C for 1 h, water quenched then aged at 595 °C for 2 h and air cooled). The variation of mechanical properties was justified by changes in the microstructure, with the as-deposited sample exhibiting prior-β grains with fine Widmanstätten-α. The mechanical properties after each heat treatment are presented in Figure 26. The stress relief heat treatment resulted in a ductility increase of 30% and a microstructure similar to that of the as-deposited Ti-6Al-4V. Hot isostatic pressing effectively removed pores, and samples had a similar increase of ductility as vacuum annealed samples (around 40% of the as-built parts), explained by the continuously coarsening of the α-phase. Solution treated samples mainly comprised of refined α-phase grains, experienced an increase of 12% in strength, but in contrast the ductility was lowered by 30%.

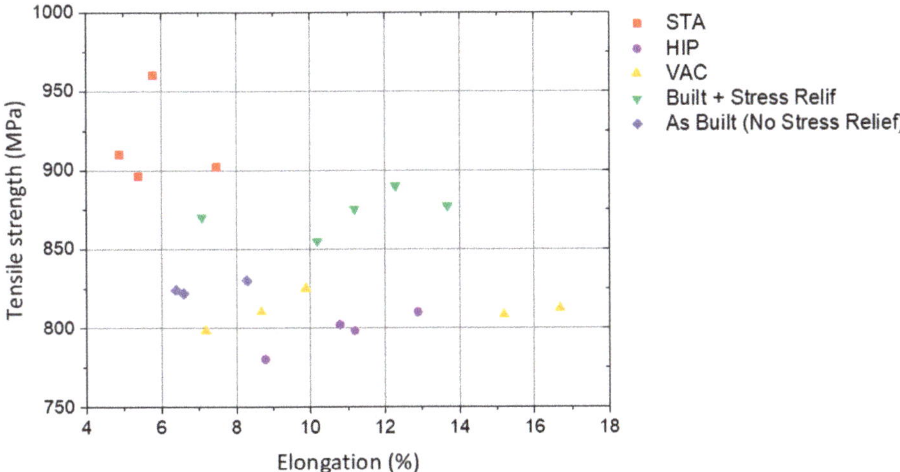

Figure 26. Mechanical properties of Ti-6Al-4V as-deposited and with post-process heat treatment (STA: solution treated plus annealing; HIP: hot isostatic pressing; VAC: vacuum annealing) (adapted from [124]).

When mixing dissimilar alloys, it is likely to obtain intermetallic compounds [125–127]. Fe_3Al is being increasingly studied with WAAM due to its unique properties [128]. In order to avoid some of the intermetallics and to reduce the anisotropy induced by the process, samples underwent homogenization (1000 °C for 7 h), homogenization plus transformation annealing in the FeAl region (850 °C for 24 h), and homogenization plus annealing in the FeAl region and a transformation treatment in the Fe_3Al region (500 °C for 120 h), in accordance to the Fe-Al binary phase diagram. Homogenization on its own eliminated Al-rich precipitates and resulted in excessive grain coarsening. With further heat treatments in the FeAl-phase region these precipitates started to randomly appear. Finally, with only a third heat treatment Al-rich precipitates and new grain boundaries formed in an equiaxed shape with a size of 150 μm. In this study, excessive time of post-heat treatments was necessary to precipitate an adequate amount of Al-rich constituents that refined grain. Thus, from a manufacturer point of view, in situ methods to perform heat treatments should be developed and industrialized to save time and reduce costs.

The effect of heat treatments on Al parts fabricated by WAAM was also studied by Qi et al. [129], where solution treatments at different temperatures plus natural aging procedures (T4 condition) were applied to AA2024 parts. The effect of those heat treatments in the mechanical properties of the parts are evidenced Figure 27. All heat treatment conditions were seen to increase the mechanical properties compared to the as-deposit samples. From this study is it clear that, depending on which property (ultimate tensile strength, yield strength, or elongation) is to be maximized, proper selection of the heat treatment conditions is fundamental.

WAAM mostly uses alloys that can be heat treatable (e.g., high strength steels, titanium alloys, and nickel superalloys, for example) or non-heat treatable (e.g., austenitic SS and 4XXX aluminum alloys). To suppress columnar grains of non-heat treatable alloys, the development of cold working variants is of major importance. Regarding heat treatable alloys, superior properties that are difficult to obtain in as-build WAAM parts are expected. Further understanding of phase transformation kinetics that occur during the process is of major importance in order to avoid or reduce the need for post-process heat treatments. For this reason, the next section will focus on modeling and simulation applied to WAAM.

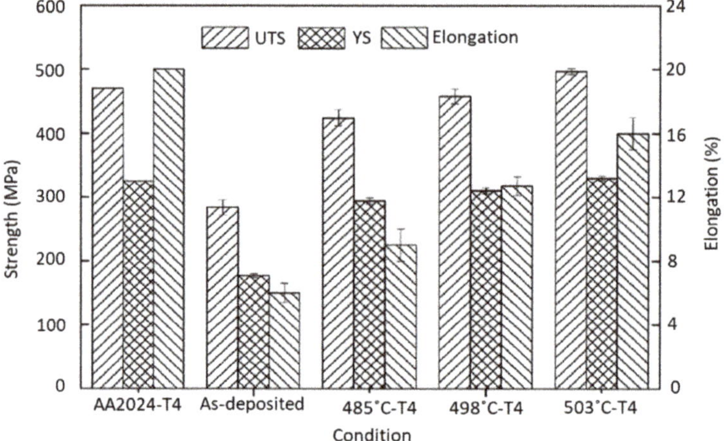

Figure 27. Tensile properties of 2024 aluminum samples [129].

7. Modeling and Simulation for WAAM

WAAM parts' microstructure and properties rely directly on process parameters, such as material, wire feed speed, heat input, and idle time between layers. The difficulties in optimizing a WAAM deposition and achieving the required characteristics, requires the execution of many experiments that can be highly costly and time consuming. Therefore, finite element analysis improvements are one of the primary goals toward a more massive adoption of WAAM. Typically, the process itself comprises melting of wire, metal transfer mode, mass transfer, gas absorption, convective flow of liquid metal, arc pressure, and solid-state phase transformations. The difficulty to model all phenomena comes from the diversity of materials and complex thermal behavior, since a portion of material is subjected to a vast number of thermal cycles of re-heating and cooling. Incorporating volumetric changes due to phase transformations can also be important [48]. Nowadays, most of the knowledge comes from trial and error experiments so if the process becomes fully modeled, accurate predictions of residual stress, distortion, and mechanical properties and microstructure can be achieved.

Chiumenti et al. [130] provided a detailed description of the formulation behind numerical simulations of WAAM processes while performing experimental validation. Their finite element model includes a heat transfer model governed by the laws of thermodynamics, Fourier's law, Stefan-Boltzmann law, considered heat dissipation by conduction and convection, phase transformation phenomena based on Scheil's equations, and the total amount of latent heat released/absorbed during phase changes. Additionally, a mechanical model which portrays a continuous damage model able to represent hot cracking singularities and porosities was included. The moving welding heat source was approximated by a double ellipsoidal power density distribution developed by Goldak et al. [131]. Montevecchi et al. [132] split the heat source into two power distribution contributions: one that characterizes the power delivered to the base material adapted from the double ellipsoid Goldak model and another constant of power distribution that should be developed to describe the energy transferred to the wire, since only 50% of the total energy was used to melt the wire [133].

These models are described as Equations (1) and (2), respectively:

$$\dot{q}_b = \frac{6\sqrt{3}\dot{Q}_b f_{f,r}}{\pi\sqrt{\pi} a_{f,r} bc} \exp\left[-3\left(\frac{x^2}{a_{f,r}^2} + \frac{y^2}{b^2} + \frac{x^2}{c^2}\right)\right] \quad (1)$$

and

$$\dot{q}_w = \frac{\dot{Q}_w}{V_{el}} \quad (2)$$

where: $\dot{Q}_b$ and $\dot{Q}_w$ are the analytic power value; coefficients a, b, and c are the semi-axis of ellipse dimensions, and V_{el} is the volume of heated elements.

Several finite element works have been conducted recently, such as evaluating the residual stress generated from clamping the substrate [134], or predicting the required idle time between layers to obtain a constant inter-layer temperature [29]. Zhao et al. [135] evaluated the stress distribution between single walls taking in consideration the deposition strategy: One wall was built always in the same direction, while the other had a deposition direction reverted in each layer. For both strategies, the stresses reduced with the height increase and it was experimentally concluded that layers in the same direction resulted in larger stresses than the part fabricated in the reverse direction. This suggests that deposition strategy is fundamental to mitigate residual stresses in WAAM parts.

Thermal models have been successfully developed to predict temperature gradients and distribution along WAAM walls, but there is a lack of models that can predict bead geometry. The difficulty comes with fully portraying heat transfer and fluid flow phenomena inside the molten pool. More recently, Bai et al. [136] succeeded in accurately predicting bead geometry by simulating heat transfer and fluid flow. Four different driving forces were considered: surface tension, Marangoni force, arc pressure, and arc shear stress. It was found out that metal liquid flow was upward inside and outward on the surface indicating that the Marangoni force dominates the fluid flow direction inside the molten pool. When depositing a high number of layers where side support is absent, the rear of the molten pool is driven by gravity and surface tension, which will determine the final bead geometry.

In another approach, the buoyancy force and frictional dissipation in the mushy zone were included [137]. The geometry of the deposit material was simulated by means of the surface energy, potential energy in the gravitational field, arc pressure, and droplet impact force. Results have shown that by mixing hot and cold liquid inside the molten pool (convection inside the molten pool) reliable predictions of bead cross section were made. Figure 28a,b depicts the experimental results and numerical estimation of bead geometry for a traveling speed of 5 mm/s and 8.3 mm/s, respectively.

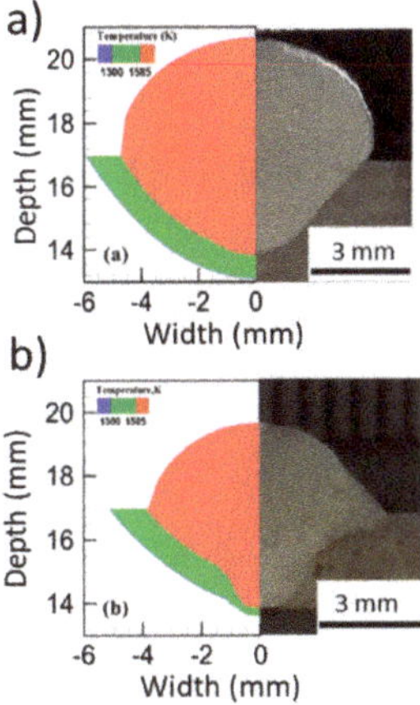

Figure 28. Comparison between predicted and experimental bead cross section for a travel speed of: (**a**) 5 mm/s and (**b**) 8.3 mm/s. (adapted from [137]).

8. Integrated Machine

Even though efforts have been made in optimizing tool path and process parameters, WAAM is sometimes referred as a near-net shape process due to its relatively poor surface finish. To answer the demands of short delivering times and complex parts, hybrid equipment which integrate additive and subtractive manufacturing technologies into one machine, arise as a promising solution to suppress WAAM's major weakness: waviness. The typical process waviness can aid crack propagation, making post machining a necessity, especially in structural applications. A schematic illustration of this hybrid solution is presented in Figure 29. The present model could be improved by the existence of online measuring equipment, which can be used to regulate parts dimensions and amend dimensions with milling [112].

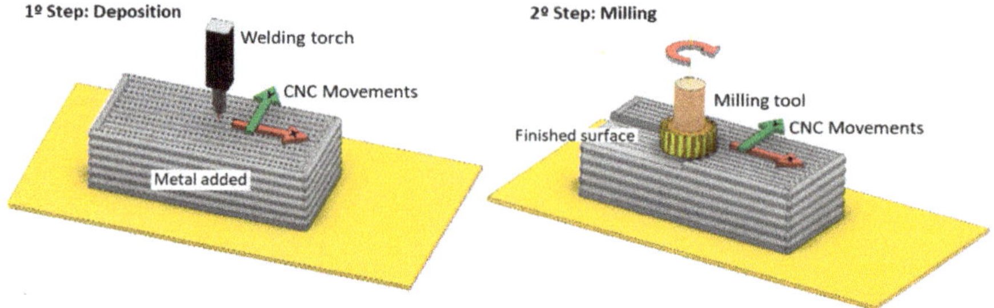

Figure 29. Steps in the hybrid process of metal additive manufacturing combining WAAM and milling (adapted from [138]).

Chen et al. [139] highlighted the difficulty of machining parts with hollow cavities, pointing out the importance of sequentially depositing material and machining it, to avoid potential problems with machining after the entire model is produced. A hybrid manufacturing system was created, aided by a software developed at IIT Bombay, named arc hybrid layer manufacturing (ArcHLM) capable of generating paths for a welding torch and a milling tool [140].

In a cost modeling and sensitivity study, the economic benefits of WAAM with subsequent machining were compared to conventional machining from a solid block. The developed model considered the prices of tools, materials, substrate, machine, software, electricity, and every consumable for each manufacturing process. Several components were analyzed such as: a wing spar, an external landing gear assembly in titanium, and a pylon mount. In all studied cases WAAM represented costs savings compared to conventional machining as shown in Table 4 [141].

Table 4. Costs of different parts per manufacturing process [141].

Designation	Process	BTF	Cost (€ × 1000)	Cost Reduction
Wing spar	Traditional	6.5	8.11	n/a
	WAAM	2.15	5.75	29%
External landing gear	Traditional	12	18.14	n/a
	WAAM	2.3	5.6	69%
Pylon mount	Traditional	5.1	2.8	n/a
	WAAM	1.5	2.68	7%

Mazak Corporation (Japan) unveiled a machine model VARIAXIS j-600AM featuring a standard WAAM head mounted on the machine headstock as well as an advanced 5-axis multi-surface subtractive capability to produce high-precision parts complete in single setups. Another advancement was made by Mutoh Industries that revealed the model Arc MA500-S1 which uses arc welding technologies. However, these machines are not versatile for research purposes, since no additional instrumentation or variation of the welding process is currently possible.

9. Defects and Non-Destructive Testing

As previously described, WAAM combined with a subtractive process is a viable solution to produce fully dense parts in an efficient way. The several process parameters of this process and complex material behavior leads to several challenges which can only be suppressed by a multi-disciplinary approach. Whilst there are no significant improvements on simulation software optimized for WAAM, integrating non-destructive testing and non-destructive evaluation sensors on equipment to evaluate parts as they are still being produced is of major importance.

9.1. Defects

WAAM process is, fundamentally, very similar to welding so defects such as, hot cracking, cold cracking, porosity, delamination, and spatter are well documented for different alloys [142–145]. Defects in WAAM can originate by poor path planning, excessive heat input, and consequently residual stresses, gas contamination, and feedstock quality. Additionally, surface finish was seen to directly affect hydrogen crack susceptibility [146].

Nevertheless, some predominant macro defects appear once layers start to be built in WAAM. These include side collapse, mainly caused by the excessive heat sink at the beginning of layers, in contrast to a low dissipation condition at the end of each layer (Figure 30a), that consequently results in unflatten top surfaces. Secondly, portions of unmelted wire can appear stuck to final parts, because of inconsistent wire stick-out on arc ignition. If the stick-out is too long the initial current will detach the wire without melting it (Figure 30b). Finally, large distortions upon unclamping as a result of the excessive heat input and consequently heat accumulation, result in a bending distortion of the component as depicted in Figure 31.

Figure 30. Image of WAAM macro defects: (**a**) side collapse; (**b**) unmelted wire (adapted from [147]).

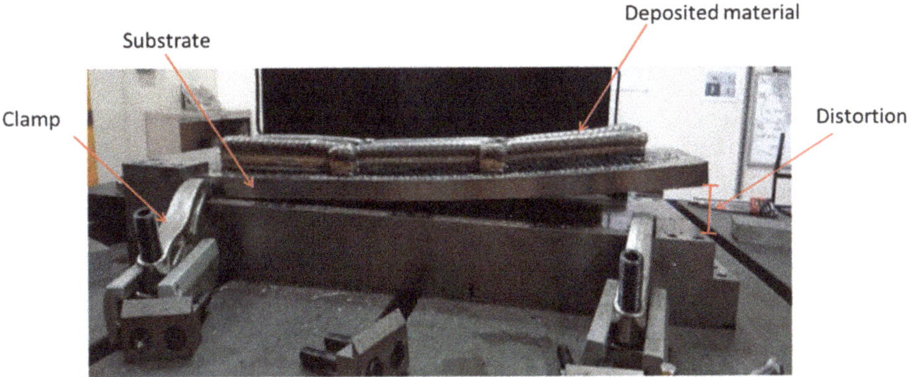

Figure 31. Distortion of a WAAM part (adapted from [148]).

The abovementioned defects can be mitigated with a correct choice of process parameters and process variants. Residual stresses that lead to distortions, and consequently to a loss of tolerance, can be relieved with post-processing heat treatments, right path planning, in situ cold/heating mechanisms, or even by the use of cold-work deformation-based techniques. Spatters are directly related with the selection of process parameters that will determine the transfer mode, which is easily overcome. Porosity is the most common defect for aluminum alloys and can be process-induced or due to poor wire batch quality. For Al-based alloys, rolling [23] and the welding mode CMT-pulse advanced [149] were seen to entirely remove pores. Side collapse and unmelted wire can be avoided by introducing sensors to assure a constant contact-tip-to-work distance, and a constant inter-layer temperature. Therefore, the following section describes the current status and difficulties on inspection and standardization of WAAM parts, covering in-situ monitoring and non-destructive testing as well as outline standardization and certification challenges.

9.2. In-Situ Monitoring

The main challenge of WAAM nowadays is developing appropriate inspection procedures. However, the increased capabilities to manufacture complex parts, decreases the ability for parts to be inspected. Therefore, using sensors to inspect parts as they are still being produced is of major importance. Everton et al. [150] reviewed in-situ monitoring and closed-loop control techniques for additive manufacturing, and some of them are yet to be exploited in WAAM.

It is well known that WAAM is very sensitive to variations of process parameters, so the implementation of an in-process monitoring device in a closed loop control system becomes important, since it would allow for rectification of parts as they are being still being built. Formation of internal defects and excessive residual stresses could be mitigated with such an approach. Traditionally, in fusion welding, monitoring is made through an evaluation of current, voltage, shielding gas flow rate, travel speed, and wire feed speed. In addition, during WAAM, sensors can be used to monitor the temperature at different regions [151], measure the size and geometry of beads [152,153], determine the weld pool characteristics, monitor the acoustic signal of deposition [154], detect electrical conductivity variations [155], and measure oxygen levels [156]. Normally, in fusion-based welding, the process parameters are held constant, but in WAAM, due to differences of thermal behavior throughout parts fabrication, geometric variations and mechanical properties are established and adjustments are necessary. Weld pool images contain abundant information that are directly related to parts quality. Therefore, the transient section of the weld pool was evaluated and monitored in [157], since defects such as cracks, delamination, and voids cause changes in heat flow and change the transient response. Thereby, on-line analysis of the arc spectrum was used to measure the weld pool characteristics by analyzing the abnormal bands according to the atomic emission spectrum, which contained necessary information about shielding gas flow rate and impurities that can influence fluidity of the weld pool and final parts quality. Spectrum analysis in combination with a CCD camera was able to detect material composition and confirm the presence of rust, oil stain, and shielding gas flow rate in the weld pool images [158]. An innovative single-neuron self-learning controller was integrated in WAAM equipment [159]. The developed controlling algorithm considered travel speed as the input control variable, and the layer width as the output to compensate layer height deviations. A non-linear Hammerstein model was used to establish a dynamic relationship between both parameters. The maximum deviations in layer width between the expected and detected layer width was 0.5 mm.

Xu et al. [148] reviewed process monitoring and control of WAAM parts and proposed a multi-sensor device to monitor each variant and output, as schematically shown in Figure 32. It considers an acoustic sensor for measuring arc pulsation and intensity, since when an irregularity occurs it will be reflected on the acoustic signal and on the signal of the current [154]. Infrared camera, thermocouples, or pyrometers would be used to monitor the molten pool and thermal cycles. In addition, it also contemplated a double profilometer and the use of add-ons, for instance, inter-layer rolling. Moreover, electrical conductivity measurements can be taken during deposition to identify the different phases that are being developed during solidification. In order to established WAAM consistency, a prior knowledge of the parts and build process is necessary to create an algorithm that will facilitate the industrial uptake of this process.

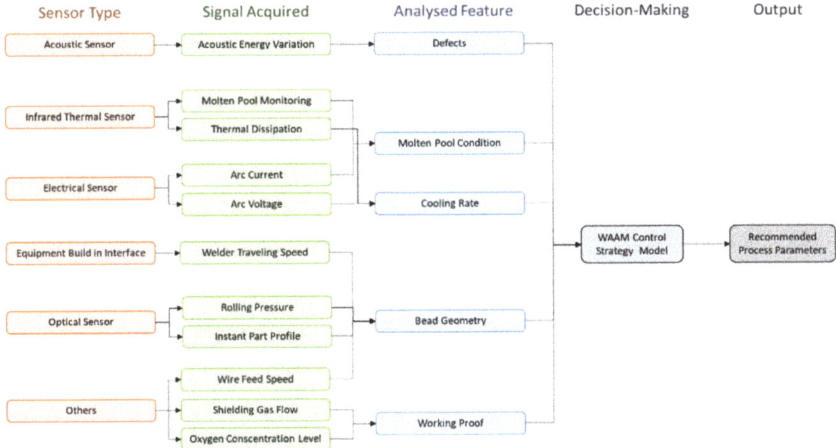

Figure 32. Schematic diagram of a full monitoring system for WAAM (adapted from [148]).

9.3. Non-Destructive Testing

Non-destructive testing (NDT) comprises a set of a high valuable technique to assess parts integrity, and its development is necessary for the rapid industrialization of WAAM. The are several techniques (i.e., radiographic testing, tomography, ultrasonic testing, eddy currents, thermography and liquid penetrants) however, some of them have limited potential to be used in WAAM, mostly due to parts waviness as described by Lopez et al. [155]. These limitations include, for example, the need to inspect high temperature surfaces, which requires that the lift-off distance between the probe and the material significantly increases. The authors experimentally tested radiography and ultrasonic testing (UT) to detect surface defects in Al WAAM parts. Conventional ultrasonic testing is limited to off-line inspection, since couplants are susceptible to high temperatures, and require machined surfaces to be applied. Nevertheless, a novel solution was to place the probe on the bottom of the substrate, yet the signal corresponding to the interface of substrate and the part must be taken into consideration. Radiography could detect lack of fusions between layers and porosities, but to fully locate them it is necessary to analyze other sections.

An ultrasonic array post-processing technique named total focusing method was seen to detect artificial holes with 3 mm, regardless of their location in part. Using conventional imaging techniques, defects near the side walls and top surface could not be detectable, and provided a non-realistic size and geometry of defects [160].

Currently, WAAM inline inspection conditions are extremely demanding for current state of the art NDT probes. The major issues include: high temperature of the last deposited layer; difficult to introduce a test medium to give a high signal-to-noise ratio; surface waviness and roughness; and small defects dimension (<1 mm). As such, the development of multiparametric non-destructive testing systems is mandatory to identify defects formation production and increase reliability in defect detection.

10. Applications

WAAM particularities makes it suitable for the creation of large-sized parts with medium complexity components, made with high-value materials. Therefore, this technology is viable to be used in industries, such as aerospace, automotive, defense, molds and dies, naval, and nuclear energy [70,161,162]. Topologically optimized structures have been increasingly being used by the aerospace and automotive industries as they reduce weight while maintaining the functionality of the part maximizing its performance. WAAM offers the potential to produce topologically optimized components, since through conventional technologies they become very expensive, with high material waste, and extensive lead times [163].

Currently, the aerospace industry focuses on the manufacture of complex-shaped components of titanium and nickel alloys, making WAAM a cost-effective method for their production, due to the difficulties associated with subtractive methods applied in these materials. Titanium represents 93% of the Lockheed SR-71 Blackbird's structural weight and during its production approximately 90% of the forging weight had to be removed by machining [164]. Norsk Titanium delivered the first titanium additively manufactured component made via WAAM approved by the federal aviation administration, as it was installed on the Boeing 787 Dreamliner. Norsk Titanium's technology allowed waste reduction, less energy consumption, and product costs reduction by up to 30% and 75% time saving, than forging with subsequent machining [165].

Ni-based alloys and stainless steels are widely used in the nuclear industry, where parts with high heat and corrosion resistance are required. WAAM is an appropriate candidate to replace some less requested portions of nickel parts to stainless steel, allowing the reduction of cost and weight of these components [99].

MX3D radically transformed the industry of construction by delivering the first additively manufactured metal bridge with a total weight of 4500 kg, 12.5 m in length, and 6.3 m wide (Figure 33a). Hirtler et al. [166] exploited one of WAAM's possible applications, that is, manufacturing from a previous semi-finished component fabricated by a conventional metal forming process. A rib was first forged and then finished by increasing its height with WAAM (Figure 33b). Magnesium alloys have been used in WAAM [167], but its application for biomedical application still does not to meet the requirements in terms of corrosion-resistance and biocompatibility. Once these are ensured, those components can be used for human vertebra prototypes, hip stem implants, and treat bone fractures [168]. Another example included an excavator arm, as depicted in Figure 33c.

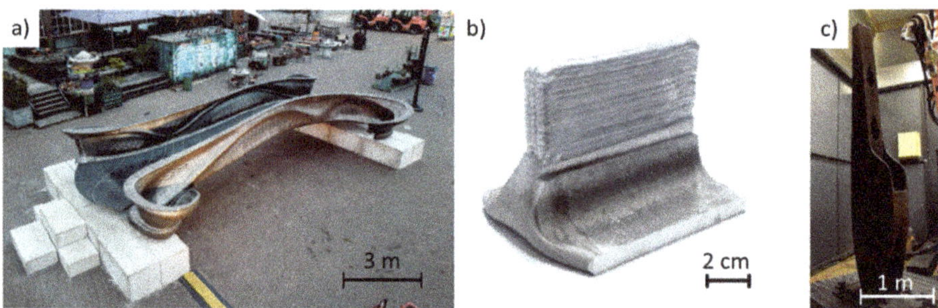

Figure 33. Various components made with WAAM: (**a**) MX3D bridge [168], (**b**) rib [166], and (**c**) excavator arm [169].

11. Summary and Future Outlook

The fabrication of complex shaped parts via WAAM is becoming well-established in both academia and the industry. Several process variants have been developed recently in order to optimize the microstructure and mechanical properties of the as-built parts. Additionally, most of the more relevant engineering alloys (titanium, aluminum, and steels, including stainless) are already used in WAAM with excellent results, proving the viability of this technique to produce custom-made large metallic parts. One aspect that is not yet well explored concerns the possibility of using WAAM for repair applications. This could greatly decrease costs associated with the need to completely renew a given structural part, since with WAAM technology it is possible to perform localized repairs.

Another key aspect that is not yet established is related to certification of WAAM parts. This step is crucial to further expand the range of applications of this technology and open the door for more demanding structural applications, where the advantages associated with WAAM can be of special interest. Concurrent to the need for certification procedures for WAAM parts, is the need to develop effective and integrated non-destructive testing systems capable of detecting defect formation during parts production. The need for the development of these in-situ monitoring methods lies in the fact that with such an approach any generated defect can be repaired right after its formation and not only at the end of production. Inspection only at the end of the parts production may lead to significant material waste and higher production times, since the location where the defect exists may be difficult to access, thus preventing their overhaul.

Some applications using WAAM-produced parts are already in the market and it is expected that industry plays a critical role in expanding the applications of WAAM parts. Combined with the development of certification procedures and effective non-destructive inline methods, it can be expected that WAAM will become one of the most used additive manufacturing technologies in the near future.

Funding: The authors acknowledge Fundação para a Ciência e a Tecnologia (FCT-MCTES) for its financial support via the project UID/EMS/00667/2019. VD acknowledges FCT-MCTES for funding the PhD grant SFRH/BD/139454/2018.

Conflicts of Interest: The authors declare no conflict of interest.

References

1. ASTM F2792-12a. *Standard Terminology for Additive Manufacturing Technologies, (Withdrawn 2015)*; ASTM International: West Conshohocken, PA, USA, 2012.
2. Kah, P.; Latifi, H.; Suoranta, R.; Martikainen, J.; Pirinen, M. Usability of arc types in industrial welding. *Int. J. Mech. Mater. Eng.* **2014**, *9*, 1–12. [CrossRef]
3. Martina, F.; Mehnen, J.; Williams, S.W.; Colegrove, P.; Wang, F. Investigation of the benefits of plasma deposition for the additive layer manufacture of Ti–6Al–4V. *J. Mater. Process. Technol.* **2012**, *212*, 1377–1386. [CrossRef]
4. Oliveira, J.P.; Crispim, B.; Zeng, Z.; Omori, T.; Fernandes, F.M.B.; Miranda, R.M. Microstructure and mechanical properties of gas tungsten arc welded Cu-Al-Mn shape memory alloy rods. *J. Mater. Process. Technol.* **2019**, *271*, 93–100. [CrossRef]
5. Ding, D.; Pan, Z.; Cuiuri, D.; Li, H. Wire-feed additive manufacturing of metal components: Technologies, developments and future interests. *Int. J. Adv. Manuf. Technol.* **2015**, *81*, 465–481. [CrossRef]
6. McAndrew, A.R.; Rosales, M.A.; Colegrove, P.A.; Hönnige, J.R.; Ho, A.; Fayolle, R.; Eyitayo, K.; Stan, I.; Sukrongpang, P.; Crochemore, A.; et al. Interpass rolling of Ti-6Al-4V wire + arc additively manufactured features for microstructural refinement. *Addit. Manuf.* **2018**, *21*, 340–349. [CrossRef]
7. Williams, S.W.; Martina, F.; Addison, A.C.; Ding, J.; Pardal, G.; Colegrove, P. Wire + Arc Additive Manufacturing. *Mater. Sci. Technol.* **2016**, *32*, 641–647. [CrossRef]
8. Ding, D.; Pan, Z.; van Duin, S.; Li, H.; Shen, C. Fabricating superior NiAl bronze components through wire arc additive manufacturing. *Materials* **2016**, *9*, 652. [CrossRef] [PubMed]
9. Martina, F.; Ding, J.; Williams, S.; Caballero, A.; Pardal, G.; Quintino, L. Tandem Metal Inert Gas process for high productivity Wire Arc Additive Manufacturing in stainless steel. *Addit. Manuf.* **2018**, *25*, 545–550. [CrossRef]
10. Oliveira, J.P.; Barbosa, D.; Fernandes, F.M.B.; Miranda, R.M. Tungsten inert gas (TIG) welding of Ni-rich NiTi plates: Functional behavior. *Smart Mater. Struct.* **2016**, *25*, 03LT01. [CrossRef]
11. Oliveira, J.P.; Fernandes, F.M.B.; Miranda, R.M.; Schell, N.; Ocaña, J.L. Effect of laser welding parameters on the austenite and martensite phase fractions of NiTi. *Mater. Charact.* **2016**, *119*, 148–151. [CrossRef]
12. Wang, F.; Williams, S.; Rush, M. Morphology investigation on direct current pulsed gas tungsten arc welded additive layer manufactured Ti6Al4V alloy. *Int. J. Adv. Manuf. Technol.* **2011**, *57*, 597–603. [CrossRef]
13. Jhavar, S.; Jain, N.K.; Paul, C.P. Development of micro-plasma transferred arc (μ-PTA) wire deposition process for additive layer manufacturing applications. *J. Mater. Process. Technol.* **2014**, *214*, 1102–1110. [CrossRef]

14. Gu, J.; Ding, J.; Williams, S.W.; Gu, H.; Ma, P.; Zhai, Y. The effect of inter-layer cold working and post-deposition heat treatment on porosity in additively manufactured aluminum alloys. *J. Mater. Process. Technol.* **2016**, *230*, 26–34. [CrossRef]
15. Cong, B.; Ding, J.; Williams, S. Effect of arc mode in cold metal transfer process on porosity of additively manufactured Al-6.3%Cu alloy. *Int. J. Adv. Manuf. Technol.* **2014**, *76*, 1593–1606. [CrossRef]
16. Almeida, P.M.S.; Williams, S. Innovative process model of Ti-6AL-4V Additive Layer Manufacturinf using Cold Metal Transfer (CMT). In Proceedings of the 21st International Solid Freeform Fabrication Symposium, Austin, TX, USA, 1–9 August 2010.
17. Shi, J.; Li, F. Effect of in-process active cooling on forming quality and efficiency of tandem GMAW–based additive manufacturing. *J. Adv. Manuf. Technol.* **2018**, 1–8. [CrossRef]
18. Martina, F.; Colegrove, P.A.; Williams, S.W.; Meyer, J. Microstructure of Interpass Rolled Wire + Arc Additive Manufacturing Ti-6Al-4V Components. *Metall. Mater. Trans. A Phys. Metall. Mater. Sci.* **2015**, *46*, 6103–6118. [CrossRef]
19. Hönnige, J.R.; Colegrove, P.A.; Ganguly, S.; Eimer, E.; Kabra, S.; Williams, S. Control of residual stress and distortion in aluminium wire + arc additive manufacture with rolling. *Addit. Manuf.* **2018**, *22*, 775–783. [CrossRef]
20. Gu, J.; Wang, X.; Bai, J.; Ding, J.; Williams, S.; Zhai, Y.; Liu, K. Deformation microstructures and strengthening mechanisms for the wire+arc additively manufactured Al-Mg4.5Mn alloy with inter-layer rolling. *Mater. Sci. Eng. A* **2018**, *712*, 292–301. [CrossRef]
21. Hönnige, J.R.; Colegrove, P.A.; Ahmad, B.; Fitzpatrick, M.E.; Ganguly, S.; Lee, T.L.; Williams, S.W. Residual stress and texture control in Ti-6Al-4V wire + arc additively manufactured intersections by stress relief and rolling. *Mater. Des.* **2018**, *150*, 193–205. [CrossRef]
22. Martina, F.; Roy, M.J.; Szost, B.A.; Terzi, S.; Colegrove, P.A.; Williams, S.W.; Withers, P.J.; Meyer, J.; Hofmann, M. Residual stress of as-deposited and rolled wire+arc additive manufacturing Ti–6Al–4V components. *Mater. Sci. Technol.* **2016**, *32*, 1439–1448. [CrossRef]
23. Gu, J.; Ding, J.; Williams, S.W.; Gu, H.; Bai, J.; Zhai, Y.; Ma, P. The strengthening effect of inter-layer cold working and post-deposition heat treatment on the additively manufactured Al–6.3Cu alloy. *Mater. Sci. Eng. A* **2016**, *651*, 18–26. [CrossRef]
24. Colegrove, P.A.; Donoghue, J.; Martina, F.; Gu, J.; Prangnell, P.; Hönnige, J. Application of bulk deformation methods for microstructural and material property improvement and residual stress and distortion control in additively manufactured components. *Scr. Mater.* **2017**, *135*, 111–118. [CrossRef]
25. Xie, Y.; Zhang, H.; Zhou, F. Improvement in Geometrical Accuracy and Mechanical Property for Arc-Based Additive Manufacturing Using Metamorphic Rolling Mechanism. *J. Manuf. Sci. Eng.* **2016**, *138*, 111002. [CrossRef]
26. Hönnige, J.R.; Colegrove, P.; Williams, S. Improvement of microstructure and mechanical properties in Wire + Arc Additively Manufactured Ti-6Al-4V with Machine Hammer Peening. *Procedia Eng.* **2017**, *216*, 8–17. [CrossRef]
27. Hönnige, J.R.; Colegrove, P.A.; Prangnell, P.; Ho, A.; Williams, S. The Effect of Thermal History on Microstructural Evolution, Cold-Work Refinement and α/\b{eta} Growth in Ti-6Al-4V Wire + Arc AM. *arXiv*, 2018; 1–25arXiv:1811.02903.
28. Sun, R.; Li, L.; Zhu, Y.; Guo, W.; Peng, P.; Cong, B.; Sun, J.; Che, Z.; Li, B.; Guo, C.; et al. Microstructure, residual stress and tensile properties control of wire-arc additive manufactured 2319 aluminum alloy with laser shock peening. *J. Alloy. Compd.* **2018**, *747*, 255–265. [CrossRef]
29. Montevecchi, F.; Venturini, G.; Grossi, N.; Scippa, A.; Campatelli, G. Idle time selection for wire-arc additive manufacturing: A finite element-based technique. *Addit. Manuf.* **2018**, *21*, 479–486. [CrossRef]
30. Geng, H.; Li, J.; Xiong, J.; Lin, X. Optimisation of interpass temperature and heat input for wire and arc additive manufacturing 5A06 aluminium alloy. *Sci. Technol. Weld. Join.* **2017**, *22*, 472–483. [CrossRef]
31. Wu, B.; Pan, Z.; Ding, D.; Cuiuri, D.; Li, H.; Fei, Z. The effects of forced interpass cooling on the material properties of wire arc additively manufactured Ti6Al4V alloy. *J. Mater. Process. Tech.* **2018**. [CrossRef]
32. Li, F.; Chen, S.; Shi, J.; Zhao, Y.; Tian, H.; Li, F.; Chen, S.; Shi, J.; Zhao, Y.; Tian, H. Thermoelectric Cooling-Aided Bead Geometry Regulation in Wire and Arc-Based Additive Manufacturing of Thin-Walled Structures. *Appl. Sci.* **2018**, *8*, 207. [CrossRef]

33. Bai, X.; Zhang, H.; Wang, G. Modeling of the moving induction heating used as secondary heat source in weld-based additive manufacturing. *Int. J. Adv. Manuf. Technol.* **2015**, *77*, 717–727. [CrossRef]
34. Li, Z.; Liu, C.; Xu, T.; Ji, L.; Wang, D.; Lu, J.; Ma, S. Reducing arc heat input and obtaining equiaxed grains by hot-wire method during arc additive manufacturing titanium alloy. *Mater. Sci. Eng. A* **2018**, *742*, 287–294. [CrossRef]
35. Alberti, E.A.; Bueno, B.M.P.; D'Oliveira, A.S.C.M. Additive manufacturing using plasma transferred arc. *Int. J. Adv. Manuf. Technol.* **2016**, *83*, 1861–1871. [CrossRef]
36. Xiong, J.; Lei, Y.; Li, R. Finite element analysis and experimental validation of thermal behavior for thin-walled parts in GMAW-based additive manufacturing with various substrate preheating temperatures. *Appl. Therm. Eng.* **2017**, *126*, 43–52. [CrossRef]
37. Lockett, H.; Ding, J.; Williams, S.; Martina, F.; Lockett, H. Design for Wire + Arc Additive Manufacture : Design rules and build orientation selection Design for Wire + Arc Additive Manufacture: Design rules and build orientation selection. *J. Eng. Des.* **2017**, *48*. [CrossRef]
38. Haselhuhn, A.S.; Wijnen, B.; Anzalone, G.C.; Sanders, P.G.; Pearce, J.M. In situ formation of substrate release mechanisms for gas metal arc weld metal 3-D printing. *J. Mater. Process. Technol.* **2015**, *226*, 50–59. [CrossRef]
39. Ding, J.; Colegrove, P.; Martina, F.; Williams, S.; Wiktorowicz, R.; Palt, M.R. Development of a laminar flow local shielding device for wire + arc additive manufacture. *J. Mater. Process. Technol.* **2015**, *226*, 99–105. [CrossRef]
40. Xu, X.; Ding, J.; Ganguly, S.; Diao, C.; Williams, S. Oxide accumulation effects on wire + arc layer-by-layer additive manufacture process. *J. Mater. Process. Technol.* **2018**, *252*, 739–750. [CrossRef]
41. Xu, X.; Ganguly, S.; Ding, J.; Dirisu, P.; Martina, F.; Liu, X.; Williams, S.W. Improving mechanical properties of wire plus arc additively manufactured maraging steel through plastic deformation enhanced aging response. *Mater. Sci. Eng. A* **2019**, *747*, 111–118. [CrossRef]
42. ASM International. Weld Solidification. *Weld Integr. Perform.* **1997**, *6*, 19.
43. Easton, M.A.; Qian, M.; Prasad, A.; StJohn, D.H. Recent advances in grain refinement of light metals and alloys, Curr. Opin. *Solid State Mater. Sci.* **2016**, *20*, 13–24. [CrossRef]
44. Oliveira, J.P.; Panton, B.; Zeng, Z.; Andrei, C.M.; Zhou, Y.; Miranda, R.M.; Fernandes, F.M.B. Laser joining of NiTi to Ti6Al4V using a Niobium interlayer. *Acta Mater.* **2016**, *105*, 9–15. [CrossRef]
45. Lütjering, G.; Williams, J.C. *Titanium*, 2nd ed.; Springer: Berlin/Heidelberg, Germany, 2007.
46. Carroll, B.E.; Palmer, T.A.; Beese, A.M. Anisotropic tensile behavior of Ti-6Al-4V components fabricated with directed energy deposition additive manufacturing. *Acta Mater.* **2015**, *87*, 309–320. [CrossRef]
47. Wang, F.; Williams, S.; Colegrove, P.; Antonysamy, A.A. Microstructure and Mechanical Properties of Wire and Arc Additive Manufactured Ti-6Al-4V. *Metall. Mater. Trans. A* **2013**, *44*, 968–977. [CrossRef]
48. DebRoy, T.; Wei, H.L.; Zuback, J.S.; Mukherjee, T.; Elmer, J.W.; Milewski, J.O.; Beese, A.M.; Wilson-Heid, A.; De, A.; Zhang, W. Additive manufacturing of metallic components—Process, structure and properties. *Prog. Mater. Sci.* **2018**, *92*, 112–224. [CrossRef]
49. Bermingham, M.J.; Mcdonald, S.D.; Dargusch, M.S.; Stjohn, D.H. Grain-refinement mechanisms in titanium alloys. *J. Mater. Res.* **2008**, *23*, 97–104. [CrossRef]
50. Kobryn, P.A.; Semiatin, S.L. Microstructure and texture evolution during solidification processing of Ti–6Al–4V. *J. Mater. Process. Technol.* **2003**, *135*, 330–339. [CrossRef]
51. Donoghue, J.; Antonysamy, A.A.; Martina, F.; Colegrove, P.A.; Williams, S.W.; Prangnell, P.B. The effectiveness of combining rolling deformation with Wire–Arc Additive Manufacture on β-grain refinement and texture modification in Ti–6Al–4V. *Mater. Charact.* **2016**, *114*, 103–114. [CrossRef]
52. Donoghue, J.M. *Hybrid Additive Manufacture and Deformation Processing for Large Scale Near-Net Shape Manufacture of Titanium Aerospace Components*; University of Manchester: Manchester, UK, 2016.
53. Bermingham, M.J.; Kent, D.; Zhan, H.; Stjohn, D.H.; Dargusch, M.S. Controlling the microstructure and properties of wire arc additive manufactured Ti-6Al-4V with trace boron additions. *Acta Mater.* **2015**, *91*, 289–303. [CrossRef]
54. Mereddy, S.; Bermingham, M.J.; Kent, D.; Dehghan-Manshadi, A.; StJohn, D.H.; Dargusch, M.S. Trace Carbon Addition to Refine Microstructure and Enhance Properties of Additive-Manufactured Ti-6Al-4V. *JOM* **2018**, *70*, 1670–1676. [CrossRef]
55. Mereddy, S.; Bermingham, M.J.; StJohn, D.H.; Dargusch, M.S. Grain refinement of wire arc additively manufactured titanium by the addition of silicon. *J. Alloy. Compd.* **2017**, *695*, 2097–2103. [CrossRef]

56. Miranda, R.M.; Assunção, E.; Silva, R.J.C.; Oliveira, J.P.; Quintino, L. Fiber laser welding of NiTi to Ti-6Al-4V. *Int. J. Adv. Manuf. Technol.* **2015**, *81*, 1533–1538. [CrossRef]
57. Karagadde, S.; Lee, P.D.; Cai, B.; Fife, J.L.; Azeem, M.A.; Kareh, K.M.; Puncreobutr, C.; Tsivoulas, D.; Connolley, T.; Atwood, R.C. Transgranular liquation cracking of grains in the semi-solid state. *Nat. Commun.* **2015**, *6*, 8300. [CrossRef]
58. Unfried-Silgado, J.; Ramirez, A.J. Modeling and characterization of as-welded microstructure of solid solution strengthened Ni-Cr-Fe alloys resistant to ductility-dip cracking Part II: Microstructure characterization. *Met. Mater. Int.* **2014**, *20*, 307–315. [CrossRef]
59. Carter, L.N.; Attallah, M.M.; Reed, R.C. Laser Powder Bed Fabrication of Nickel-Base Superalloys: Influence of Parameters ; Characterisation, Quantification and Mitigation of Cracking. *Superalloys* **2012**, 577–586. [CrossRef]
60. Xing, X.; Di, X.; Wang, B. The effect of post-weld heat treatment temperature on the microstructure of Inconel 625 deposited metal. *J. Alloys Compd.* **2014**, *593*, 110–116. [CrossRef]
61. Wang, J.F.; Sun, Q.J.; Wang, H.; Liu, J.P.; Feng, J.C. Effect of location on microstructure and mechanical properties of additive layer manufactured Inconel 625 using gas tungsten arc welding. *Mater. Sci. Eng. A* **2016**, *676*, 395–405. [CrossRef]
62. Jurić, I.; Garašić, I.; Bušić, M.; Kožuh, Z. Influence of Shielding Gas Composition on Structure and Mechanical Properties of Wire and Arc Additive Manufactured Inconel 625. *JOM* **2018**, 71. [CrossRef]
63. Xu, X.; Ganguly, S.; Ding, J.; Seow, C.E.; Williams, S. Enhancing mechanical properties of wire + arc additively manufactured INCONEL 718 superalloy through in-process thermomechanical processing. *Mater. Des.* **2018**, *160*, 1042–1051. [CrossRef]
64. Xu, X.; Ding, J.; Ganguly, S.; Williams, S. Investigation of process factors affecting mechanical properties of INCONEL 718 superalloy in wire + arc additive manufacture process. *J. Mater. Process. Technol.* **2018**, *265*, 201–209. [CrossRef]
65. Molins, R.; Hochstetter, G.; Chassaigne, J.C.; Andrieu, E. Oxidation effects on the fatigue crack growth behaviour of alloy 718 at high temperature. *Acta Mater.* **1997**, *45*, 663–674. [CrossRef]
66. Xiao, H.; Li, S.; Han, X.; Mazumder, J.; Song, L. Laves phase control of Inconel 718 alloy using quasi-continuous-wave laser additive manufacturing. *Mater. Des.* **2017**, *122*, 330–339. [CrossRef]
67. Wu, B.; Pan, Z.; Ding, D.; Cuiuri, D.; Li, H.; Xu, J.; Norrish, J. A review of the wire arc additive manufacturing of metals: Properties, defects and quality improvement. *J. Manuf. Process.* **2018**, *35*, 127–139. [CrossRef]
68. Haden, C.V.; Zeng, G.; Iii, F.M.C.; Ruhl, C.; Krick, B.A.; Harlow, D.G. Wire and arc additive manufactured steel: Tensile and wear properties. *Addit. Manuf.* **2017**, *16*, 115–123. [CrossRef]
69. Yilmaz, O.; Ugla, A.A. Microstructure characterization of SS308LSi components manufactured by GTAW-based additive manufacturing : Shaped metal deposition using pulsed current arc. *Int. J. Adv. Manuf. Technol.* **2016**. [CrossRef]
70. Queguineur, A.; Rückert, G.; Cortial, F.; Hascoët, J.Y. Evaluation of wire arc additive manufacturing for large-sized components in naval applications. *Weld. World.* **2018**, *62*, 259–266. [CrossRef]
71. Wang, L.; Xue, J.; Wang, Q. Correlation between arc mode, microstructure, and mechanical properties during wire arc additive manufacturing of 316L stainless steel. *Mater. Sci. Eng. A* **2019**, *751*, 183–190. [CrossRef]
72. Ge, J.; Lin, J.; Chen, Y.; Lei, Y.; Fu, H. Characterization of wire arc additive manufacturing 2Cr13 part: Process stability, microstructural evolution, and tensile properties. *J. Alloy. Compd.* **2018**, *748*, 911–921. [CrossRef]
73. Xiaoyong, S.W.; Zhou, Q.; Wang, K.; Peng, Y.; Ding, J.; Kong, J. Study on microstructure and tensile properties of high nitrogen Cr-Mn steel processed by CMT wire and arc additive manufacturing. *Mater. Des.* **2019**, *166*, 107611. [CrossRef]
74. Wang, T.; Zhang, Y.; Wu, Z.; Shi, C. Microstructure and properties of die steel fabricated by WAAM using H13 wire. *Vacuum* **2018**, *149*, 185–189. [CrossRef]
75. Xu, X.; Ganguly, S.; Ding, J.; Guo, S.; Williams, S.; Martina, F. Microstructural evolution and mechanical properties of maraging steel produced by wire+arc additive manufacture process. *Mater. Charact.* **2017**, *143*, 152–162. [CrossRef]
76. Rodrigues, T.A.; Duarte, V.; Avila, J.A.; Santos, T.G.; Miranda, R.M.; Oliveira, J.P. Wire and arc additive manufacturing of HSLA steel: Effect of Thermal Cycles on Microstructure and Mechanical Properties. *Addit. Manuf.* **2019**, *27*, 440–450. [CrossRef]

77. Wang, H.; Jiang, W.; Ouyang, J.; Kovacevic, R. Rapid prototyping of 4043 Al-alloy parts by VP-GTAW. *J. Mater. Process. Technol.* **2004**, *148*, 93–102. [CrossRef]
78. Ding, Y.; Muñiz-Lerma, J.A.; Trask, M.; Chou, S.; Walker, A.; Brochu, M. Microstructure and mechanical property considerations in additive manufacturing of aluminum alloys. *MRS Bull.* **2016**, *41*, 745–751. [CrossRef]
79. Gu, J.L.; Ding, J.L.; Cong, B.Q.; Bai, J.; Gu, H.M.; Williams, S.W.; Zhai, Y.C. The Influence of Wire Properties on the Quality and Performance of Wire+Arc Additive Manufactured Aluminium Parts. *Adv. Mater. Res.* **2014**, *1081*, 210–214. [CrossRef]
80. Fixter, J.; Gu, J.; Ding, J.; Williams, S.W.; Prangnell, P.B. Preliminary Investigation into the Suitability of 2xxx Alloys for Wire-Arc Additive Manufacturing. *Mater. Sci. Forum.* **2016**, *877*, 611–616. [CrossRef]
81. Zhang, C.; Li, Y.; Gao, M.; Zeng, X. Wire arc additive manufacturing of Al-6Mg alloy using variable polarity cold metal transfer arc as power source. *Mater. Sci. Eng. A* **2018**, *711*, 415–423. [CrossRef]
82. Geng, H.; Li, J.; Xiong, J.; Lin, X.; Zhang, F. Optimization of wire feed for GTAW based additive manufacturing. *J. Mater. Process. Tech.* **2017**, *243*, 40–47. [CrossRef]
83. Geng, F.Z.H.; Li, J.; Xiong, J.; Lin, X.; Huang, D. Formation and improvement of surface waviness for additive manufacturing 5A06 aluminium alloy component with GTAW system. *Rapid Prototyp. J.* **2018**, *24*, 342–350. [CrossRef]
84. Ortega, A.G.; Galvan, L.C.; Deschaux-Beaume, F.; Mezrag, B.; Rouquette, S. Effect of process parameters on the quality of aluminium alloy Al5Si deposits in wire and arc additive manufacturing using a cold metal transfer process. *Sci. Technol. Weld. Join.* **2018**, *23*, 316–332. [CrossRef]
85. Horgar, A.; Fostervoll, H.; Nyhus, B.; Ren, X.; Eriksson, M.; Akselsen, O.M. Additive manufacturing using WAAM with AA5183 wire. *J. Mater. Process. Technol.* **2018**, *259*, 68–74. [CrossRef]
86. Wang, L.; Suo, Y.; Liang, Z.; Wang, D.; Wang, Q. Effect of titanium powder on microstructure and mechanical properties of wire + arc additively manufactured Al-Mg alloy. *Mater. Lett.* **2019**. [CrossRef]
87. Sales, A.; Ricketts, N.J. Effect of Scandium on Wire Arc Additive Manufacturing of 5 Series Aluminium Alloys. In *Light Metals 2019*; Springer: Berlin, Germany, 2019; pp. 1455–1461.
88. Guo, J.; Zhou, Y.; Liu, C.; Wu, Q.; Chen, X.; Lu, J. Wire arc additive manufacturing of AZ31 magnesium alloy: Grain refinement by adjusting pulse frequency. *Materials* **2016**, *9*, 823. [CrossRef]
89. *ASTM B91-12. Standard Specification for Magnesium-Alloy Forgings*; ASTM International: West Conshohocken, PA, USA, 2012.
90. Han, S.; Zielewski, M.; Holguin, D.M.; Parra, M.M.; Kim, N. Optimization of AZ91D Process and Corrosion Resistance Using Wire Arc Additive Manufacturing. *Appl. Sci.* **2018**, *8*, 1306. [CrossRef]
91. Tammas-Williams, S.; Todd, I. Design for additive manufacturing with site-specific properties in metals and alloys. *Scr. Mater.* **2017**, *135*, 105–110. [CrossRef]
92. Oliveira, J.P.; Cavaleiro, A.J.; Schell, N.; Stark, A.; Miranda, R.M.; Ocana, J.L.; Fernandes, F.M.B. Effects of laser processing on the transformation characteristics of NiTi: A contribute to additive manufacturing. *Scr. Mater.* **2018**, *152*, 122–126. [CrossRef]
93. Qi, Z.; Cong, B.; Qi, B.; Sun, H.; Zhao, G.; Ding, J. Microstructure and mechanical properties of double-wire + arc additively manufactured Al-Cu-Mg alloys. *J. Mater. Process. Tech.* **2018**, *255*, 347–353. [CrossRef]
94. Somashekara, M.A.; Suryakumar, S. Studies on Dissimilar Twin-Wire Weld-Deposition for Additive Manufacturing Applications. *Trans. Indian Inst. Met.* **2017**, *70*, 2123–2135. [CrossRef]
95. Suryakumar, S.; Somashekara, M.A. Manufacture of functionally gradient materials using weld-deposition. In Proceedings of the 24th International SFF Symposium—An Additive Manufacturing Conference, SFF 2013, Austin, TX, USA, 12–14 August 2013; pp. 939–949. [CrossRef]
96. Shen, C.; Pan, Z.; Cuiuri, D.; Roberts, J.; Li, H. Fabrication of Fe-FeAl Functionally Graded Material Using the Wire-Arc Additive Manufacturing Process, Metall. *Mater. Trans. B* **2016**, *47*, 763–772. [CrossRef]
97. Wei, P.S. The Physics of Weld Bead Defects. In *Welding Processes*; Kovacevic, R., Ed.; InTech: Rijeka, Czech Republic, 2012; ISBN 978-953-51-0854-2.
98. Bobbio, L.D.; Bocklund, B.; Otis, R.; Borgonia, J.P.; Dillon, R.P.; Shapiro, A.A.; McEnerney, B.; Liu, Z.-K.; Beese, A.M. Characterization of a functionally graded material of Ti-6Al-4V to 304L stainless steel with an intermediate V section. *J. Alloys Compd.* **2018**. [CrossRef]
99. Abe, T.; Sasahara, H. Dissimilar metal deposition with a stainless steel and nickel-based alloy using wire and arc-based additive manufacturing. *Precis. Eng.* **2016**, *45*, 387–395. [CrossRef]

100. Shen, C.; Pan, Z.; Ding, D.; Yuan, L.; Nie, N.; Wang, Y.; Luo, D.; Cuiuri, D.; van Duin, S.; Li, H. The in fl uence of post-production heat treatment on the multi-directional properties of nickel-aluminum bronze alloy fabricated using wire-arc additive manufacturing process. *Addit. Manuf.* **2018**, *23*, 411–421. [CrossRef]
101. Kazanas, P.; Deherkar, P.; Almeida, P.; Lockett, H.; Williams, S. Fabrication of geometrical features using wire and arc additive manufacture. *Proc. Inst. Mech. Eng. Part B J. Eng. Manuf.* **2012**, *226*, 1042–1051. [CrossRef]
102. Zhang, Y.M.; Chen, Y.; Li, P.; Male, A.T. Weld deposition-based rapid prototyping: A preliminary study. *J. Mater. Process. Technol.* **2003**, *135*, 347–357. [CrossRef]
103. Hu, Z.; Qin, X.; Shao, T.; Liu, H. Understanding and overcoming of abnormity at start and end of the weld bead in additive manufacturing with GMAW. *Int. J. Adv. Manuf. Technol.* **2018**, *95*, 2357–2368. [CrossRef]
104. Ding, D.; Pan, Z.; Cuiuri, D.; Li, H. A multi-bead overlapping model for robotic wire and arc additive manufacturing (WAAM). *Robot. Comput. Integr. Manuf.* **2015**, *31*, 101–110. [CrossRef]
105. Ding, D.; Pan, Z.; Cuiuri, D.; Li, H. A tool-path generation strategy for wire and arc additive manufacturing. *Int. J. Adv. Manuf. Technol.* **2014**, *73*, 173–183. [CrossRef]
106. Ding, D.; Shen, C.; Pan, Z.; Cuiuri, D.; Li, H.; Larkin, N.; van Duin, S. Towards an automated robotic arc-welding-based additive manufacturing system from CAD to finished part. *CAD Comput. Aided Des.* **2016**, *73*, 66–75. [CrossRef]
107. Blum, H. A transformation for extracting new descriptors of shape. *Model. Percept. Speech Vis. Form.* **1967**, *19*, 362–380.
108. Kao, J.-H. Process Planning for Additive/Subtractive Solid Freeform Fabrication Using Medial Axis Transform. Ph.D. Thesis, Stanford University Stanford, Stanford, CA, USA, June 1999; p. 173.
109. Ding, D.; Pan, Z.; Cuiuri, D.; Li, H. A practical path planning methodology for wire and arc additive manufacturing of thin-walled structures. *Robot. Comput. Integr. Manuf.* **2015**, *34*, 8–19. [CrossRef]
110. Venturini, G.; Montevecchi, F.; Scippa, A.; Campatelli, G. Optimization of WAAM Deposition Patterns for T-crossing Features. *Procedia CIRP.* **2016**, *55*, 95–100. [CrossRef]
111. Xu, X.; Ding, J.; Ganguly, S.; Diao, C.; Williams, S. Preliminary Investigation of Building Strategies of Maraging Steel Bulk Material Using Wire + Arc Additive Manufacture. *J. Mater. Eng. Perform.* **2018**, 5–11. [CrossRef]
112. Ma, G.; Zhao, G.; Li, Z.; Yang, M.; Xiao, W. Optimization strategies for robotic additive and subtractive manufacturing of large and high thin-walled aluminum structures. *Int. J. Adv. Manuf. Technol.* **2018**, 1–18. [CrossRef]
113. Nguyen, L.; Buhl, J.; Bambach, M. Decomposition Algorithm for Tool Path Planning for Wire-Arc Additive Manufacturing. *J. Mach. Eng.* **2018**, *18*, 96–107. [CrossRef]
114. Nilsiam, Y.; Sanders, P.; Pearce, J.M. Slicer and process improvements for open-source GMAW-based metal 3-D printing. *Addit. Manuf.* **2017**, *18*, 110–120. [CrossRef]
115. Oliveira, J.P.; Fernandes, F.M.B.; Miranda, R.M.; Schell, N.; Ocaña, J.L. Residual stress analysis in laser welded NiTi sheets using synchrotron X-ray diffraction. *Mater. Des.* **2016**, *100*, 180–187. [CrossRef]
116. Attallah, M.M.; Jennings, R.; Wang, X.; Carter, L.N. Additive manufacturing of Ni-based superalloys: The outstanding issues. *MRS Bull.* **2016**, *41*, 758–764. [CrossRef]
117. Withers, P.J.; Bhadeshia, H.K.D.H. Residual stress. Part 1–measurement techniques. *Mater. Sci. Technol.* **2001**, *17*, 355–365. [CrossRef]
118. Colegrove, P.A.; Coules, H.E.; Fairman, J.; Martina, F.; Kashoob, T.; Mamash, H.; Cozzolino, L.D. Microstructure and residual stress improvement in wire and arc additively manufactured parts through high-pressure rolling. *J. Mater. Process. Technol.* **2013**, *213*, 1782–1791. [CrossRef]
119. Mughal, M.P.; Fawad, H.; Mufti, R.A.; Siddique, M. Deformation modelling in layered manufacturing of metallic parts using gas metal arc welding: Effect of process parameters. *Model. Simul. Mater. Sci. Eng.* **2005**, *13*, 1187–1204. [CrossRef]
120. Denlinger, E.R.; Heigel, J.C.; Michaleris, P.; Palmer, T.A. Effect of inter-layer dwell time on distortion and residual stress in additive manufacturing of titanium and nickel alloys. *J. Mater. Process. Technol.* **2015**, *215*, 123–131. [CrossRef]
121. Yang, Y.; Jin, X.; Liu, C.; Xiao, M.; Lu, J.; Fan, H.; Ma, S. Residual Stress, Mechanical Properties, and Grain Morphology of Ti-6Al-4V Alloy Produced by Ultrasonic Impact Treatment Assisted Wire and Arc Additive Manufacturing. *Metals* **2018**, *8*, 934. [CrossRef]

122. Wu, B.; Ding, D.; Pan, Z.; Cuiuri, D.; Li, H.; Han, J.; Fei, Z. Effects of heat accumulation on the arc characteristics and metal transfer behavior in Wire Arc Additive Manufacturing of Ti6Al4V. *J. Mater. Process. Technol.* **2017**, *250*, 304–312. [CrossRef]
123. Wu, B.; Pan, Z.; Ding, D.; Cuiuri, D.; Li, H. Effects of heat accumulation on microstructure and mechanical properties of Ti6Al4V alloy deposited by wire arc additive manufacturing. *Addit. Manuf.* **2018**, *23*, 151–160. [CrossRef]
124. Bermingham, M.J.; Nicastro, L.; Kent, D.; Chen, Y.; Dargusch, M.S. Optimising the mechanical properties of Ti-6Al-4V components produced by wire + arc additive manufacturing with post-process heat treatments. *J. Alloys Compd.* **2018**, *753*, 247–255. [CrossRef]
125. Oliveira, J.P.; Zeng, Z.; Andrei, C.; Fernandes, F.M.B.; Miranda, R.M.; Ramirez, A.J.; Omori, T.; Zhou, N. Dissimilar laser welding of superelastic NiTi and CuAlMn shape memory alloys. *Mater. Des.* **2017**, *128*, 166–175. [CrossRef]
126. Zeng, Z.; Oliveira, J.P.; Yang, M.; Song, D.; Peng, B. Functional fatigue behavior of NiTi-Cu dissimilar laser welds. *Mater. Des.* **2017**, *114*, 282–287. [CrossRef]
127. Zeng, Z.; Panton, B.; Oliveira, J.P.; Han, A.; Zhou, Y.N. Dissimilar laser welding of NiTi shape memory alloy and copper. *Smart Mater. Struct.* **2015**, *24*, 125036. [CrossRef]
128. Shen, C.; Pan, Z.; Cuiuri, D.; van Duin, S.; Luo, D.; Dong, B.; Li, H. Influences of postproduction heat treatment on Fe3Al-based iron aluminide fabricated using the wire-arc additive manufacturing process. *J. Adv. Manuf. Technol.* **2018**, *97*, 335–344. [CrossRef]
129. Qi, Z.; Cong, B.; Qi, B.; Zhao, G.; Ding, J. Properties of wire + arc additively manufactured 2024 aluminum alloy with different solution treatment temperature. *Mater. Lett.* **2018**, *230*, 275–278. [CrossRef]
130. Chiumenti, M.; Cervera, M.; Salmi, A.; de Saracibar, C.A.; Dialami, N.; Matsui, K. Finite element modeling of multi-pass welding and shaped metal deposition processes. *Comput. Methods Appl. Mech. Eng.* **2010**, *199*, 2343–2359. [CrossRef]
131. Goldak, J.; Bibby, M.; Moore, J.; House, R.; Patel, B. Computer modeling of heat flow in welds. *Metall. Trans. B* **1986**, *17*, 587–600. [CrossRef]
132. Montevecchi, F.; Venturini, G.; Scippa, A.; Campatelli, G. Finite Element Modelling of Wire-arc-additive-manufacturing Process. *Procedia CIRP* **2016**, *55*, 109–114. [CrossRef]
133. Dupont, J.N.; Marder, A.R. Thermal Efficiency of Arc Welding Processes. *Weld. Res. Suppl.* **1995**, 406s–416s. [CrossRef]
134. Wang, X.; Wang, A. Finite element analysis of clamping form in wire and arc additive manufacturing, 2017 7th Int. Conf. Model. Simulation. *Appl. Optim. ICMSAO* **2017**, 1–5. [CrossRef]
135. Zhao, H.; Zhang, G.; Yin, Z.; Wu, L. Three-dimensional finite element analysis of thermal stress in single-pass multi-layer weld-based rapid prototyping. *J. Mater. Process. Technol.* **2012**, *212*, 276–285. [CrossRef]
136. Bai, X.; Colegrove, P.; Ding, J.; Zhou, X.; Diao, C.; Bridgeman, P.; Hönnige, J.; Zhang, H.; Williams, S. Numerical analysis of heat transfer and fluid flow in multilayer deposition of PAW-based wire and arc additive manufacturing. *Int. J. Heat Mass Transf.* **2018**, *124*, 504–516. [CrossRef]
137. Ou, W.; Mukherjee, T.; Knapp, G.L.; Wei, Y.; Debroy, T. Fusion zone geometries, cooling rates and solidification parameters during wire arc additive manufacturing. *Int. J. Heat Mass Transf.* **2018**, *127*, 1084–1094. [CrossRef]
138. Diéguez, J.L.; Santana, A.; Afonso, P.; Zanin, A.; Wernke, R. Preliminary development of a Wire and Arc Additive Manufacturing system. *Procedia Manuf.* **2017**, *13*, 895–902. [CrossRef]
139. Chen, Y.H.; Song, Y. Development of a layer based machining system. *CAD Comput. Aided Des.* **2001**, *33*, 331–342. [CrossRef]
140. Karunakaran, K.P.; Suryakumar, S.; Pushpa, V.; Akula, S. Low cost integration of additive and subtractive processes for hybrid layered manufacturing. *Robot. Comput. Integr. Manuf.* **2010**, *26*, 490–499. [CrossRef]
141. Filomeno, M.; Williams, S. Wire + Arc Additive Manufacturing vs. Traditional Machining from Solid: A Cost Comparison'. 2015. Available online: http://waammat.com/documents/waam-vs-machining-from-solid-a-cost-comparison (accessed on 25 February 2019).
142. Murti, V.S.R.; Srinivas, P.D.; Banadeki, G.H.D.; Raju, K.S. Effect of heat input on the metallurgical properties of HSLA steel in multi-pass MIG welding. *J. Mater. Process. Tech.* **1993**, *37*, 723–729. [CrossRef]
143. Amer Welding Society (AWS). *Welding Handbook—Metals and Their Weldability–Vol 4*, 7th ed.; Amer. Welding Society: Miami, FL, USA, 1982.
144. Zong, R.; Chen, J.; Wu, C.; Padhy, G.K. Influence of shielding gas on undercutting formation in gas metal arc welding. *J. Mater. Process. Technol.* **2016**, *234*, 169–176. [CrossRef]

145. Olson, D.L.; Siewert, T.A.; Liu, S. (Eds.) *ASM Handbook Volume 6: Welding, Brazing, and Soldering*; ASM International: Almere, The Netherlands, 1993. [CrossRef]
146. Ryan, E.M.; Sabin, T.J.; Watts, J.F.; Whiting, M.J. The in fl uence of build parameters and wire batch on porosity of wire and arc additive manufactured aluminium alloy 2319. *J. Mater. Process. Tech.* **2018**, *262*, 577–584. [CrossRef]
147. Id, L.J.; Lu, J.; Tang, S.; Wu, Q.; Wang, J.; Ma, S.; Fan, H.; Liu, C. Research on Mechanisms and Controlling Methods of Additive Manufacturing. *Materials* **2018**, *11*, 1104. [CrossRef]
148. Xu, F.; Dhokia, V.; Colegrove, P.; Mcandrew, A.; Henstridge, A.; Newman, S.T.; Xu, F.; Dhokia, V.; Colegrove, P.; Mcandrew, A. Realisation of a multi-sensor framework for process monitoring of the wire arc additive manufacturing in producing Ti-6Al-4V parts. *Int. J. Comput. Integr. Manuf.* **2018**, 1–14. [CrossRef]
149. Cong, B.; Ouyang, R.; Qi, B.; Ding, J. Influence of Cold Metal Transfer Process and Its Heat Input on Weld Bead Geometry and Porosity of Aluminum-Copper Alloy Welds. *Rare Met. Mater. Eng.* **2016**, *45*, 606–611. [CrossRef]
150. Everton, A.T.C.S.K.; Hirsch, M.; Stravroulakis, P.; Leach, R.K. Review of in situ process monitoring and in situ metrology for metal additive manufacturing. *Mater. Des.* **2016**, *95*, 431–445. [CrossRef]
151. Yang, D.; Wang, G.; Zhang, G. Thermal analysis for single-pass multi-layer GMAW based additive manufacturing using infrared thermography. *J. Mater. Process. Technol.* **2017**, *244*, 215–224. [CrossRef]
152. Xiong, J.; Zhang, G. Online measurement of bead geometry in GMAW-based additive manufacturing. *Meas. Sci. Technol.* **2013**, *24*. [CrossRef]
153. Font, T.; Diao, C.; Ding, J.; Williams, S.; Zhao, Y. A passive imaging system for geometry measurement for the plasma arc welding process. *IEEE Trans. Ind. Electron.* **2017**, *64*, 7201–7209. [CrossRef]
154. Polajnar, I.; Bergant, Z.; Grum, J. Arc Welding Process Monitoring By Audible Sound. *Appl. Contemp. Non-Destr. Test. Eng.* **2013**. [CrossRef]
155. Lopez, L.Q.A.; Bacelar, R.; Pires, I.; Santos, T.G.; Sousa, J.P. Non-destructive testing application of radiography and ultrasound for wire and arc additive manufacturing. *Addit. Manuf.* **2018**, *21*, 298–306. [CrossRef]
156. Artaza, T.; Alberdi, A.; Murua, M.; Gorrotxategi, J.; Frías, J.; Puertas, G.; Melchor, M.A.; Mugica, D.; Suárez, A. Design and integration of WAAM technology and in situ monitoring system in a gantry machine. *Procedia Manuf.* **2017**, *13*, 778–785. [CrossRef]
157. Zalameda, J.N.; Burke, E.R.; Hafley, R.A.; Taminger, K.M.; Domack, C.S.; Brewer, A.; Martin, R.E. Thermal imaging for assessment of electron-beam freeform fabrication (EBF^3) additive manufacturing deposits. *SPIE Digit. Libr.* **2013**, *8705*, 87050M. [CrossRef]
158. Zhao, Z.; Guo, Y.; Bai, L.; Wang, K.; Han, J. Quality monitoring in wire-arc additive manufacturing based on cooperative awareness of spectrum and vision. *Optik* **2019**, *181*, 351–360. [CrossRef]
159. Xiong, J.; Yin, Z.; Zhang, W. Closed-loop control of variable layer width for thin-walled parts in wire and arc additive manufacturing. *J. Mater. Process. Technol.* **2016**, *233*, 100–106. [CrossRef]
160. Norman, C.; Javadi, Y.; Macleod, C.N.; Pierce, S.G.; Gachagan, A. Ultrasonic phased array inspection of wire plus arc additive manufacture (WAAM) samples using conventional and total focusing method (TFM) imaging approaches. In Proceedings of the 57th British Institue of Non-destructive Testing Annual Conference, Nottingham, UK, 10–12 September 2018.
161. Brandl, E.; Baufeld, B.; Leyens, C.; Gault, R. Additive manufactured Ti-6A1-4V using welding wire: Comparison of laser and arc beam deposition and evaluation with respect to aerospace material specifications. *Phys. Procedia* **2010**, *5*, 595–606. [CrossRef]
162. Busachi, A.; Erkoyuncu, J.; Colegrove, P.; Martina, F.; Ding, J. Designing a WAAM based manufacturing system for defence applications. *Procedia CIRP* **2015**, *37*, 48–53. [CrossRef]
163. Thompson, M.K.; Moroni, G.; Vaneker, T.; Fadel, G.; Campbell, R.I.; Gibson, I.; Bernard, A.; Schulz, J.; Graf, P.; Ahuja, B.; et al. Design for Additive Manufacturing: Trends, opportunities, considerations, and constraints. *CIRP Ann.-Manuf. Technol.* **2016**, *65*, 737–760. [CrossRef]
164. Merlin, P. Design and Development of the Blackbird: Challenges and Lessons Learned. In Proceedings of the 47th AIAA Aerospace Sciences Meeting including The New Horizons Forum and Aerospace Exposition, Aerospace Sciences Meetings, Orlando, FL, USA, 5–8 January 2009; pp. 1–38. [CrossRef]
165. Scott, A. Printed Titanium Parts Expected to Save Millions in Boeing Dreamliner Costs. Available online: https://www.reuters.com/article/us-norsk-boeing/printed-titanium-parts-expected-to-save-millions-in-boeing-dreamliner-costs-idUSKBN17C264 (accessed on 26 February 2019).

166. Hirtler, M.; Jedynak, A.; Sydow, B.; Sviridov, A.; Bambach, M. Investigation of microstructure and hardness of a rib geometry produced by metal forming and wire-arc additive manufacturing. In Proceedings of the 5th International Conference on New Forming Technology (ICNFT 2018), Bremen, Germany, 18–21 September 2018; pp. 1–6.
167. Takagi, H.; Sasahara, H.; Abe, T.; Sannomiya, H.; Nishiyama, S.; Ohta, S.; Nakamura, K. Material-property evaluation of magnesium alloys fabricated using wire-and-arc-based additive manufacturing. *Addit. Manuf.* **2018**, *24*, 498–507. [CrossRef]
168. Buchanan, C.; Gardner, L. Metal 3D printing in construction: A review of methods, research, applications, opportunities and challenges. *Eng. Struct.* **2019**, *180*, 332–348. [CrossRef]
169. Greer, A.C.; Nycz, A.; Noakes, M.; Post, B.; Kurfess, T.; Love, L.; Greer, C.; Nycz, A.; Noakes, M.; Richardson, B.; et al. Introduction to the Design Rules for Metal Big Area Additive Manufacturing. *Addit. Manuf.* **2019**, *27*, 159–166. [CrossRef]

Article

Temperature and Microstructure Evolution in Gas Tungsten Arc Welding Wire Feed Additive Manufacturing of Ti-6Al-4V

Corinne Charles Murgau [1], Andreas Lundbäck [2,*], Pia Åkerfeldt [3] and Robert Pederson [1,4]

1 Department of Engineering Science, University West, 46129 Trollhättan, Sweden; corinne.charlesmurgau@gmail.com (C.C.M.); robert.pederson@hv.se (R.P.)
2 Division of Mechanics of Solid Materials, Luleå University of Technology, 97181 Luleå, Sweden
3 Division of Materials Science, Luleå University of Technology, 97181 Luleå, Sweden; pia.akerfeldt@ltu.se
4 GKN Aerospace Engine Systems, 46181 Trollhättan, Sweden
* Correspondence: andreas.lundback@ltu.se; Tel.: +46-920-492-856

Received: 4 October 2019; Accepted: 24 October 2019; Published: 28 October 2019

Abstract: In the present study, the gas tungsten arc welding wire feed additive manufacturing process is simulated and its final microstructure predicted by microstructural modelling, which is validated by microstructural characterization. The Finite Element Method is used to solve the temperature field and microstructural evolution during a gas tungsten arc welding wire feed additive manufacturing process. The microstructure of titanium alloy Ti-6Al-4V is computed based on the temperature evolution in a density-based approach and coupled to a model that predicts the thickness of the α lath morphology. The work presented herein includes the first coupling of the process simulation and microstructural modelling, which have been studied separately in previous work by the authors. In addition, the results from simulations are presented and validated with qualitative and quantitative microstructural analyses. The coupling of the process simulation and microstructural modeling indicate promising results, since the microstructural analysis shows good agreement with the predicted alpha lath size.

Keywords: additive manufacturing; titanium; Ti-6Al-4V; microstructural modeling; metal deposition; finite element method

1. Introduction

The Gas Tungsten Arc Welding (GTAW) wire feed additive manufacturing (AM) process involves the deposition of metal material using a tungsten arc as an energy source. GTAW belongs to the group of Direct Energy Deposit (DED) methods in the family of AM processes. A solid metal wire is fed through a conventional wire feeder and deposited layer-by-layer onto a substrate. The energy source is electrical heating, which melts the metal wire as well as the base material. The addition of multiple layers produces a dense near-net-shape part. GTAW wire feed additive manufacturing has a high deposition rate [1]. Flexibility and cost savings in manufacturing are important driving forces in the development of such techniques [2].

The titanium alloy, Ti-6Al-4V, is extensively used in aerospace applications because of its high strength to weight ratio [3]. With an appropriate protective environment to minimize the oxygen content, this alloy shows good weldability and has also been shown to be suitable for additive manufacturing.

In additive manufacturing, multiple layers are melted on top of each other in a layer-by-layer manner, inducing a number of heating-up/cooling-down cycles in the manufactured material. Since Ti-6Al-4V is a two-phase alloy consisting of 90–95% of hcp-α phase and 5–10% bcc-β phase at room temperature, the final microstructure of the material after being additive manufactured can exhibit a variety of microstructures, depending on the conditions of the thermal cycles. For the industry to fully adopt additive manufacturing, and to be able to qualify titanium parts for aerospace applications, a complete understanding of the mechanical behavior and control of the resulting properties are prerequisites.

Process simulations provide information on how to manufacture components in order to achieve the required properties and support the development and understanding of the manufacturing process itself. Finite Element (FE) simulations, a conventional method used in the modeling of welding processes, particularly for larger components, are applied on the macro scale [4–7]. In the present work, a validated FE model presented by Lundbäck and Lindgren [4] is used to predict the thermal field that, in turn, determines the microstructural evolution. A number of different strategies exist regarding how to include detailed and explicit modeling on a microscopic scale in a macroscopic simulation. One such strategy is to use submeshes located at the nodes of the FE model [8], while another is to use dual-mesh methods that place microstructural domains at the nodes of a macroscopic FE calculation [9,10]. However, with these strategies, the calculations become very cumbersome if a large component is to be simulated. Therefore, for industrial needs, it is more pragmatic to use a density type of model, also called the internal state variable approach by Grong and Shercliff [11]. In this work, a density-type approach is used in order to model the microstructure on a larger scale. Moreover, it facilitates the future combination of the microstructural model with a mechanical model to compute material properties.

The aim of the present study has been to, for the first time, couple the process simulation model, developed by the authors [4], with a microstructural model, which has also been developed by the authors [12], to predict the alpha lath size in the GTAW wire feed AM process. The results were validated by building a well-defined, 10-layer-high weld sequence, in which, at specific positions, its microstructural features have been characterized and compared to the predicted results.

2. Process Description and Microstructural Characterization

2.1. Experimental Setup

The Ti-6Al-4V was wire-feed deposited on a 3.25 mm-thick plate using a tungsten arc weld heat source. Four weld sequences with a height of 10 layers were continuously added onto the plate, as presented in Figure 1. Each layer is approximately 0.7 mm in height. Continuous addition means that no waiting time between each deposited layer was applied, except for the short time during welding torch movement to the new starting position. In order to avoid oxidation and alpha case formation during the building process, the oxygen level was kept below 10 ppm in the building chamber. This was achieved by applying an over-pressurized argon gas flowing through the chamber. The weld passes are numbered 1 to 4; Figure 1 indicates each starting point. The building sequence is such that walls 1 through 4 are deposited sequentially for the current layer and the structure is built in a layer-by-layer fashion. In total, ten layers were deposited for each wall.

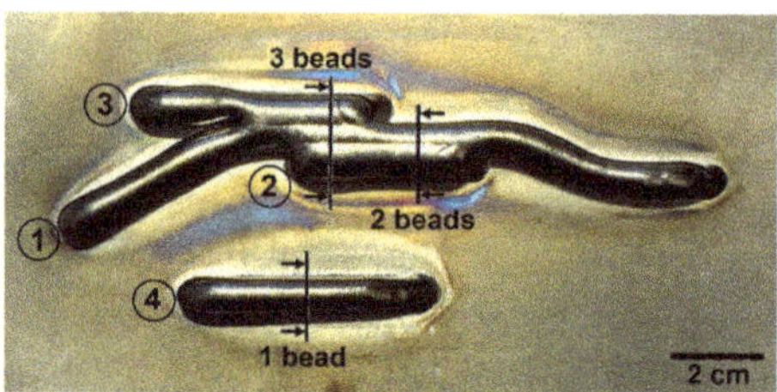

Figure 1. The built feature consisting of the four sequences of ten layers in height, including the starting positions and sequence order. The arrows indicate the position for microstructural characterization of each cross-section.

2.2. Microstructural Characterization

The microstructures of three selected cross-sections, as indicated in Figure 1, were characterized. The location of each cross-section was selected so that the microstructure of different widths (one, two, and three weld beads) of the built material could be characterized. Each sample was mounted, ground, and polished using conventional methods for titanium alloys, and finally, etched with Kroll's reagent (1 mL HF, 2 mL HNO_3, and 100 mL distilled H_2O) to reveal the microstructure.

The microstructural characterization was carried out in a light optical microscope (NIKON eclipse MA200) equipped with image analysis software (NIS Elements Basic Research). Large-area mapping was first carried out to capture overview images of the cross-sections. Thereafter, the microstructural characterization was planned in detail and the areas to be analyzed were decided. In each area, ten images at 1000-times magnification were captured and, in each image, the α lath thickness was measured at 25 randomly selected sites, yielding 250 α lath measurements in total. In addition, the fraction of grain boundary α was assessed for the single weld bead cross-section through manual measurements. The working sequence was as follows: first, large-area mapping (approximately 110 images) of the selected region at 500-times magnification was carried out at the highest resolution (2560 × 1920), and then the grain boundary α was carefully marked and colored red using an image editing software (Adobe Photoshop CC 2015) and digital zoom. The fraction of grain boundary α was subsequently calculated by comparing the total number of pixels to the number of red pixels in the images.

3. Process Modeling

The similarities between the GTAW process and multipass welding permit the application of welding simulation techniques. Computational Welding Mechanics (CWM) establishes methods and models that are applicable for the control of welding processes to obtain optimal mechanical performance. The book by Lindgren [13] describes different modeling options and strategies.

3.1. Thermal Model

For the thermal model, the weld pool details are replaced by a heat input model. The modeling is thus considerably simplified, but is still able to create a model fit for the purpose. The implemented logic and model is thoroughly described in [4]. Some noteworthy clarifications to the process model are briefly mentioned below.

The heat input model is the commonly-used double-ellipsoid with Gaussian distribution, as proposed by Goldak et al. [14]. An adaptive rescaling of the heat input through the efficiency factor is, when needed, used to control the variation of the heat input function due to the rather coarse mesh used for the discretization of the model. The parameters for the heat input and heat dissipation are the same as presented in [4]. For convenience, they are listed in Table 1, where Q is the nominal effect of the heat source, η is the efficiency factor, h is the heat transfer coefficient, and e is the emissivity. The a, b, c_f, and c_r parameters define the geometry of the double ellipsoid. The model was validated thermally in [4], and found to be in good agreement with the measurements.

Table 1. Parameters for the heat input and heat dissipation.

Q (W)	η	a (m)	b (m)	c_f (m)	c_r (m)	h (W/m^{2}K)	e
896	0.58	0.004	0.0012	0.004	0.006	18	0.05

The material data that has been used in the thermal model is from [15]. The thermal conductivity and the specific heat capacity can be seen in Figure 2. Whenever the temperature in the model is outside the given data, a cut-off value is used, i.e., the last known value. A latent heat of 290 J/kg is applied in the transition between the solidus and liquidus temperatures, which are set to 1604 °C and 1700 °C, respectively.

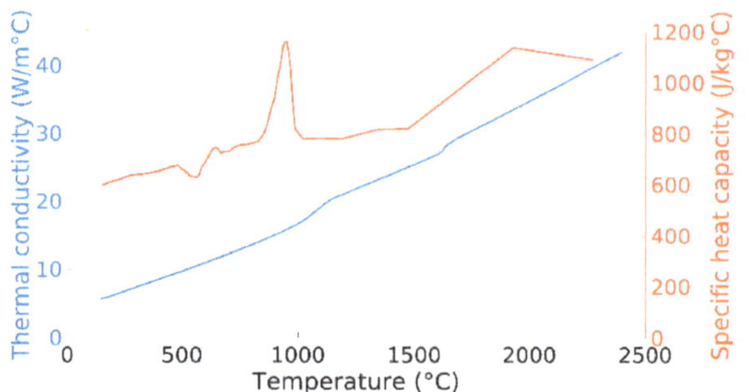

Figure 2. Material data used in the thermal model.

The number of nodes and elements in this model are approximately 19,000 and 13,000, respectively. The element type is 8-noded, fully integrated, hexahedral elements. The FE-software (version, company, city, country) used in this work, MSC.Marc (version, company, city, country), and its pre- and post-processor Mentat, have a number of interfaces for user-defined subroutines. The heat input is defined via user subroutines. The microstructural model that will be described in the next section is also defined via these user subroutines.

3.2. Microstructure Model

The phase evolution is computed during heating, cooling, and during repeated re-heating and cooling. During heating, when the temperature exceeds about 700 °C, the α phase starts to transform to the β phase [16]. Normally, during AM as well as welding, the heating rates are too fast for an equilibrium to exist. This means that $X\beta < X\beta\text{-eq}$, even at temperatures exceeding the so-called β-transus of the alloy. Above the β-transus temperature at equilibrium, only the β phase exists. During cooling, existing β phase transforms into α phase. For slow to moderate cooling rates, the transformation is diffusion controlled. The initial α phase normally nucleates at the prior β grain boundaries and continues to grow along these grain boundaries before starting to grow into the prior β grains, in a lamellar morphology. This lamellar type of microstructure is called the Widmanstätten microstructure here. For very high cooling rates, non-diffusional transformation of $\beta \rightarrow \alpha$ phase takes place; this type of α phase is, therefore, defined as martensitic α, i.e., the microstructure will be martensitic. For moderate to high cooling rates, a massive transformation of $\beta \rightarrow \alpha$ can occur [17]. This transformation is not included in the model. Solid–solid phase changes, on heating as well as on cooling, are mostly characterized by transformation mechanisms of nucleation and growth processes. The β phase decomposition to α Widmanstätten and α at prior β grain boundaries, as well as the β formation during heating, are all diffusion-controlled processes [18]. In contrast, α martensite formation from the β phase is a diffusionless transformation.

3.2.1. Solid–Solid Phase Change Model

The diffusional phase transformations of β to α and α to β are evaluated using a modified Johnson–Mehl–Avrami theory. This equation is strictly valid only under isothermal transformation; therefore, the additivity principle has been adopted, and discretization into a series of smaller isothermal steps is used during temperature variations. The rule of additivity [16] is commonly used to calculate non-isothermal transformation from isothermal transformation data using simple rate laws. It should be noted that the additivity rule should be applied only under certain conditions in which the reaction is additive [10,16,18]. Grong and Shercliff [11] compared the additivity approach to numerical solutions when modeling microstructural state variables, with a focus on applications in heat treatments and welding. Even though discrepancies are observed, the approximation is evaluated to be sufficient for many problems, particularly when the constant data is estimated against experimental data for microstructure. The generalization steps of the Johnson–Mehl–Avrami equation used in this work are thoroughly detailed in [12]. The incomplete transformations toward equilibrium at the current temperature are circumvented by normalizing the equations. The interaction between simultaneous transformations is handled by assuming that the current fraction of the resultant phase is taken relative to the total content of the transforming phase.

The diffusionless martensite formation was chosen to be modeled by the classical Koistinen–Marburger equation, for which a direct incremental formulation of the equation, shown to be simpler and equally well accurate [19], was chosen. The overall equations for the model and their discretization are detailed in [12].

3.2.2. Morphology Parameter: Alpha Lath Thickness

The morphology size parameter associated with Widmanstätten α phase, i.e., the α lath thickness parameter, was modeled by a simplified energy model approach [20]. The α phase formation temperature is here, as a first approximation, considered dominant in determining the α lath thickness. The empirical Arrhenius equation is used to express the temperature dependence of the α lath thickness. A first approximation for kinetic parameters was used in [20]; the proposed values seemed to be outside of the expected parameters' dimensions. Irwin et al. [21] updated the values for the parameters after additional optimization supported by a new set of experimental results. The new suggested values, Arrhenius prefactor k = 1.42 μm and activation temperature R = 294 K, appear to have appropriate dimensions, and are thus used in this work. The equation and its explicit form used in the model can be found in [20].

3.2.3. Implementation Strategy

The development of the microstructural model is based on the finite element method, and is supported by the features in the software MSC.Marc. The microstructure is homogeneously described by state variables associated at each of the integration points of the finite element mesh. This approach means that Representative Volume Elements (RVE) (see Figure 3) are considered at each integration point of the elements. The calculated value corresponds to an average behavior over this domain. For example, the phase fraction in an integration point then corresponds to the fraction of the phase in the RVE connected to this particular integration point.

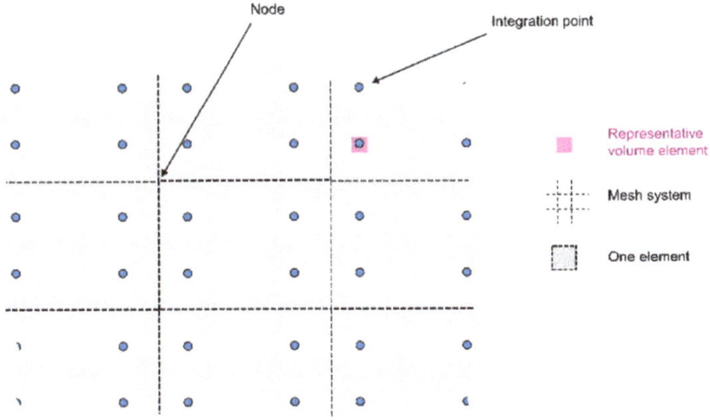

Figure 3. Schematic illustration of density type of model.

Four state variables are used in the model to represent the fraction of microstructural constituents. One more state variable, denoted as the size parameter, is used for the α lath thickness. The state variable denominations and the size parameter can be seen in Table 2. The α phase can form a number of different types of microstructure, but for modeling purposes, a choice to approximate the diffusional-formed α phase into a twofold-different microstructure is made, namely (i) grain boundary α (α_{gb}), and (ii) Widmanstätten (α_w) microstructure. The grain boundary α (α_{gb}) is the α phase that is formed in the prior β grain boundaries. The Widmanstätten (α_w) structure represents the α phase that forms inside the prior β grains (which is here considered to include both colony type and basket weave microstructures). The α martensite (α_m) structure and the β phase are also represented.

Table 2. Microstructural parameters and morphology description used in this work.

Phase Constituents		**Type**	**State Variable**	**Size Parameters**
α	**Diffusional α**	Grain boundary	$X_{\alpha gb}$	-
		Intergranular, Basket-weave, Colony	$X_{\alpha w}$	Lath thickness, $t_{\alpha\text{-lath}}$
	Non-diffusional α	Martensite	$X_{\alpha m}$	-
β		-	X_{β}	-

The interactions between the constituents are schematically presented in the upper-right square of Figure 4. In this work, four different diffusional transformations and one non-diffusional transformation are implemented. The transformation processes and microstructural constituent interactions that are implemented in the model are shown in the table in Figure 4. A detailed description of the logic and the overall model can be found in [12].

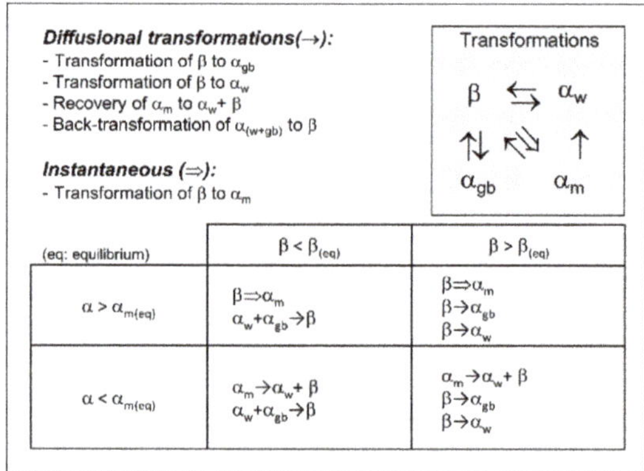

Figure 4. Transformation process and constituent interactions.

3.3. Adaptive Time Sub-stepping

To optimize the solution routine and reduce the computational time, an adaptive sub-stepping to calculate the microstructural model was adopted. To improve the accuracy, a refined time stepping is used when the temperature is in the phase transformation region. When necessary, each thermal step is thus divided into several smaller thermal sub-steps, assuming a linear temperature variation inside the original time step.

4. Results and Discussion

4.1. Microstructural Analysis

The microstructural analysis was performed on the areas highlighted in Figure 5, denoted A, B, C, D, and E. In general, the microstructure of the GTAW wire feed built Ti-6Al-4V consists of large, columnar, prior β grains that grow in the temperature gradient direction, through several layers. The directions of the columnar prior β grains can be seen in Figure 5. The prior β grains, seen as large areas of different color and/or contrast in Figure 5, are increasingly oriented towards the sides of the cross-section because of the temperature gradient. Within the prior β grain, fine α laths are observed in the form of either basketweave α structure (see Figure 6a) or colony α structure (see Figure 6b). However, only small regions of colony α are observed for the GTAW wire feed built Ti-6Al-4V, with the main part of the α laths being in the form of the basketweave α structure. The prior β grain boundaries are decorated by grain boundary α, as shown in Figure 6c. It is noteworthy that the thickness and prevalence of grain boundary α varies in different cross-sections. In some regions, the grain boundary α is continuous, as it is in Figure 6c, while it is absent or discontinuous in other grain boundaries.

Table 3. The result of the quantitative material characterization of the three cross-sections containing widths of one, two, and three beads, respectively.

Cross Section	Area	Average α Lath Thickness (μm)	Grain Boundary α Phase Fraction (%)
One bead	A	1.1 ± 0.4	0.21
	B	1.0 ± 0.3	0.11
	C	0.9 ± 0.3	0.05
	All	1.0 ± 0.3	-
Two beads	A	1.1 ± 0.4	-
	B	1.0 ± 0.4	-
	C	1.0 ± 0.3	-
	D	1.0 ± 0.3	-
	E	1.0 ± 0.4	-
	All	1.0 ± 0.4	-
Three beads	A	1.0 ± 0.3	-
	B	1.1 ± 0.5	-
	C	1.3 ± 0.5	-
	D	0.9 ± 0.3	-
	E	1.0 ± 0.3	-
	All	1.1 ± 0.4	-

The result of the quantitative microstructural characterization is summarized in Table 3. In general, only a small difference is observed when comparing the different areas. The variation is within the range of the standard deviation. However, one tendency is that slightly thicker α laths form with increasing thickness of deposited material (number of beads). This could be explained by the increased number of heating cycles due to the additional beads that allow the diffusional growth of the α laths to continue for a longer time. The α lath measurements are consistent with those previously reported [22] (0.8 ± 0.2 mm) for GTAW wire feed built Ti-6Al-4V, not for the same, but for similar process conditions. In contrast to the α lath thickness, a large variation of the grain boundary α fraction is observed. The fraction of grain boundary α depends on the location of the prior β grains and, furthermore, on the size of the prior β grains. As seen in Figure 5, the widths of the prior β grains vary significantly within the cross-sections, making the fraction estimation highly sensitive to the location of the evaluated area. Moreover, because of its limited thickness, the grain boundary α could be difficult to discern, which may have been the case for some limited regions of the areas investigated in the present study. For future work, it is therefore recommended that the validation be carried out of the grain boundary α on a deposited material with a slower cooling rate, for which a thicker grain boundary α is expected to be better defined, and thus, easier to measure.

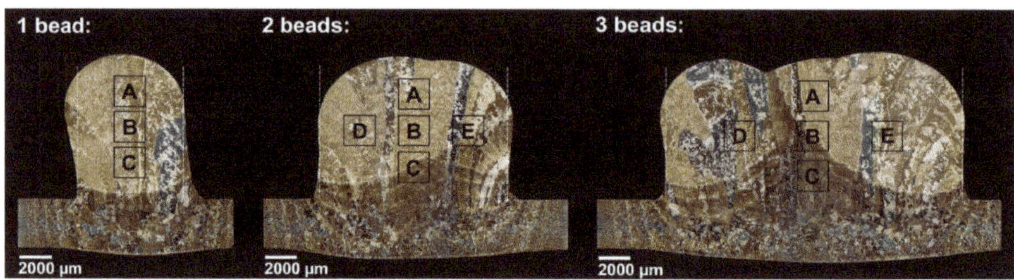

Figure 5. Light optical micrographs showing the characterized cross-sections indicated in Figure 1. The different areas, denoted as A, B, C, D, and E, correspond to the location of the measurements presented in Table 3.

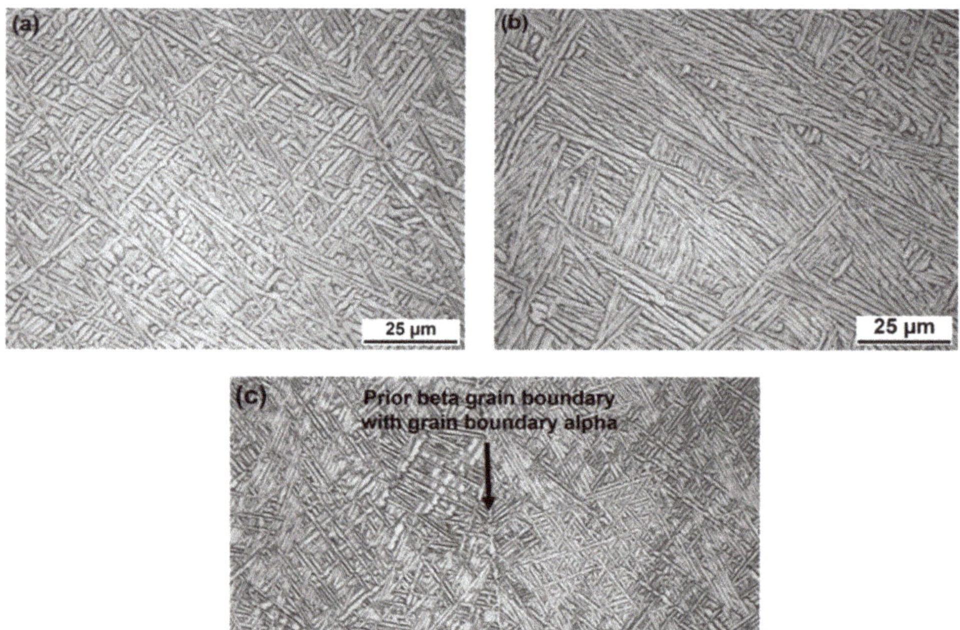

Figure 6. Light optical micrographs showing the microstructure of the GTAW wire feed built Ti-6Al-4V consists of fine α laths in the form of (**a**) basketweave α or (**b**) colony α structure. The prior β grain boundary is decorated with grain boundary α (**c**).

4.2. Microstructural Simulation

The additive manufacturing process is characterized by cyclic temperature variations, leading to repetitive phase transformations and microstructural changes in the deposited material and substrate. Temperature history is thus the main factor when modeling the microstructure. Continuous microstructural modeling allows for the microstructural changes that occur during processing to be followed. While microstructural analysis provides information about the results after the additive manufacturing process, the microstructural model, by following the entire deposition process, gives continuous information about the microstructure changes during the additive manufacturing process. The temperature history experienced at a selected point during additive manufacturing (see Figure 7a) is representative of a typical temperature profile experienced by the deposited material. The location of this point is in area B in the single weld bead sample in Figure 5. In Figure 7b,c, the simulated phase transformations versus time and temperature, respectively, can be seen. As explained earlier, during cooling, the β phase transforms into a mixture of α_{gb} and α_{wid}. The phase fraction of each microconstituent depends on the cooling rate, i.e., a faster cooling rate promotes more α_{wid} than α_{gb}, and vice versa. During the following heating sequence, the α_{gb} and α_{wid} transform to the β phase, which cyclically transforms again to α_{gb} and α_{wid} over the temperature history. It is important for the numerics in the microstructural model that the temperature change during a time step is not too large. Therefore, an internal time sub-stepping logic was developed to control this. In the current work, a maximum temperature change within the phase transformation region of 5 °C is allowed. If the temperature change in the thermal solution exceeds this value, the time step will be subdivided internally. The authors have observed that, for this specific set-up, the model yields a slightly different result when using the sub-stepping logic during the heating cycle. However, during the cooling phase, they are nearly identical. For another type of AM process, e.g., powder bed, the sub-stepping logic is probably of more importance, as the cooling ratios are much higher.

Figures 8 and 9 show the simulated results of the α lath thickness. Despite the simplicity of the chosen model, the simulation results agree well with the experimental measurements. The simulated thickness, evaluated to be approximately 1.1 µm, is in agreement with the 1.0 to 1.1 µm obtained by microstructural analysis. It is also interesting to note that the α lath thickness shows a tendency to increase with increasing wall width, i.e., when built with more weld beads. A similar trend is found in the experimental evaluation.

The model predicts 3–6% α_{gb} in the deposited material, as seen in Figure 10, and the experimental measurements indicate α_{gb} fractions between 0.05–0.2%, as seen in Table 3. The small amount of α_{gb} phase and the large variation of the α_{gb} amount that was observed experimentally result in some uncertainty. However, it is obvious that the model overpredicts the amount of α_{gb}. As already stated in the microstructural analysis section, a more accurate validation case could be achieved by using deposited material containing a thicker grain boundary α, which could be obtained in slowly-cooled deposited material.

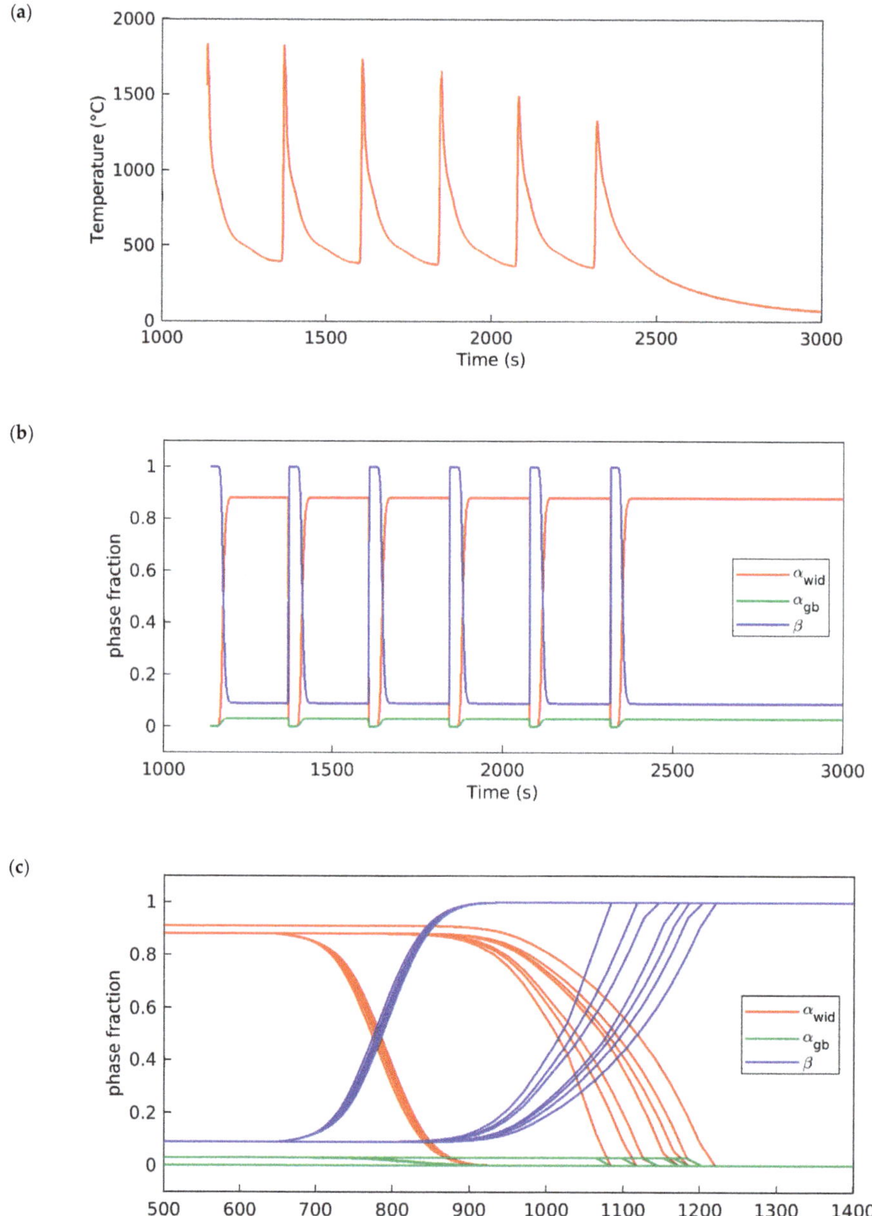

Figure 7. Simulation results at a node positioned in one weld bead cross-section positioned in the B area (Figure 5). (**a**) Temperature variations vs time. (**b**) Corresponding simulated α_{gb}, α_{wid}, and β phase fractions vs time. (**c**) Corresponding simulated α_{gb}, α_{wid}, and β phase fractions vs. temperature.

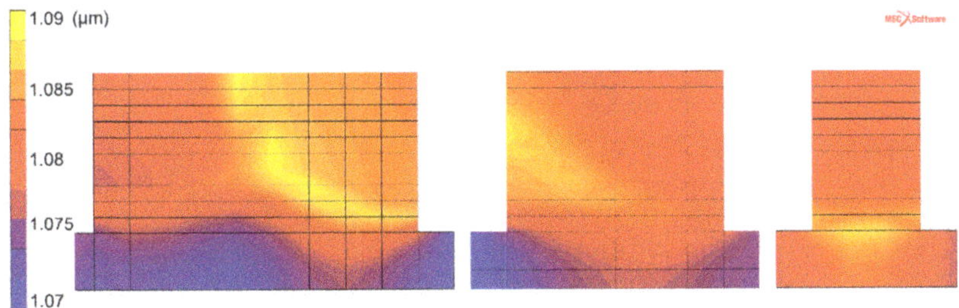

Figure 8. Simulated α lath thickness (µm), cross-sections with walls of widths of one, two, and three weld beads.

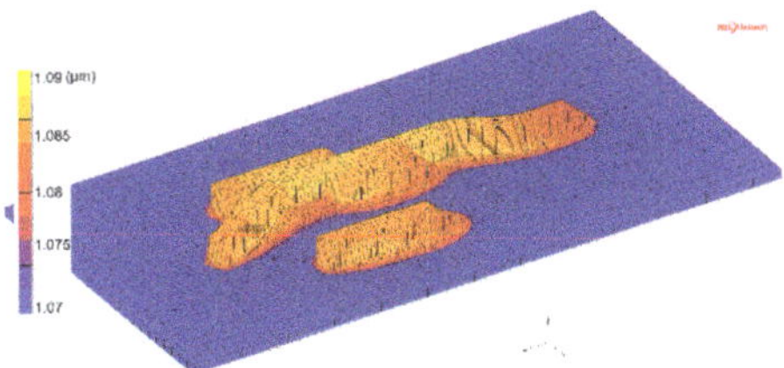

Figure 9. Simulated α lath thickness (µm) variation in the additive manufactured part.

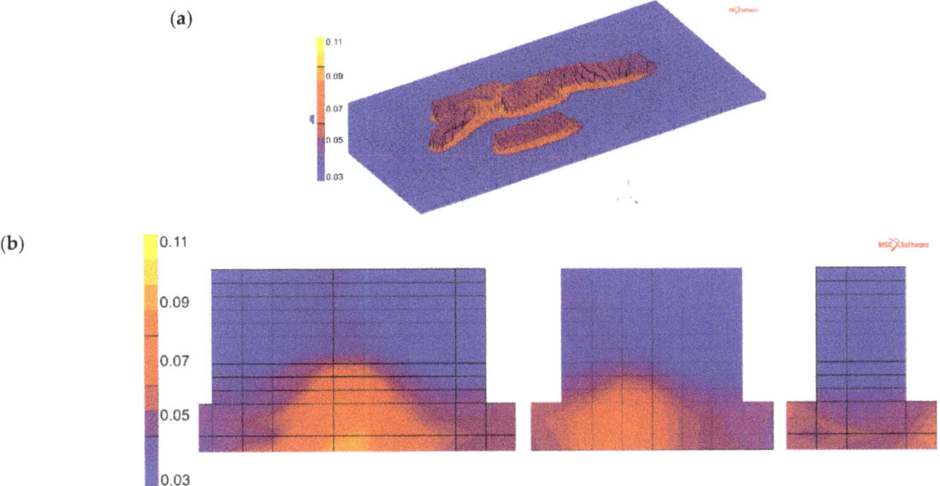

Figure 10. Simulated αgb phase fraction after wire feed additive manufacturing. (**a**) Complete view of the sample. (**b**) Cross-sections of one, two, and three weld-beads-wide walls.

An example of the results that can be obtained during the additive manufacturing process by using simulations is shown in Figure 11. The effect of the newly-deposited layer on the previously deposited layers is presented here. The simulated β phase fraction illustrates the ongoing phase transformations that are taking place while depositing the consecutive metal layers. The model, as well as the microstructure characterization, show no martensitic areas in the microstructure for the current manufacturing parameters.

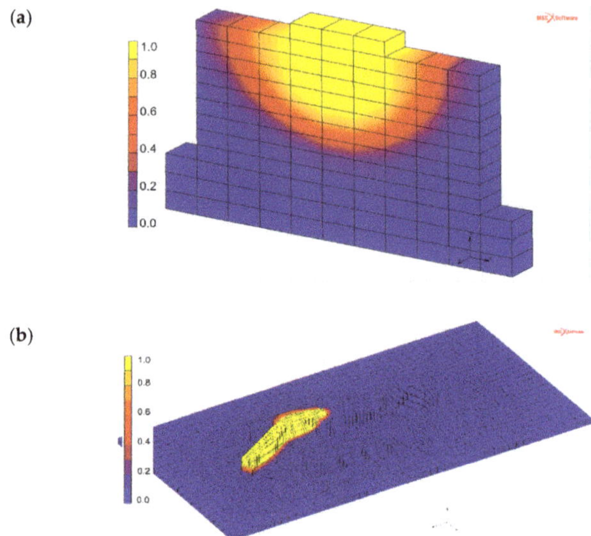

Figure 11. Simulated β phase fraction during wire feed additive manufacturing. (**a**) Cross-section of the three-weld beads wall. (**b**) Complete view of the sample.

The presented method for the modeling of wire feed additive manufacturing has been implemented into subroutines that can be evoked using commercial, general-purpose, finite element software. Moreover, microstructural modeling increases our understanding of the microstructural evolution, not only for post building, but also during the additive manufacturing process. The α lath thickness has been successfully predicted. The β phase transformation model (equivalent to the complementary total α phase transformation model) was validated in a previous publication [12] and used by others, e.g., [21,23]. However, the α_{gb} phase transformation part of the model is yet to be validated.

In addition, the microstructural model is an important tool that enables continued additive manufacturing process development to obtain improved material properties. Because physical and mechanical properties are dependent on the microstructure, it is also interesting to consider the microstructural development during the AM process. Then, the model can be coupled with a flow stress model needed in a thermo-mechanical analysis. The plastic properties will then depend on the current temperature and microstructure. The microstructural model is coupled with the dislocation density-based plasticity model presented in [24]. This type of constitutive model has a natural handshake with microstructural models; the coupling and application of these two models are presented in [25].

5. Conclusions

In this work, a process simulation model was coupled to a microstructural model. A density type of microstructural model was chosen in the current work. The main reason for this was that it should be possible to apply the model to industrial applications on a macroscopic level. The study presented herein shows promising results, since the size of the predicted alpha laths correlated very well with the microstructural analysis. Other types of models, such as sub-models or dual-mesh models, can give more detailed information about the microstructure, particularly the morphology; however, they typically become very computationally demanding, and are suitable for the whole field in an industrial application. In addition, the density type facilitates the combination of the microstructural model with a mechanical model to compute material properties, which is also shown in [25].

Author Contributions: All co-authors contributed to the planning of the work and preparation of the manuscript. C.C.M. developed the phase transformation model. C.C.M. and A.L. implemented the model and preformed the simulations. P.Å. carried out and planned the microstructural characterization. R.P. was involved in initiating the project that financed this work and was also involved in discussing the results and conclusions of this work.

Funding: This research was funded by the European 6th Framework Program through the research project VERDI (Virtual Engineering for Robust manufacturing with Design Integration). The authors also acknowledge the financial support from VINNOVA foundation (Swedish Governmental Agency for Innovation Systems) through the NFFP program (Swedish National Aviation Engineering Research Program). The authors are also grateful for the financial support from the Swedish National Space Board through the NRFP2 program (Swedish National Space Research Program). The Graduate School in Space Technology for supporting and encouraging the project.

Acknowledgments: The authors would like to express their gratitude to Lars-Erik Lindgren at the Division of Mechanics of Solid Materials, Luleå University of Technology, for his valuable discussions about the microstructure model and to Mats Högström at University West, Sweden for performing the additive manufacturing trials. The close collaboration with GKN Aerospace Engine Systems in supporting this work is highly recognized.

Conflicts of Interest: The authors declare no conflict of interest.

References

1. Ding, D.; Pan, Z.; Cuiuri, D.; Li, H. Wire-feed additive manufacturing of metal components: Technologies, developments and future interests. *Int. J. Adv. Manuf. Technol.* **2015**, *81*, 465–481. [CrossRef]
2. Brice, C. Net shape processing of titanium alloys for enhanced performance and improved affordability. In Proceedings of the 12th World Conference on Titanium, Bejing, China, 19–24 June 2011; Yufeng, N., Bin, G., Eds.; pp. 1697–1703.
3. Peters, M.; Leyens, C. *Titanium and Titanium Alloys: Fundamentals and Applications*; Wiley-VCH: Weinheim, Germany, 2003; ISBN 3527305343.
4. Lundbäck, A.; Lindgren, L.-E. Modelling of metal deposition. *Finite Elem. Anal. Des.* **2011**, *47*, 1169–1177. [CrossRef]
5. Agelet de Saracibar, C.; Lundbäck, A.; Chiumenti, M.; Cervera, M. Shaped Metal Deposition Processes. In *Encyclopedia of Thermal Stresses*; Hetnarski, R.B., Ed.; Springer: Dordrecht, The Netherlands, 2014; pp. 4346–4355.
6. Chiumenti, M.; Cervera, M.; Salmi, A.; Agelet de Saracibar, C.; Dialami, N.; Matsui, K. Finite element modeling of multi-pass welding and shaped metal deposition processes. *Comput. Methods Appl. Mech. Eng.* **2010**, *199*, 2343–2359. [CrossRef]
7. Graf, M.; Hälsig, A.; Höfer, K.; Awiszus, B.; Mayr, P. Thermo-Mechanical Modelling of Wire-Arc Additive Manufacturing (WAAM) of Semi-Finished Products. *Metals* **2018**, *8*, 1009. [CrossRef]
8. Gandin, C.-A.; Rappaz, M. A 3D Cellular Automaton algorithm for the prediction of dendritic grain growth. *Acta Mater.* **1997**, *45*, 2187–2195. [CrossRef]
9. Katzarov, I.; Malinov, S.; Sha, W. Finite element modeling of the morphology of β to α phase transformation in Ti-6Al-4V alloy. *Metall. Mater. Trans. A* **2002**, *33*, 1027–1040. [CrossRef]
10. Thiessen, R.G.; Richardson, I.M.; Sietsma, J. A Dual-Mesh Strategy for Microstructure Development in a Macroscopic Heat Affected Zone: Studies on AISI316L and AISI1005. In Proceedings of the Trends in Welding Research: Proceedings of the 7th International Conference, Pine Mountain, GA, USA, 16–20 May 2005; David, S.A., Ed.; ASM International: Pine Mountain, GA, USA, 2006; pp. 993–1000.
11. Grong, Ø.; Shercliff, H.R. Microstructural modelling in metals processing. *Prog. Mater. Sci.* **2002**, *47*, 163–282. [CrossRef]
12. Charles Murgau, C.; Pederson, R.; Lindgren, L.-E. A model for Ti–6Al–4V microstructure evolution for arbitrary temperature changes. *Model. Simul. Mater. Sci. Eng.* **2012**, *20*, 055006. [CrossRef]
13. Lindgren, L.-E. *Computational Welding Mechanics: Thermomechanical and Microstructural Simulations*; CRC Press: Boca Raton, FL, USA, 2007; ISBN 9781845693558.
14. Goldak, J.; Chakravarti, A.; Bibby, M. A new finite element model for welding heat sources. *Metall. Trans. B* **1984**, *15*, 299–305. [CrossRef]
15. Boivineau, M.; Cagran, C.; Doytier, D.; Eyraud, V.; Nadal, M.-H.; Wilthan, B.; Pottlacher, G. Thermophysical Properties of Solid and Liquid Ti-6Al-4V (TA6V) Alloy. *Int. J. Thermophys.* **2006**, *27*, 507–529. [CrossRef]
16. Cahn, J.W. Transformation kinetics during continuous cooling. *Acta Metall.* **1956**, *4*, 572–575. [CrossRef]

17. Xu, W.; Brandt, M.; Sun, S.; Elambasseril, J.; Liu, Q.; Latham, K.; Xia, K.; Qian, M. Additive manufacturing of strong and ductile Ti–6Al–4V by selective laser melting via in situ martensite decomposition. *Acta Mater.* **2015**, *85*, 74–84. [CrossRef]
18. Christian, J.W. *The Theory of Transformations in Metals and Alloys: Part. I + II*; Elsevier: Amsterdam, The Netherlands, 2002; ISBN 9780080440194.
19. Back Gyhlesten, J.; Lindgren, L.-E. Simplified Implementation of the Koistinen-Marburger Model for Use in Finite Element Simulations. In Proceedings of the 11th International Congress on Thermal Stresses, Salerno, Italy, 5–9 June 2016.
20. Charles, C.; Järvstråt, N. Modelling Ti-6Al-4V microstructure by evolution laws implemented as finite element subroutines: Application to TIG metal deposition. In Proceedings of the 8 th International Conference on Trends in Welding Research, Pine Mountain, GA, USA, 1–6 June 2008; pp. 477–485.
21. Irwin, J.; Reutzel, E.W.; Michaleris, P.; Keist, J.; Nassar, A.R. Predicting Microstructure From Thermal History During Additive Manufacturing for Ti-6Al-4V. *J. Manuf. Sci. Eng.* **2016**, *138*, 111007. [CrossRef]
22. Neikter, M.; Åkerfeldt, P.; Pederson, R.; Antti, M.-L.; Sandell, V. Microstructural characterization and comparison of Ti-6Al-4V manufactured with different additive manufacturing processes. *Mater. Charact.* **2018**, *143*, 68–75. [CrossRef]
23. Salsi, E.; Chiumenti, M.; Cervera, M.; Salsi, E.; Chiumenti, M.; Cervera, M. Modeling of Microstructure Evolution of Ti6Al4V for Additive Manufacturing. *Metals* **2018**, *8*, 633. [CrossRef]
24. Babu, B.; Lindgren, L.-E. Dislocation density based model for plastic deformation and globularization of Ti-6Al-4V. *Int. J. Plast.* **2013**, *50*, 94–108. [CrossRef]
25. Babu, B.; Lundbäck, A.; Lindgren, L.-E. Simulation of additive manufacturing of Ti-6Al-4V using a coupled physically-based flow stress and metallurgical model. *Materials* **2019**. Submitted.

Article

Characterisation of a High-Performance Al–Zn–Mg–Cu Alloy Designed for Wire Arc Additive Manufacturing

Paulo J. Morais [1,*], Bianca Gomes [1], Pedro Santos [1], Manuel Gomes [1], Rudolf Gradinger [2], Martin Schnall [2], Salar Bozorgi [2], Thomas Klein [2], Dominik Fleischhacker [3], Piotr Warczok [4], Ahmad Falahati [4,5] and Ernst Kozeschnik [4,5]

1 Instituto de Soldadura e Qualidade, Av. Prof. Dr. Cavaco Silva, 33, 2740-120 Porto Salvo, Portugal; bfgomes@isq.pt (B.G.); pmsantos@isq.pt (P.S.); magomes@isq.pt (M.G.);
2 LKR Light Metals Technologies Ranshofen, Austrian Institute of Technology, Lamprechtshausenerstraße 61, 5282 Ranshofen-Braunau, Austria; rudolf.gradinger@ait.ac.at (R.G.); martin.schnall@ait.ac.at (M.S.); salar.bozorgi@ait.ac.at (S.B.); thomas.klein@ait.ac.at (T.K.)
3 SinusPro GmbH, Conrad-von-Hötzendorf-Straße 127, 8010 Graz, Austria; dominik.fleischhacker@sinuspro.at
4 MatCalc Engineering GmbH, Gumpendorfer Strasse 21, 1060 Vienna, Austria; piotr.warczok@mceng.at (P.W.); ahmad.falahati@tuwien.ac.at (A.F.); ernst.kozeschnik@tuwien.ac.at (E.K.)
5 Institute of Materials Science and Technology, TU Wien, Getreidemarkt 9/E308, 1060 Vienna, Austria
* Correspondence: pjmorais@isq.pt

Received: 5 March 2020; Accepted: 30 March 2020; Published: 1 April 2020

Abstract: Ever-increasing demands of industrial manufacturing regarding mechanical properties require the development of novel alloys designed towards the respective manufacturing process. Here, we consider wire arc additive manufacturing. To this end, Al alloys with additions of Zn, Mg and Cu have been designed considering the requirements of good mechanical properties and limited hot cracking susceptibility. The samples were produced using the cold metal transfer pulse advanced (CMT-PADV) technique, known for its ability to produce lower porosity parts with smaller grain size. After material simulations to determine the optimal heat treatment, the samples were solution heat treated, quenched and aged to enhance their mechanical performance. Chemical analysis, mechanical properties and microstructure evolution were evaluated using optical light microscopy, scanning electron microscopy, energy dispersive X-ray spectroscopy, X-ray fluorescence analysis and X-ray radiography, as well as tensile, fatigue and hardness tests. The objective of this research was to evaluate in detail the mechanical properties and microstructure of the newly designed high-performance Al–Zn-based alloy before and after ageing heat treatment. The only defects found in the parts built under optimised conditions were small dispersed porosities, without any visible cracks or lack of fusion. Furthermore, the mechanical properties are superior to those of commercial 7xxx alloys and remarkably independent of the testing direction (parallel or perpendicular to the deposit beads). The presented analyses are very promising regarding additive manufacturing of high-strength aluminium alloys.

Keywords: wire arc additive manufacturing; precipitation hardening; Al–Zn–Mg–Cu alloys; microstructure characterisation; mechanical properties

1. Introduction

Additive manufacturing (AM) is a manufacturing technology which has been evolving at an enormous rate in the last 30 years. AM uses 3D CAD models to build parts by adding material successively, layer by layer, allowing the production of complex geometric shapes and topology optimised parts. The additive manufacturing of metals, which has become an area of extensive research work, is transforming many industrial sectors by reducing component lead time and material waste. Many different industries using metallic materials, such as aerospace, automotive, tooling, and health care, among others, are already taking advantage of AM [1,2]. Nevertheless, metallic parts produced by AM are susceptible to various defects, property degradation and compositional changes that need to be addressed [3].

Due to the high strength-to-weight ratio and ductility at room temperature, Al–Zn–Mg–Cu alloys are used in the automotive, aerospace, aircraft and competitive sport industries [4–7]. In these industries, new materials with enhanced mechanical properties are always needed to satisfy the ever-increasing demands for the weight reduction of structures and components. The strength of Al–Zn–Mg–Cu alloys is mainly controlled by the ageing process through the precipitation and growth of very fine precipitates of the η'-phase (a semi-coherent, metastable precursor of the equilibrium $MgZn_2$ phase) evolving from the Zn and Mg Guinier–Preston (GP) zones [8–10]. The purpose of Cu in these alloys is to increase the ageing rate by increasing the degree of super-saturation, possibly, through nucleation of Al_2CuMg (S-Phase) [11–13].

The ageing of Al–Zn–Mg–Cu alloys from room temperature to relatively low ageing temperatures is accompanied by the generation of GP zones with an approximately spherical shape [14]. With increasing ageing time, the GP zones increase in size and the strength of the alloy increases. Extended ageing at temperatures above room temperature transforms the GP zones in alloys with relatively high Zn and Mg ratios into the transition precipitates known as η', T′ or S′, the precursor of the equilibrium $MgZn_2$, T ($Al_2Mg_3Zn_3$) and S (Al_2CuMg)-phase precipitates [15–17].

The effect of alloying elements on welding additive processability and mechanical properties of Al–Zn–Mg–Cu alloys have been studied in different research works [18–20]. For example, the addition of Mn to Fe-containing Al–Zn–Mg–Cu alloys can promote the formation of the α-phase ($Al_{15}(Fe,Mn)_3Si_2$) instead of the more harmful β-phases (Al_5FeSi) [11,21–23]. Al–Zn–Mg alloys are known as materials difficult to cast and weld, due to subsequent hot cracking failures. This feature clearly creates a challenge to design alloys suitable for the wire arc additive manufacturing process. Analysis of the available hot tearing diagrams for this system indicated that alloys with a Mg:Zn ratio >1 are expected to have acceptable weldability [24]. Furthermore, alloys with this solute ratio might exhibit satisfactory mechanical properties [25,26]. The alloy compositions were designed based on intensive literature studies and theoretical thermodynamic considerations.

Among the manifold additive manufacturing techniques, technologies using metal powder can produce parts with high accuracy and comparably low deposition rates, which renders them more suitable to produce small parts with complex geometry [27]. When the powder is replaced by wire, higher deposition rates can be achieved, the costs with raw material are reduced, and the process is safer. Due to these benefits, the wire arc additive manufacturing process has been considered and many studies have been conducted using this technology [28–32]. Using this process, deposition rates up to 13-fold higher than in powder-based techniques can be achieved [33] and the investment costs in equipment are reduced when compared to, e.g., laser-based AM [34]. Wire arc additive manufacturing parts, however, present higher surface roughness, in the range of approximately 500 μm [27], which could be reduced by process control. Cold metal transfer (CMT) is a modified gas metal arc welding (GMAW) process developed by Fronius. Characteristic for the CMT technology is a low heat input and splash-free welding. Porosity can stem from wire or substrate contaminants or from volatile alloying elements. These contaminants are cracked by the high energy of the electric arc, resulting in gas porosity that is trapped upon solidification and can hardly be reduced during processing [35]. The characteristic curve used (arc) is a mix of CMT pulses and alternating voltage (CMT-PADV), which has been shown to be well suited for wire arc additive manufacturing of aluminium [36].

The aim of the present work was to understand the influence of the wire arc process and new alloying system on the mechanical properties of the produced parts before and after the heat treatment, including detailed microstructural characterisation.

2. Materials and Methods

2.1. Alloy Composition and Material Processing

The experimental alloy composition was cast using vertical continuous casting. The resultant chemical composition is given in Table 1. Subsequently, cylindrical preforms with a diameter of 35 mm and a length of 100 mm were machined and heated in a furnace to 435 °C. These preforms were then extruded to wires with a diameter of 1.6 mm and a length of approximately 4 m. The resultant segments were then joined and, after visual inspection, coiled, thus enabling wire arc experiments with novel wires with specified non-commercial chemical compositions and a small lot size.

Table 1. Chemical composition of the alloy in the as-cast condition measured by optical emission spectroscopy (OES) in weight percent.

As-cast (OES)	Al	Mg	Zn	Mn	Si	Cu	Fe	Cr	Ti	Zr
wt.%	89.30	5.87	3.58	0.49	0.07	0.33	0.11	0.04	0.05	0.12

The additive manufactured samples were produced with a CMT Advanced 4000 R (Fronius International GmbH, Wels, Austria) and a robot arm (ABB, Zurich, Switzerland) using CMT pulse advanced (PADV). Rectangular-shaped parts of approximately 170 × 30 × 120 mm^3 (l × w × h) and an R equal to 5 mm for the corners were manufactured (Figure 1). A water-cooled base plate was used. The cleaning of the substrate before the deposition of the first layer and between layers was conducted with a stainless steel brush. The shielding gas used was argon.

For further analyses, three different material conditions were investigated in detail: (i) the as-built (AB) material condition; (ii) the T6 material condition, which was obtained by a solution heat treatment at 470 °C for 5 h, followed by rapid cooling in water and a subsequent ageing treatment at 120 °C for 24 h; and (iii) the T73 material condition, which was obtained by a solution heat treatment at 470 °C for 5 h, followed by rapid cooling in water and a subsequent two-stage ageing treatment at 120 °C for 24 h, and then 160 °C for 24 h.

2.2. Characterisation Techniques

Tensile test specimens were extracted from the samples in the three different material conditions, in both transverse and longitudinal directions (Figure 1(b) and (c)). Standard specimens, according to ISO 6892-1:2016, were used to determine the average mechanical properties of the alloys in the different heat treatment conditions and direction (Figure 2a). Subsized specimens (Figure 2b) were used to probe local properties in the sample different regions (bottom, middle or top of the sample), and thus, to evaluate the homogeneity of the mechanical properties along the height and width of the produced parts. Three to five standard size specimens were tested for each material, heat treatment and conditions, whereas six to nine subsized specimens scattered over the whole face of the sample were tested for local properties determination (Figure 1b,c).

Fatigue test specimens were extracted from the T73 tempered material condition in the longitudinal and transverse direction. The test conditions and specimen geometry (Figure 3) were in accordance with ASTM E466 standard. The specimen gauge length was further polished down to grade 1000 emery paper prior to fatigue testing. The tests were performed on a servo hydraulic machine under load control (Instron 8502, Norwood, Massachusetts, USA). Based on the tensile test results obtained for this material, a yield stress of 330 MPa was assumed and the maximum stress for fatigue tests set up in the range of 95% to 45% of the yield. All tests were performed with a stress ratio R = 0.1, commonly used in aeronautical and structural testing, a sinusoidal wave form and a frequency of 20 Hz. All tests were run up to failure—after which, the fracture surface was observed using a scanning electron microscope (JEOL JSM–6500F, Tokyo, Japan) to identify any features associated with the fracture initiation and propagation processes.

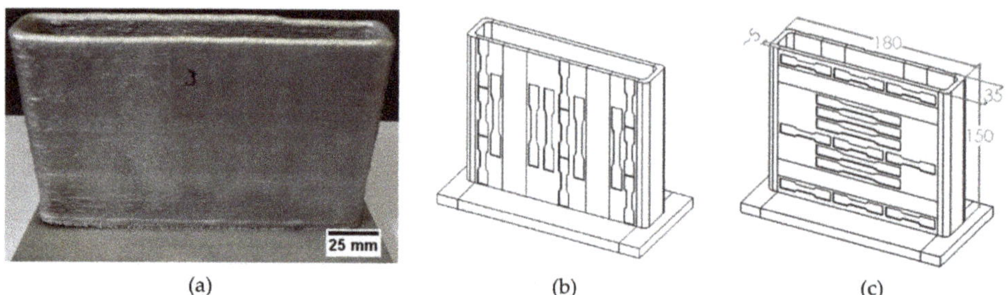

Figure 1. (**a**) Rectangular-shaped specimen produced by wire arc additive manufacturing on a flat substrate; (**b**) transverse tensile specimens; (**c**) longitudinal tensile specimens. All dimensions in mm.

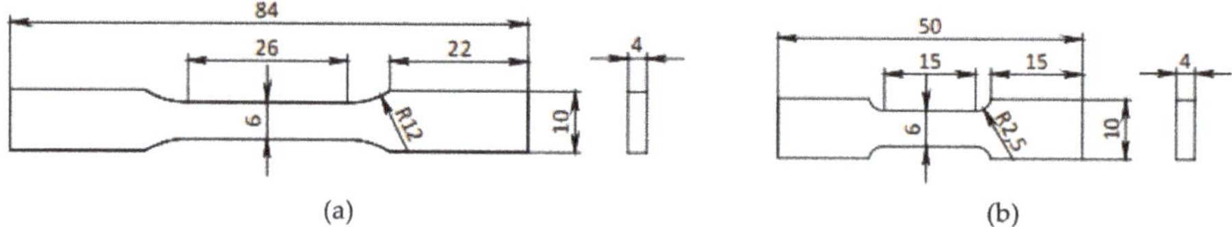

Figure 2. Tensile specimens: (**a**) regular size; (**b**) subsize. All dimensions in mm.

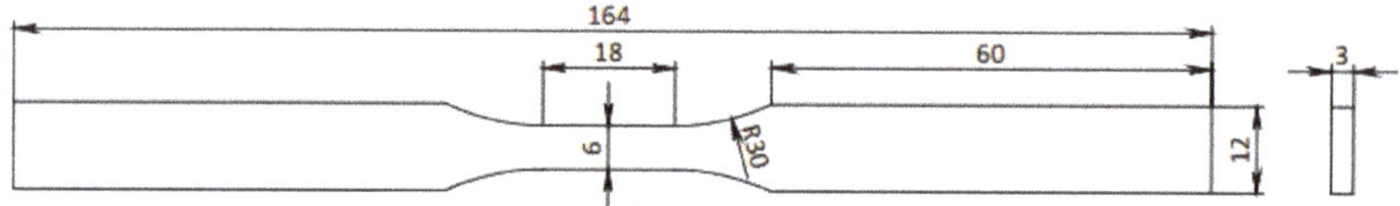

Figure 3. Fatigue specimen. All dimensions in mm.

The transverse parts in the various conditions mentioned above were cold mounted, grinded with SiC paper and polished with diamond paste. Microhardness measurements were performed with a hardness tester (Mitutoyo Akashi AVK-C0, Kawasaki, Japan), using the Vickers method, with a load of 9.81 N and indentation time of 15 s. These measurements were performed in the centreline of the specimens along its height (one specimen along the height of the produced walls per condition was tested). Microstructure analysis was performed by optical light microscopy (Carl Zeiss Axiotech 100HD-3D, Oberkochen, Germany) and stereoscopy (Olympus SZX7, Tokyo, Japan). Specimens were electrochemically etched with Barkers Reagent (5 mL HBF4 (48%) + 200 mL water) using 15 V for 90–240 s. Scanning electron microscopy (SEM) and energy dispersive X-ray spectroscopy (EDS) results were obtained from the samples in as polished condition using a JSM–6500F microscope (JOEL, Tokyo, Japan) fitted with an EDS detector (Oxford Instruments X-Max^N, Abingdon, UK).

Porosity analysis was performed in the as-polished samples, considering the three different regions and conditions presented above. Five images were obtained randomly for each region and evaluated using ImageJ software (National Institutes of Health, Bethesda, Maryland, USA) [37]. First, the scale was set in the software and, then, the image was thresholded carefully to ensure all the porosities remained in the image. Thus, all the particles with a minimum of 0.5 circularity were detected and measured by the software. A second analysis was performed to evaluate whether "non-pores" were measured and to measure the pores not detected. The porosity diameter considered was the average of width and height values of the smallest rectangle enclosing the selected pore. In addition, porosity was also analysed by X-ray radiography with X-ray equipment (YXLON ANDREX RIX-02, Hudson, Ohio, USA). The X-ray inspection followed the procedure prescribed by ISO 10675-2 standard.

In the grain measurement, the standard ASTM E112–2013 was used as a reference to establish the procedure. The number of grains in the specimens and respective areas were calculated using ImageJ software based on the visual counting performed in the etched micrographs. The images evaluated were of 200x or 500x magnification depending on the grain size. The ratio of the grains per area were then calculated based on the numbers of grain counted in a known area. To compare all the results, the average number of grains per 10^6 μm^2 was considered. In most cases, five images were evaluated for each condition.

2.3. Simulation of the Ageing Response

MatCalc software (MatCalc Engineering, Vienna, Austria) version 6.02 with the corresponding thermodynamic, mobility and physical databases for Al alloys was used for material simulations. The precipitation sequence during ageing was modelled in a simplified manner by considering Cu–Mg clusters and AlMgZn GP zones, nucleating directly from the FCC Al precipitation domain (bulk nucleation sites), while T-phase and S-phase precipitates were formed by a transformation of clusters and GP zones. Ageing treatments with various heat treatments were simulated, taking the annealed material at 500 °C as the initial material condition for precipitation kinetics simulations. For the validation of the results observed, experimentally, samples from the wire arc deposition process were solution heat treated to a peak temperature of 500 °C for 20 min. After quenching, three samples were subjected to an ageing process: (i) at 90 °C for 24 h, (ii) at 140 °C for 24 h and (iii) pre-aged at 90 °C for 24 h and aged at 140 °C for 24 h afterwards. Brinell hardness measurements were conducted in various time intervals during ageing. The chemical composition considered for the simulation was based on the chemical analysis of the alloy obtained after deposition.

3. Results and Discussion

3.1. Wire Arc Deposition

In many instances, the limiting factor with regard to manufacturing is the capability of the alloys to be shaped into components. Consequently, there is a continuous demand to improve existing alloys, as well as to design new ones, in order to meet the requirements of various processing technologies. In the design of the alloy investigated in the present work, engineering properties such as weldability, hot-tearing susceptibility as well as resulting mechanical properties were in focus as detailed by Schnall et al. [38]. The chemical composition of the Al–Zn–Mg–Cu alloy was, thereby, iteratively adapted using a set of thermodynamic calculations, experimental determination of thermophysical properties and engineering tests of the hot-tearing susceptibility.

The absence of visible cracking, major porosity and no macroscopic shrinkage is a good indication of the process stability and the apparent good weldability of the newly designed alloy. Visual inspection of the specimens' surfaces (Figure 1) suggests a low level of surface waviness, which is beneficent if the final surface needs to be machined, i.e., a lower buy-to-fly ratio is achieved or it lowers possible notch effects at unmachined surfaces of the final component. It is noted that the geometric features of the wire arc additive manufacturing specimens were geometrically consistent during consecutive manufacturing of several samples, suggesting uniform deposition conditions with a high reproducibility.

The arc's colour during deposition appeared cyan, qualitatively suggesting the burn off of volatile chemical species. In Table 2, the chemical composition of the specimen after wire arc deposition also evidences the burn off of Mg and Zn during the additive manufacturing process. Hence, it is concluded that when aiming at a specified chemical composition of the final component, the wire needs to be over-alloyed with all volatile chemical elements.

Table 2. Chemical composition of the deposited alloy in the as-built condition measured by OES in weight percent.

As-built (OES)	Al	Mg	Zn	Mn	Si	Cu	Fe	Cr	Ti	Zr
wt.%	90.00	5.33	3.44	0.49	0.07	0.31	0.11	0.04	0.06	0.12

Based on the macrographic results (Figure 4), it can be seen that the middle and top layers exhibited lower waviness, which may be explained by the higher temperature of the material in the deposition of these layers [39]. A higher temperature, together with the concomitant reduced viscosity, results in a higher flow-capability of the melt. Furthermore, the only discontinuities found in the micrographic images are pores. X-ray radiography indicated few dispersed porosities smaller than 1 mm in diameter. The general results presented so far demonstrate that modest adaptations in alloy composition enable improved weldability, allowing for crack-free fabrication of samples by wire arc additive manufacturing, potentially also enabling conventional welding of 7xxx alloy sheets, which will be the subject of further research.

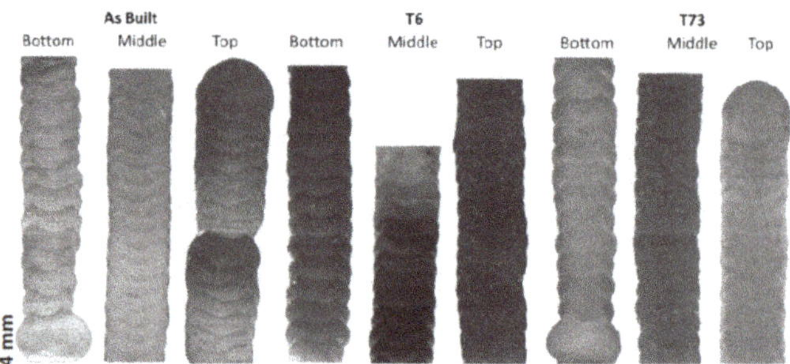

Figure 4. Macrographs of the transverse parts extracted from the samples in various material conditions.

A constant level of hardness is seen along the height of the walls, where the average values found for samples in the AB, T6 and T73 conditions are 103, 131 and 143 HV, respectively (Figure 5). The constant hardness in each material condition suggests sufficient chemical homogeneity after wire arc processing, thus enabling a homogeneous hardening precipitate structure. It is noted that homogeneous mechanical properties counteract strain localisation upon loading, which is beneficial in case of highly loaded structural applications.

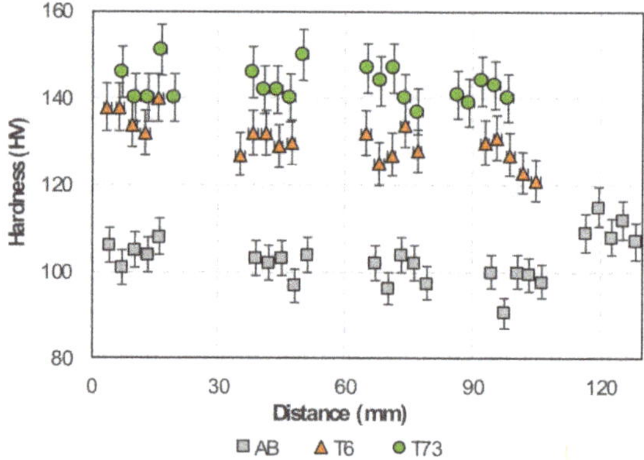

Figure 5. Evolution of Vickers hardness from the bottom (first layer) to the top of the wall, for three different material conditions (as-built, T6 and T73).

3.2. Microstructure Characterisation

3.2.1. Microstructure Evolution

When evaluating the micrographs of the deposited samples (Figure 6), it can be seen in the bottom region of the as-built condition that a dendritic structure is predominant, and no clear grain structure is revealed. In all the other conditions, a mixture of equiaxed and elongated grains are seen. It can be noted that a smaller equiaxed grain structure is discernible in the interlayer region, followed by elongated grains and a zone of coarser grains with mostly equiaxed grains. This is a common feature of aluminium processed by directed energy deposition due to solidification conditions within a pronounced thermal gradient [40].

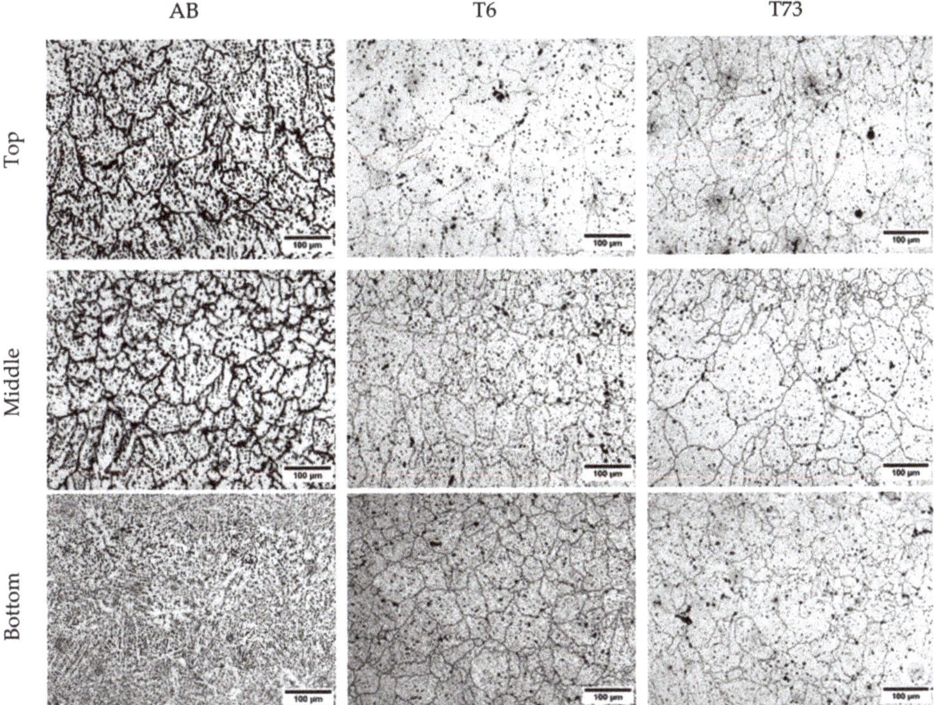

Figure 6. Micrograph of the samples in the AB, T6 and T73 conditions prepared using Barkers reagent.

3.2.2. Porosity Evaluation

The pore volumetric fraction analysis shows that this parameter increases along the samples' height from the bottom to the top (Figure 7). This result might be related to the thermal gradient that is created during the wire arc deposition process. As the specimen's height increases, so does the mean temperature, which affects the metallurgical processes occurring in the melt zone to a yet unknown extent. The smallest average pore diameter found was 6 µm and the highest was 10 µm. The variations among the three types of conditions (as-built, T6 and T73) are expected, since the heat treatments usually promote not only the increase in secondary porosity volume but also the coalescence of small micropores [41]. According to Gu et al. [42], some mechanisms responsible for micropores growth in Al alloys are Ostwald ripening [43], nucleation of new pores and growth of secondary pores [44]. When correlating these results to the results from the tensile test, performed with the subsize specimens extracted from the transverse direction, no visible relation is clearly identified

(Figure 12), suggesting that the pores observed are sufficiently small and few as to not to interfere within the elastic deformation regime and do not cause premature failure upon plastic deformation (see Section 3.4).

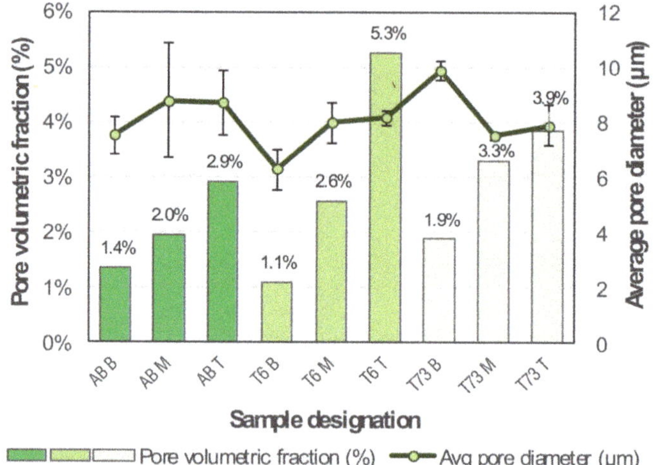

Figure 7. Pore volumetric fraction and average diameter of pores of the samples in the AB, T6 and T73 conditions.

3.2.3. Grain Number

When evaluating the number of grains in the different regions (bottom, middle and top) of the sample in the T6 and T73 conditions (Figure 8), it can be seen that the grains are smaller (higher in quantity) in the bottom region (it was not possible to measure any grain structure in the sample in the as-built condition in the bottom region due to this regions particular microstructure as visible in Figure 6). This observation underpins our previous argumentation: as the specimen height increases, the mean specimen temperature is increased. Thus, thermally controlled coarsening reactions prevail and dominate the resultant grain size via grain growth.

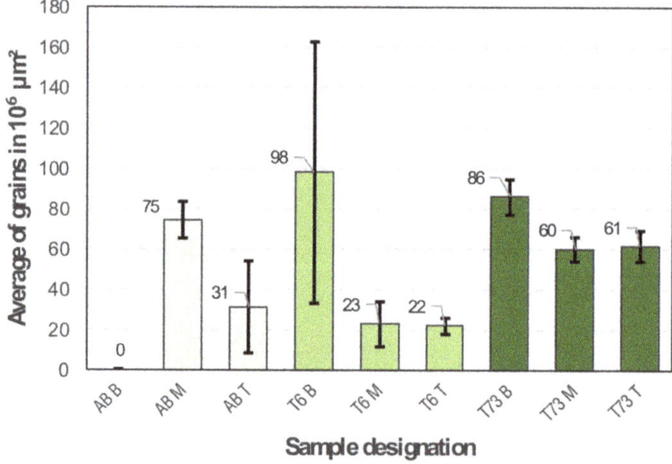

Figure 8. Average of grains in 10^6 μm^2 of the samples in the AB, T6 and T73 conditions.

3.3. Heat Treatment Simulation

In order to estimate the effects of the various heat treatments on the mechanical properties, MatCalc simulations were performed as described in subchapter 2. The maximum yield strength within the simulated time frame of 500 h was observed for ageing at 90 °C, when the material approaches a value of 380 MPa, as can be seen in Figure 9a. Application of lower temperatures resulted in lower yield strength values within the simulated time frame of 500 h. Application of higher temperatures resulted in earlier occurrence of the peak ageing response but in all cases below the strength value found for ageing at 90 °C. The application of a two-stage ageing treatment consisting of 24 h ageing at 90 °C followed by 32 h ageing at 140 °C resulted in the replication of the strength peak at 380 MPa, as shown in Figure 9b. It was found that the application of higher temperatures for the second ageing step resulted in faster coarsening of the initial precipitates, leading to lower strength peaks and faster over-ageing. Application of lower temperatures for the second ageing (i.e., 90 °C) did not increase the simulated strength peak, resulting only in its delayed occurrence.

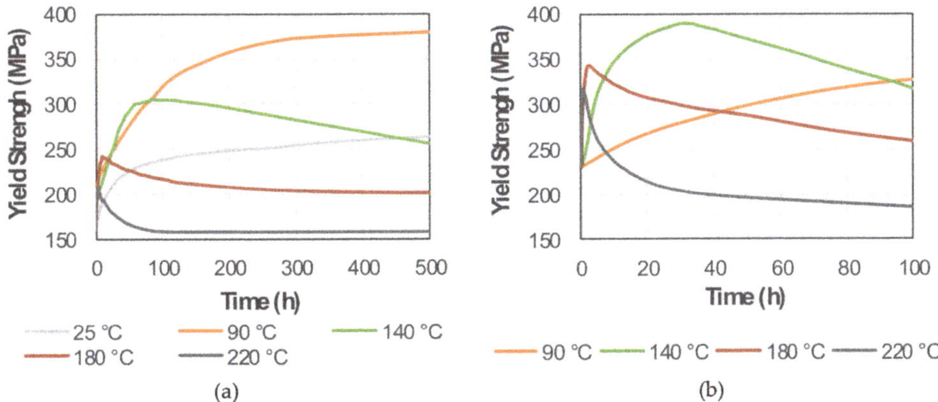

Figure 9. Simulated yield strength for various ageing treatments at various temperatures of: (**a**) single-stage treatment; (**b**) second treatment of double-stage treatment after ageing at 90 °C for 24 h.

For validation of the simulation results, Brinell hardness tests were conducted and compared to the simulated yield strength for three different conditions (Figure 10). For the sample aged at 90 °C, the hardness increases slowly and reaches a plateau at 145–150 HB after ~275 h without any indication of over-ageing. A hardness peak at 130 HB was found for the sample aged at 140 °C after 125 h of ageing, showing a slow hardness decrease for longer ageing duration. The sample subjected to two-stage ageing treatment, held first at 90 °C for 24 h, followed by ageing at 140 °C, exhibited a notable hardness increase within 10 h of the second ageing treatment reaching the value ~155 HB. With in 100 h of the second stage ageing, the sample hardness remained above 150 HB, decreasing slowly afterwards. It is concluded that the experimentally observed trends satisfactorily underpin the simulation results and fit well to results of a recent work [9].

3.4. Mechanical Properties

Concerning the deposition direction in the AB, T6 and T73 conditions, the tensile strength and elongation are lower in transverse specimens, indicating a certain degree of anisotropy (Figure 11). It should be noted that the yield strength is nearly isotropic, which is highly important for practical applications. It can be seen that the ageing treatment increases yield strength, tensile strength and elongation. The T73 condition presents the highest yield strength, while T6 presented a good combination of strength and ductility. When comparing the results obtained in this research with the designed alloy to various Al alloys used in the selective laser melting (SLM) the good mechanical properties achieved are noticeable [45].

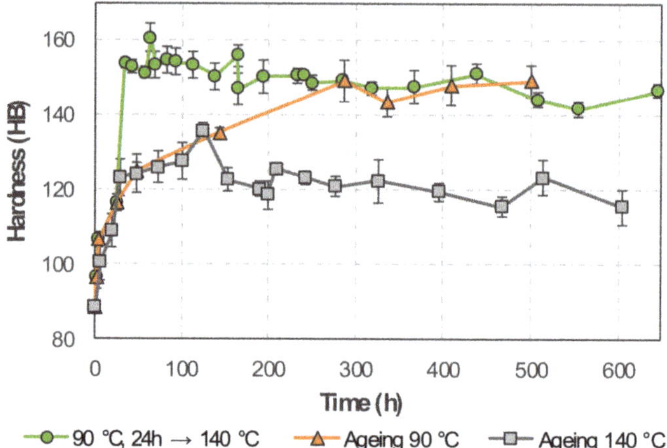

Figure 10. Brinell hardness evolution in samples subjected to various ageing treatments.

In the AB, T6 and T73 conditions, the tensile strength and the fracture elongation for standard and subsize specimens are similar, indicating homogenous mechanical properties along the height of the produced samples, as indicated by the mechanical characteristics of the analysed specimens taken in transverse direction (Figure 12).

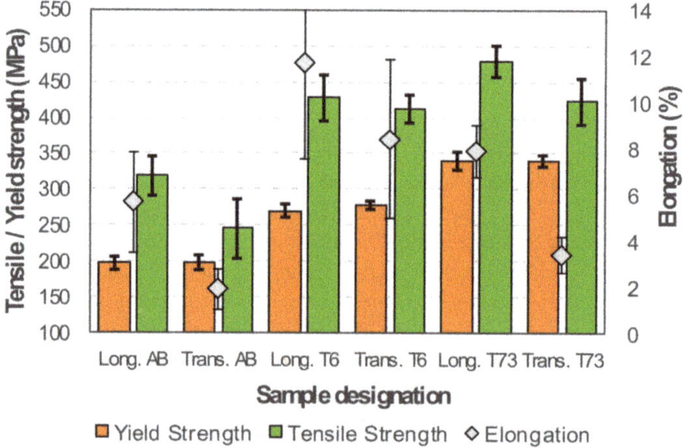

Figure 11. Average yield strength, tensile strength and elongation of samples in the AB, T6 and T73 conditions.

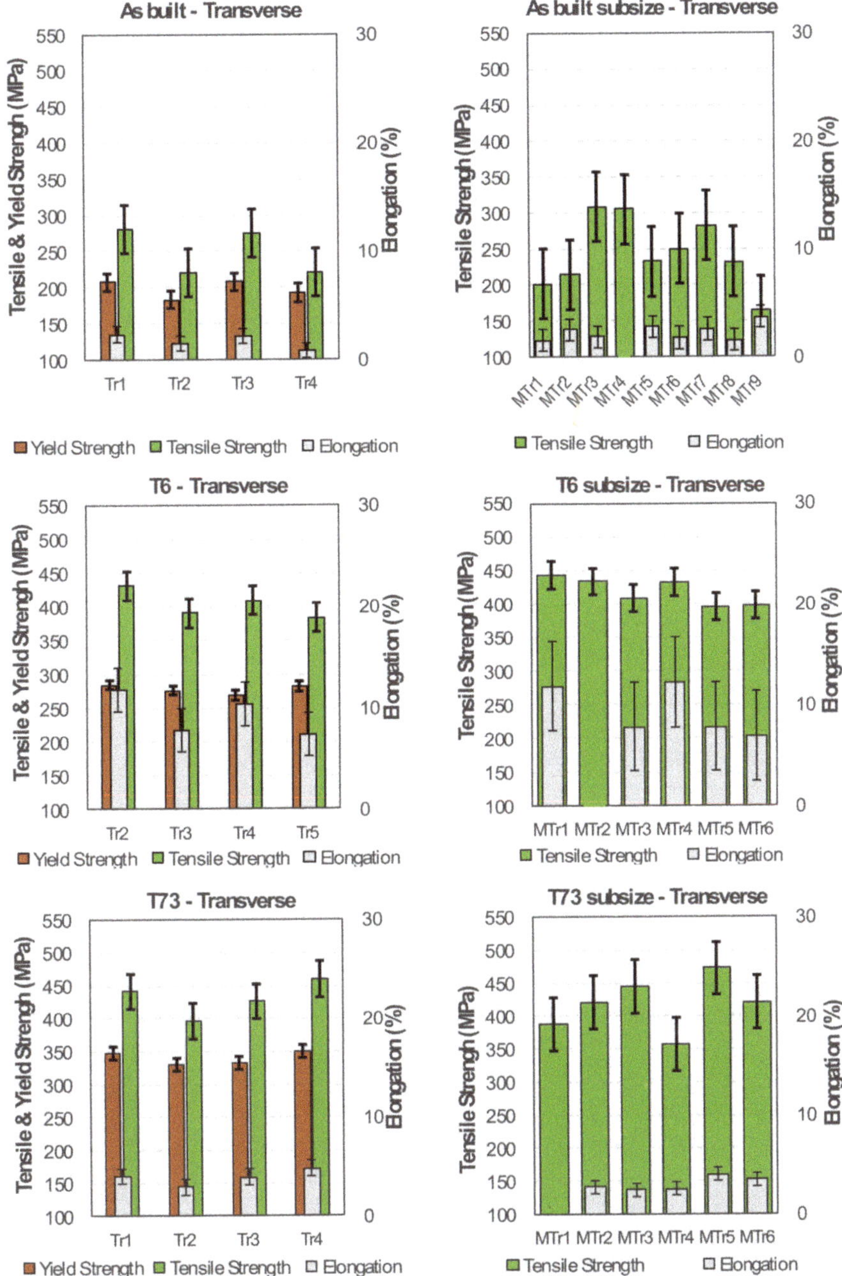

Figure 12. Yield strength, tensile strength and fracture elongation of the regular and subsize specimens in the AB, T6 and T73 conditions.

Due to the limited amount of material available, fatigue testing was performed on specimens in the T73 condition only. This testing was intended to compare the material behaviour in the transverse and longitudinal directions, thus to obtain some indications on the performance of the manufacturing process, rather than fully characterise the fatigue behaviour of the material. The fatigue test results exhibited a considerable scatter for both deposition directions (Figure 13), particularly at higher stress levels. This scatter can presumably be explained by material structural heterogeneities and residual porosity.

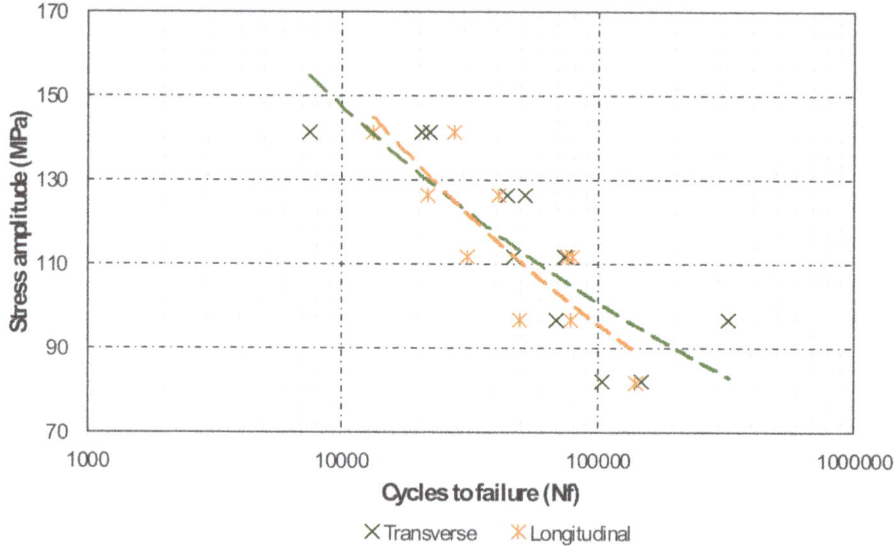

Figure 13. Wöhler curves of longitudinal and transverse specimens in the T73 condition parts.

The main fracture initiation site feature evidenced in almost all specimens analysed was porosity (Figure 14). Post-test analysis of the fracture surfaces revealed that fracture often initiated from porosities at or close to the specimen surface. In some instances, these defects were previously internal or subsurface defects that emerged near the surface as a result of specimen machining and polishing (Figure 15).

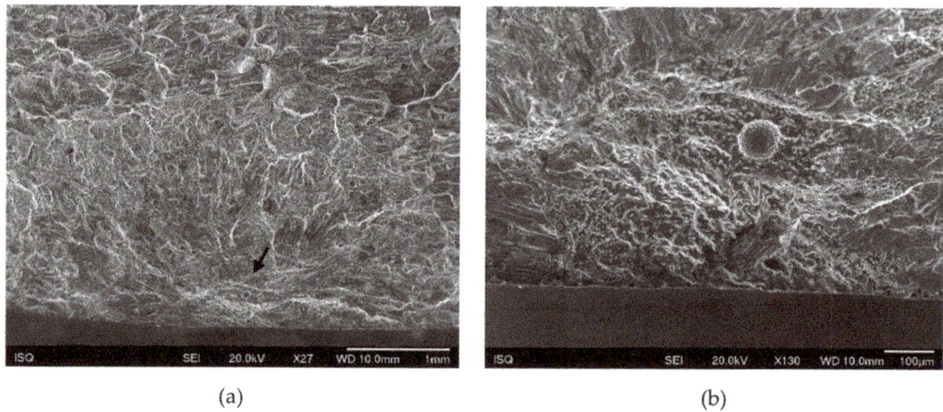

Figure 14. (**a**) A small pore in the initiation site of a longitudinal specimen; (**b**) detail of (a).

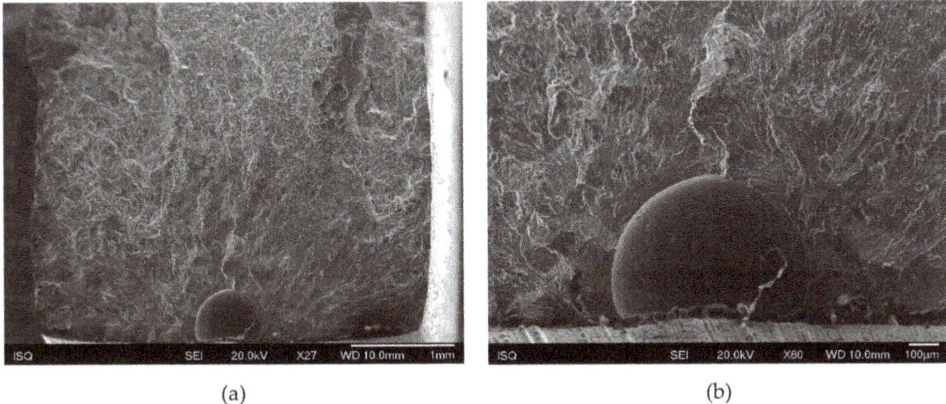

Figure 15. (**a**) A large subsurface pore in the initiation site of a perpendicular specimen; (**b**) detail of (a).

Some intergranular brittleness was also observed, possibly as a result of minor oscillations of deposition conditions leading to some fatigue endurance reduction. This intergranular fracture mode seemed more frequent in perpendicular specimens than in longitudinal ones (Figure 16).

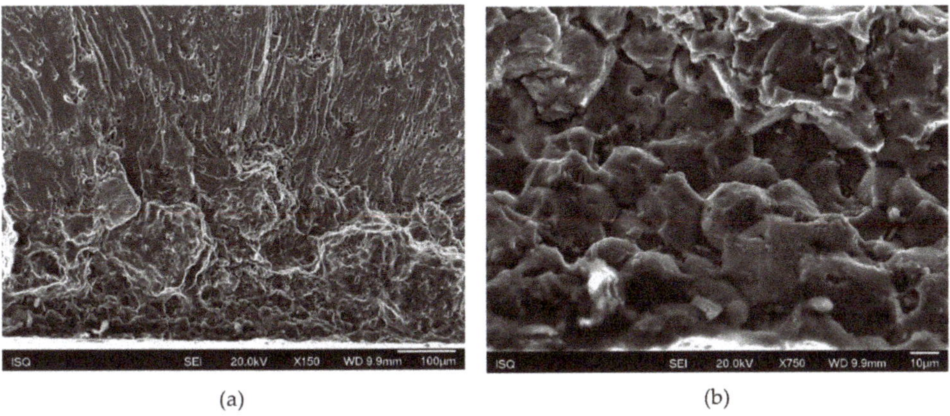

Figure 16. (**a**) Initiation site of a perpendicular specimen; (**b**) detail of the region between the pore and the surface showing intergranular decohesion.

Besides intergranular brittleness, the initiation region and earlier crack growth was often associated with cleavage fracture of grains shifting in a later propagation stage and final unstable fracture to dimpled ductile fracture. Again, porosity was often observed in the propagation path as well as intergranular decohesion, more frequent in the perpendicular specimens (Figure 17).

Despite the differences in the fracture behaviour, the material shows an almost identical behaviour in both longitudinal and transverse directions, which precludes potential for future use in mechanical parts. Comparison of fatigue data is difficult given the large variation of heat treatments, materials and testing conditions. The small number of fatigue tests performed in the present work also contribute to impair such comparisons. Nevertheless, the present results evidence the processability of 7xxx aluminium alloys by wire arc additive manufacturing with remarkable mechanical properties [46–48].

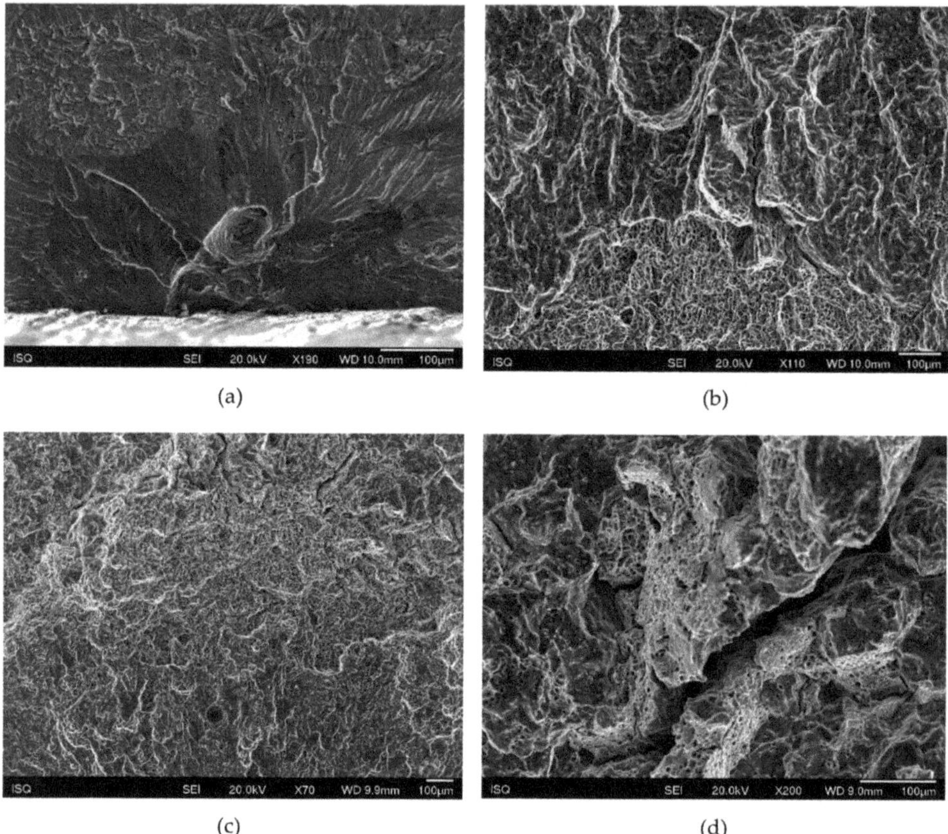

Figure 17. (**a**) Cleavage associated with porosity at the initiation site; (**b**) transition region; (**c**) propagation path with dimpled structure and pores; (**d**) propagation path with intergranular decohesion and dimpled structure.

4. Conclusions

From the performed analyses of the novel Al–Zn–Mg–Cu alloy with improved processability, the following major conclusions can be drawn:

- Wire arc additive manufacturing of the Al5–Mg3–Zn–Cu alloy resulted in a component with exceptionally high mechanical strength when compared to other Al alloys. In fact, the achieved mechanical properties are even superior to the values available for many commercial 7xxx alloys. Furthermore, the heat treatment simulations indicate that even higher mechanical properties can be achieved using optimised ageing treatments.
- Concerning defects, the manufactured parts only exhibit few dispersed porosities and volumetric pore fraction changes along the height of the sample. The volumetric pore fraction is smaller in the bottom and bigger in the top region, while the grains are smaller in the bottom and bigger in the upper region. Both observations are in line with an increasing mean component temperature with an increasing height. Equiaxed and elongated grains are seen along the sample—a typical feature of wire arc processed aluminium. It can be noted that smaller equiaxed grains are seen in the interlayer region, followed by elongated grains and a zone of coarser grains with mostly equiaxed grains.

- The fatigue results showed a high scatter caused mainly by residual porosity, which was the main feature associated with failure initiation and eventual risk of intergranular brittleness.
- The high strength reached under 60 h of ageing time can be attributed to the two-stage ageing treatment, which promoted full formation of precipitation and growth of very fine precipitates of the η′ or T′-phase (semi-coherent, metastable precursors of the equilibrium $MgZn_2$ or T ($Al_2Mg_3Zn_3$) phases, respectively) from GP zones.
- The hardness measurements performed on the aged samples deposited by the wire arc process confirmed the simulation trends and the advantages of the two-stage ageing treatment as a processing method for this material.

This study demonstrates that the wire arc additive manufacturing of a novel 7xxx alloy is not only feasible but results in very good mechanical properties following ensuing heat treatments.

Author Contributions: Conceptualization, P.J.M., B.G., and T.K.; methodology, P.J.M., B.G., M.S., and S.B.; software, D.F., P.W., A.F., and E.K.; validation, P.J.M., B.G., M.S., and S.B.; formal analysis, P.J.M., B.G., M.S., S.B., and T.K.; investigation, B.G., P.S., M.G., M.S., and S.B.; resources, R.G. and P.J.M.; data curation, P.J.M., M.S., and T.K.; writing—original draft preparation, P.J.M., B.G., T.K.; writing—review and editing, P.J.M., B.G., S.B., and T.K.; supervision, R.G. and P.J.M.; project administration, M.S., R.G., and P.J.M.; funding acquisition, R.G. and P.J.M.. All authors have read and agreed to the published version of the manuscript.

Funding: This research was funded by the Austrian Ministry for Transport, Innovation and Technology (bmvit), the Austrian Research Promotion Agency (FFG) and Fundação para a Ciência e a Tecnologia (FCT, Portugal) in the frame of M-era.Net (call 2016, grant n°. 859833).

Conflicts of Interest: The authors declare no conflict of interest.

References

1. Azam, F.I.; Rani, A.M.A.; Altaf, K.; Rao, T.V.V.L.N.; Zaharin, H.A. An In-Depth Review on Direct Additive Manufacturing of Metals. *IOP Conf. Ser. Mater. Sci. Eng.* **2018**, *328*, 012005. [CrossRef]
2. Abdulhameed, O.; Al-Ahmari, A.; Ameen, W.; Mian, S.H. Additive manufacturing: Challenges, trends, and applications. *Adv. Mech. Eng.* **2019**, *11*, 1687814018822880. [CrossRef]
3. Mukherjee, T.; Zuback, J.S.; De, A.; DebRoy, T. Printability of alloys for additive manufacturing. *Sci. Rep.* **2016**, *6*, 19717. [CrossRef] [PubMed]
4. Hirsch, J. Aluminium in Innovative Light-Weight Car Design. *Mater. Trans.* **2011**, *52*, 818–824. [CrossRef]
5. Gloria, A.; Montanari, R.; Richetta, M.; Varone, A. Alloys for aeronautic applications: State of the art and perspectives. *Metals* **2019**, *9*, 662. [CrossRef]
6. Staley, J.; Lege, D. Advances in aluminium alloy products for structural applications in transportation. *J. Phys. IV* **1993**, *3*, C7–C179. [CrossRef]
7. Lumley, R. *Fundamentals of Aluminium Metallurgy: Recent Advances*; Woodhead Publishing: Sawston, UK, 2018; ISBN 0-08-102064-3.
8. Ringer, S.; Hono, K. Microstructural evolution and age hardening in aluminium alloys: Atom probe field-ion microscopy and transmission electron microscopy studies. *Mater. Charact.* **2000**, *44*, 101–131. [CrossRef]
9. Stemper, L.; Mitas, B.; Kremmer, T.; Otterbach, S.; Uggowitzer, P.; Pogatscher, S. Age-hardening of high pressure die casting AlMg alloys with Zn and combined Zn and Cu additions. *Mater. Des.* **2019**, *181*, 107927. [CrossRef]
10. Ditta, A.; Wei, L.; Xu, Y.; Wu, S. Microstructural characteristics and properties of spray formed Zn-rich Al-Zn-Mg-Cu alloy under various aging conditions. *Mater. Charact.* **2020**, 110133. [CrossRef]
11. Mondolfo, L. *Aluminum Alloys Structure and Properties*; Elsevier: Amsterdam, The Netherlands, 1976.
12. Strawbridge, D.; Hume-Rothery, W.; Little, A. The Constitution of Al-Mg-Zn Alloys at 460 °C. *J. Inst. Met.* **1948**, *74*, 191–225.
13. Polmear, I.; StJohn, D.; Nie, J.-F.; Qian, M. *Light Alloys: Metallurgy of the Light Metals*; Butterworth-Heinemann: Oxford, UK, 2017; ISBN 0-08-099430-X.
14. Davis, J.R. *Alloying: Understanding the basics*; ASM international: Materials Park, OH, USA, 2001; ISBN 1-61503-063-8.
15. Ostermann, F. *Anwendungstechnologie Aluminium*; Springer: Berlin/Heidelberg, Germany, 2007.

16. Murayama, M.; Hono, K. Three dimensional atom probe analysis of pre-precipitate clustering in an Al-Cu-Mg-Ag alloy. *Scr. Mater.* **1998**, *38*, 1315. [CrossRef]
17. Dos Santos, J.F.; Staron, P.; Fischer, T.; Robson, J.D.; Kostka, A.; Colegrove, P.; Wang, H.; Hilgert, J.; Bergmann, L.; Hütsch, L.L. Understanding precipitate evolution during friction stir welding of Al-Zn-Mg-Cu alloy through in-situ measurement coupled with simulation. *Acta Mater.* **2018**, *148*, 163–172. [CrossRef]
18. Martin, J.H.; Yahata, B.D.; Hundley, J.M.; Mayer, J.A.; Schaedler, T.A.; Pollock, T.M. 3D printing of high-strength aluminium alloys. *Nature* **2017**, *549*, 365. [CrossRef]
19. Montero-Sistiaga, M.L.; Mertens, R.; Vrancken, B.; Wang, X.; Van Hooreweder, B.; Kruth, J.-P.; Van Humbeeck, J. Changing the alloy composition of Al7075 for better processability by selective laser melting. *J. Mater. Process. Technol.* **2016**, *238*, 437–445. [CrossRef]
20. Casati, R.; Coduri, M.; Riccio, M.; Rizzi, A.; Vedani, M. Development of a high strength Al–Zn–Si–Mg–Cu alloy for selective laser melting. *J. Alloys Compd.* **2019**, *801*, 243–253. [CrossRef]
21. Taylor, J.A. Iron-containing intermetallic phases in Al-Si based casting alloys. *Procedia Mater. Sci.* **2012**, *1*, 19–33. [CrossRef]
22. Couture, A. Iron in aluminum casting alloys-a literature survey. *Int. Cast Met. J.* **1981**, *6*, 9–17.
23. Dinnis, C.; Taylor, J. Manganese as a" neutraliser" of iron-related porosity in Al-Si foundry alloys. *Appl. Catal. B Environ.* **2007**, *101*, 388–396.
24. Eskin, D.G.; Suyitno; Katgerman, L. Mechanical properties in the semi-solid state and hot tearing of aluminium alloys. *Prog. Mater. Sci.* **2004**, *49*, 629–711. [CrossRef]
25. Fukuda, T. Weldability of 7000 series aluminium alloy materials. *Weld. Int.* **2012**, *26*, 256–269. [CrossRef]
26. Cao, C.; Zhang, D.; Wang, X.; Ma, Q.; Zhuang, L.; Zhang, J. Effects of Cu addition on the precipitation hardening response and intergranular corrosion of Al-5.2 Mg-2.0 Zn (wt.%) alloy. *Mater. Charact.* **2016**, *122*, 177–182. [CrossRef]
27. Williams, S.W.; Martina, F.; Addison, A.C.; Ding, J.; Pardal, G.; Colegrove, P. Wire+ arc additive manufacturing. *Mater. Sci. Technol.* **2016**, *32*, 641–647. [CrossRef]
28. Almeida, P.S. Process Control and Development in Wire and Arc Additive Manufacturing. PhD Thesis, Cranfield University, Cranfield, UK, 2012.
29. Schmidt, M.; Merklein, M.; Bourell, D.; Dimitrov, D.; Hausotte, T.; Wegener, K.; Overmeyer, L.; Vollertsen, F.; Levy, G.N. Laser based additive manufacturing in industry and academia. *CIRP Ann.* **2017**, *66*, 561–583. [CrossRef]
30. Plangger, J.; Schabhüttl, P.; Vuherer, T.; Enzinger, N. CMT additive manufacturing of a high strength steel alloy for application in crane construction. *Metals* **2019**, *9*, 650. [CrossRef]
31. Rodrigues, T.A.; Duarte, V.; Miranda, R.; Santos, T.G.; Oliveira, J. Current status and perspectives on wire and arc additive manufacturing (WAAM). *Materials* **2019**, *12*, 1121. [CrossRef]
32. Köhler, M.; Fiebig, S.; Hensel, J.; Dilger, K. Wire and Arc Additive Manufacturing of Aluminum Components. *Metals* **2019**, *9*, 608. [CrossRef]
33. Ding, D.; Pan, Z.; Cuiuri, D.; Li, H. A multi-bead overlapping model for robotic wire and arc additive manufacturing (WAAM). *Robot. Comput.-Integr. Manuf.* **2015**, *31*, 101–110. [CrossRef]
34. Cunningham, C.; Wikshåland, S.; Xu, F.; Kemakolam, N.; Shokrani, A.; Dhokia, V.; Newman, S. Cost modelling and sensitivity analysis of wire and arc additive manufacturing. *Procedia Manuf.* **2017**, *11*, 650–657. [CrossRef]
35. Ryan, E.; Sabin, T.; Watts, J.; Whiting, M. The influence of build parameters and wire batch on porosity of wire and arc additive manufactured aluminium alloy 2319. *J. Mater. Process. Technol.* **2018**, *262*, 577–584. [CrossRef]
36. Cong, B.; Qi, Z.; Qi, B.; Sun, H.; Zhao, G.; Ding, J. A comparative study of additively manufactured thin wall and block structure with Al-6.3% Cu alloy using cold metal transfer process. *Appl. Sci.* **2017**, *7*, 275. [CrossRef]
37. Schneider, C.A.; Rasband, W.S.; Eliceiri, K.W. NIH Image to ImageJ: 25 years of image analysis. *Nat. Methods* **2012**, *9*, 671–675. [CrossRef] [PubMed]
38. Schnall, M.; Bozorgi, S.; Klein, T.; Gradinger, R.; Birgmann, A.; Morais, P.; Warzok, P. Sonder-Aluminium-Schweißzusätze optimiert für die Verarbeitung im Wire-Arc Additive Manufacturing-Prozess. In *DVS Congress*; DVS Media GmbH: Dusseldorf, Germany, 2019.
39. Geng, H.; Li, J.; Xiong, J.; Lin, X. Optimisation of interpass temperature and heat input for wire and arc additive manufacturing 5A06 aluminium alloy. *Sci. Technol. Weld. Join.* **2016**, *22*, 1–12. [CrossRef]

40. Froend, M.; Ventzke, V.; Dorn, F.; Kashaev, N.; Klusemann, B.; Enz, J. Microstructure by design: An approach of grain refinement and isotropy improvement in multi-layer wire-based laser metal deposition. *Mater. Sci. Eng. A* **2020**, *772*, 138635. [CrossRef]
41. Talbot, D.; Granger, D. Secondary hydrogen porosity in aluminium. *J. Inst. Met.* **1964**, *92*, 290.
42. Gu, J.; Yang, S.; Gao, M.; Bai, J.; Zhai, Y.; Ding, J. Micropore evolution in additively manufactured aluminum alloys under heat treatment and inter-layer rolling. *Mater. Des.* **2020**, *186*, 108288. [CrossRef]
43. Toda, H.; Hidaka, T.; Kobayashi, M.; Uesugi, K.; Takeuchi, A.; Horikawa, K. Growth behavior of hydrogen micropores in aluminum alloys during high-temperature exposure. *Acta Mater.* **2009**, *57*, 2277–2290. [CrossRef]
44. Anyalebechi, P.N.; Hogarth, J. Effect of supereutectic homogenization on incidence of porosity in aluminum alloy 2014 ingot. *Metall. Mater. Trans. B* **1994**, *25*, 111–122. [CrossRef]
45. Aboulkhair, N.T.; Simonelli, M.; Parry, L.; Ashcroft, I.; Tuck, C.; Hague, R. 3D printing of Aluminium alloys: Additive Manufacturing of Aluminium alloys using selective laser melting. *Prog. Mater. Sci.* **2019**, *106*, 100578. [CrossRef]
46. Zhao, T.; Jiang, Y. Fatigue of 7075-T651 aluminum alloy. *Int. J. Fatigue* **2008**, *30*, 834–849. [CrossRef]
47. Vu, C.; Wern, C.; Kim, B.H.; Kim, S.K.; Choi, H.J.; Yi, S. Fatigue Behaviour of New Eco-7175 Extruded Aluminium Alloy. In Proceedings of the 8th European Conference for Aeronautics and Space Sciences (EUCASS), Madrid, Spain, 1–4 July 2019. [CrossRef]
48. Ozdes, H. The Relationship between High-Cycle Fatigue and Tensile Properties in Cast Aluminum Alloys. Master's Thesis, University of North Florida, Jacksonville, FL, USA, 2016.

Article

Reduction of Energy Input in Wire Arc Additive Manufacturing (WAAM) with Gas Metal Arc Welding (GMAW)

Philipp Henckell *, Maximilian Gierth, Yarop Ali, Jan Reimann and Jean Pierre Bergmann

Production Technology Group, Technische Universität Ilmenau, D-98693 Ilmenau, Germany; maximilian.gierth@tu-ilmenau.de (M.G.); yarop.ali@tu-ilmenau.de (Y.A.); jan.reimann@tu-ilmenau.de (J.R.); jeanpierre.bergmann@tu-ilmenau.de (J.P.B.)

* Correspondence: philipp.henckell@tu-ilmenau.de; Tel.: +49-3677-692762

Received: 15 April 2020; Accepted: 25 May 2020; Published: 29 May 2020

Abstract: Wire arc additive manufacturing (WAAM) by gas metal arc welding (GMAW) is a suitable option for the production of large volume metal parts. The main challenge is the high and periodic heat input of the arc on the generated layers, which directly affects geometrical features of the layers such as height and width as well as metallurgical properties such as grain size, solidification or material hardness. Therefore, processing with reduced energy input is necessary. This can be implemented with short arc welding regimes and respectively energy reduced welding processes. A highly efficient strategy for further energy reduction is the adjustment of contact tube to work piece distance (CTWD) during the welding process. Based on the current controlled GMAW process an increase of CTWD leads to a reduction of the welding current due to increased resistivity in the extended electrode and constant voltage of the power source. This study shows the results of systematically adjusted CTWD during WAAM of low-alloyed steel. Thereby, an energy reduction of up to 40% could be implemented leading to an adaptation of geometrical and microstructural features of additively manufactured work pieces.

Keywords: additive manufacturing; wire arc additive manufacturing; WAAM; GMAW; energy input per unit length; processing strategy; contact tip to work piece distance; electrical stickout

---

1. Introduction and State of the Art

The industrial application of additive manufacturing processes for metallic parts is growing continuously. A market growth of 41.9% was found for metallic parts in 2018 [1]. This includes technologies such as powder bed fusion (PBF) as well as direct energy deposition (DED) processes. Thereby, a differentiation is made between the applied power source (e.g., laser, electron beam and arc) and the deployed material (e.g., powder and wire) [2]. Herein, differentiations can be adhered in terms of productivity and buildup rates as well as near net shape and surface roughness of manufactured parts.

When using arc-based processes like gas metal arc welding (GMAW), deposition rates of up to 8 kg/h can be achieved [3,4]. Thereby, the buildup volume is only limited by the handling system so that manufacturing of large-volume components can be carried out [4,5]. Though, the dimensional accuracy or surface quality is limited due to comparatively large melt pool sizes. A subtractive post manufacturing process, such as milling is necessary to meet the required tolerances [3,6,7].

In recent years, this technology has been used for the fabrication of high-performance parts in aerospace industry or the energy sector. Therefore, materials such as titanium alloys [8,9] or nickel-base alloys [10,11] have gained increasing interest in the last years. The development of cost efficient production systems [12] with high deposition rates makes this technology accessible for industries such as architecture or construction engineering. Moreover, the freedom of design for complex 3D-structures surpasses the limits of fabrication with conventional methods. Thereby, wire arc additive manufacturing (WAAM) of steel parts is mainly addressed. Recent investigations on low-alloyed steel [13–16] or high-alloyed steel [17,18] demonstrate the potential for these sectors.

The main challenge in WAAM is the high, periodic heat input due to the arc welding process. Thus, geometrical features of the deposited layers are influenced by large melt pool sizes as well as microstructural properties such as grain size. The analysis and reduction of heat input during WAAM is a key factor for the application of this technology. Therefore, investigations on energy reduced welding regimes were carried out [19,20]. Moreover, cooling strategies were developed to reduce interlayer temperatures and enhance processing times [21,22]. However, the described methods of energy reduction are commonly coupled to specific hardware such as welding power sources with energy reduced processes or additional equipment, e.g., fluid bath, cooling clamps, etc. An applicable approach for the reduction of energy input during WAAM by gas metal arc welding is the extension of the free wire length during processing. Though, electrical properties such as welding current, respectively welding power can be reduced as described below.

In GMAW, a determining factor for the energy input, the process stability and the formation of the arc is the free wire length l_{FW}, respectively electrical stickout, which is defined as the distance between the contacting point of the fed wire with the contact tip and the attaching point of the arc on the wire [23,24]. Due to an altering arc length, especially in the short arc welding regime, the electrical stickout is difficult to determine. Therefore, the definition of the contact tube to work piece distance (CTWD) has become established [25–27]. In Figure 1a the definition of CTWD and electrical stickout is shown schematically.

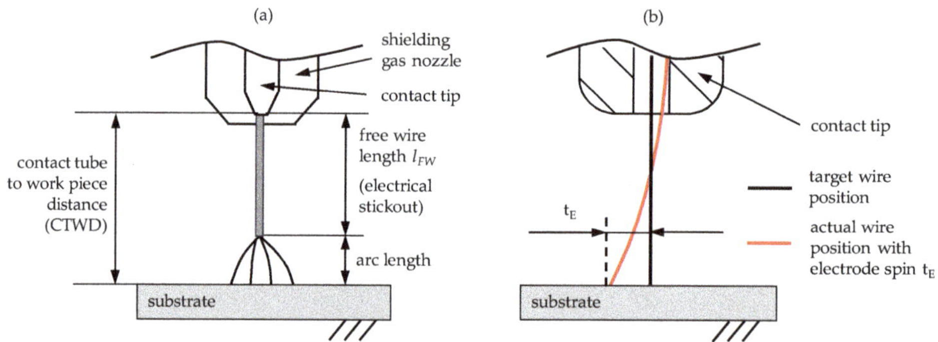

Figure 1. (**a**) Definition of contact tube to work piece distance (CTWD) and (**b**) effect of the electrode spin on wire positioning in gas metal arc welding [26–28].

The increase of CTWD during the welding process is limited, due to the spin of the electrode resulting from the wire spooling (compare Figure 1b). Subsequently, an offset of the wire tip to the desired welding position t_E may occur with increased CTWD. The consequences are unstable arc behavior, increased spattering or weld seam irregularities such as lack of fusion [29,30]. In WAAM the electrode offset affects the software supported path planning with the risk of dimensional deviations from the CAD model and the work piece [28].

However, the adjustment of the CTWD is a simple and highly efficient strategy to affect energy input in gas metal arc welding, respectively WAAM. This can be demonstrated throughout the following Equations (1)–(4). It has to be stated that a change of CTWD has no influence on the wire feeding speed and almost no influence on the arc voltage [31].

In GMAW, the welding current I_W is conducted over the free wire length l_{FW} to the arc [32]. With an extended free wire length, the electrical resistance R_{FW} of the electrode increases [27] (compare Equation (1)). Thereby, ρ_W describes the specific resistance of the wire material and A_{FW} represents the cross sectional area of the wire. For metallic wires, the specific resistance ρ_W is temperature-dependent and can be described by Equation (2). Thereby, α is the temperature coefficient, T the temperature and T_0 a temperature at which the specific resistance $\rho_W(T_0)$ is known, e.g., T_0 = 293.15 K = 20 °C. Equation (2) shows a linear correlation of temperature and specific resistance [33]. Thus, increasing temperature leads to increasing specific resistance for metals and hence affects the electrical resistance of the free wire length R_{FW}.

$$R_{FW} = \rho_W \times \frac{l_{FW}}{A_{FW}} \tag{1}$$

$$\rho_W(T) = \rho_W(T_0) \times (1 + \alpha \cdot (T - T_0)) \tag{2}$$

In addition to the increased resistance of the free wire length R_{FW}, the set welding current I_W and the resistance of the arc R_{Arc} influence the welding power P_W. Thereby, the current load of the electrode can be decreased with constantly set welding power (Equation (3)) [24,34]. According to the first Joule's law, this leads to increased resistance heating of the free wire Q_{FW} proportional to the square of the welding current I_W, the resistance of the wire R_{FW} and the welding time t_W (compare Equation (4)).

$$P_W = I_W^2 \times (R_{FW} + R_{Arc}) \tag{3}$$

$$Q_{FW} = I_W^2 \times R_{FW} \times t_W \tag{4}$$

As a result, the preheating of the free wire length leads to reduced welding power for melting the electrode. Thereby, the same deposition rate can be achieved [24,27]. A characteristic feature of the resulting material transfer is the formation of larger droplets. This can be described by the softening of the wire throughout the resistance heating as well as a reduction of the electromagnetic pinch-force, due to the reduced welding current. The decreased welding current leads to decreased penetration depth and increased weld seam height in joint welding and cladding [25,32].

The described correlations show a reduction of the welding current at a constant deposition rate, respectively a reduced heat input during the welding process. This makes the process strategy particularly suitable for additive manufacturing using WAAM. So far, the described advantages of CTWD extension have not been investigated in the context of wire arc additive manufacturing and exhibit large potential for the affection of geometrical, microstructural and mechanical properties of additively manufactured work pieces.

2. Scope of the Investigations

The objective of this study was the reduction of energy input during WAAM with gas metal arc welding throughout the systematic extension of the contact tube to work piece distance. Thereby, effects on droplet transfer such as resulting droplet diameter or detaching frequency were analyzed by high velocity video recording. Moreover, measurements of welding current and voltage were analyzed to record the resulting reduction of energy input during additive manufacturing. Moreover, investigations on the microstructure and grain size as well as mechanical properties such as material hardness were carried out on low-alloyed steel. Finally, the analysis of cooling rates of the set layers enabled the correlation of interactions between the process strategy and mechanical properties of the built structures.

3. Experimental Methods

Additive manufacturing was carried out with a gas metal arc welding process. Therefore, a welding power source of the type Alpha Q 552 Puls (EWM AG, Mündersbach, Germany) and a water-cooled welding torch WH W 500 (Alexander Binzel Schweißtechnik GmbH & Co. KG, Buseck, Germany) was used. The reproducible guidance of the welding torch was realized by a 6-axis industrial robot KR15-2 (KUKA AG, Augsburg, Germany). The GMAW process was carried out in short arc welding regime throughout the present case study. During WAAM, a temperature controlled strategy with flexible dwell time was implemented. Therefore, the welded layers were naturally cooling to 100 °C before the next layer was applied. The process was monitored with IR-camera of the type ImageR 8320 (InfraTec GmbH, Dresden, Germany). This strategy prevents excessive heat input and heat accumulation in the structure and enables comparable cooling conditions for the analysis of microstructural evolution. Therefore, cooling rates were simultaneously recorded by IR-camera in the range between 900 and 400 °C with a frequency of 50 Hz. The camera was positioned horizontally to the manufactured wall structures with a distance of 600 mm. The analysis of cooling rates was carried out with the software IRBIS 3 plus (InfraTec GmbH, Dresden, Germany). The experimental setup is shown in Figure 2.

High velocity camera imaging was carried out with HV-camera of the type CR2000x2 (Optronis GmbH, Kehl, Germany) to analyze arc behavior and droplet transfer during GMAW. The frame rate was set to 2500 fps at a resolution of 800 pixels × 600 pixels. The shutter speed was set to 1/10.000 s and the aperture adjusted manually to fade-out the arc partially in order to visualize droplet transfer.

Measurement data was recorded by a Dewetron DEWE-PCI 16 measuring system (version DEWE-800, Dewetron GmbH, Grambach, Austria) with data acquisition of current, voltage and welding power. The generated wall structures were separated from the substrate by wire-electro discharge machining (EDM). Afterwards the samples were prepared for metallographic analysis including grinding, polishing and etching with alcoholic Nital solution based on 3% nitric acid. The hardness measurements were carried out using a DuraScan 70 machine (Struers GmbH, Willich, Germany) using the Vickers testing method according to [35] with a force of 9.807 N (HV1).

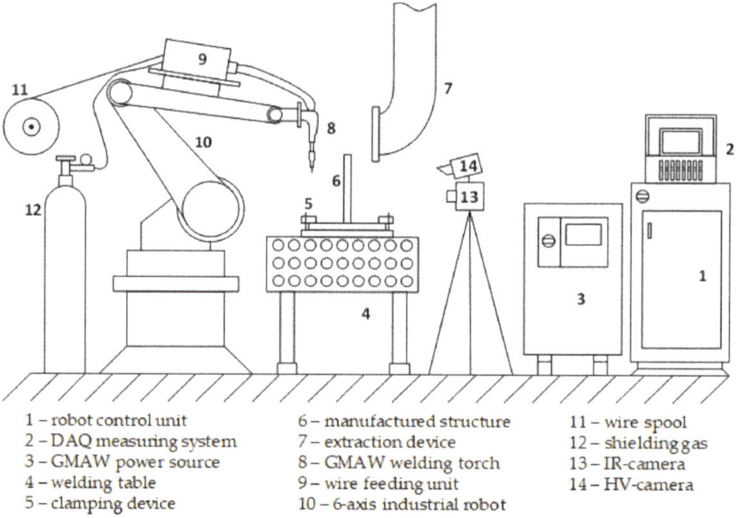

Figure 2. Experimental setup of the wire arc additive manufacturing (WAAM) process.

Low-alloyed steel S355 J2 + N (1.0570) was used as substrate material with a thickness of 20 mm. A solid wire of low-alloyed steel G4Si1/SG3 (1.5130, Ø = 1.0 mm) was used as filler material in the GMAW process. Herein, the silicon percentage was increased by 0.3% by the manufacturer to enhance the binding of oxygen and reduce the formation of porosity in the weld bead. Prospectively, the chosen filler material is of high interest for industrial sectors such as construction engineering and architecture. The following Table 1 shows the chemical composition of the substrate material and filler wire for WAAM. During additive manufacturing, 98%Ar/2%CO_2 was used as shielding gas for the GMAW process.

Table 1. Chemical composition of substrate and welding wire (%).

Function	Material	C_{max}	Si_{max}	Mn_{max}	Cu_{max}	Fe_{max}
substrate	S355J2 + N (1.0570)	0.20	0.55	1.60	0.55	balance
welding wire	G4Si1/SG3 (1.5130)	0.07	1.00	1.64	0.05	balance

During experimental trials, the contact tube to work piece distance was systematically increased in steps of 8 mm. The starting value was set to CTWD = 12 mm, which is the standard distance based on the formula CTWD = 10 … 12 × $Ø_{wire}$. The maximum extension was set to CTWD = 52 mm. Referring to the state of the art, a spin of the welding electrode t_E could be observed resulting from the wire spooling. The maximum measured value of t_E is shown in Figure 3a at the maximum extension of CTWD = 52 mm. Hereby, the electrode spin was measured at a constant value of t_E = 1.8 mm. A commonly displaced position of the electrode to one direction does not affect the additive manufacturing process and can be compensated throughout the programming of an offset in the handling system.

In order to provide constant gas shielding conditions on the weld bead, a constructional adaptation of the shielding gas nozzles was applied for the experimental trials with increased CTWD. Herein, the increased distance between substrate and contact tube requires a focused gas flow to the weld bead equally to the initial setup. Figure 3b shows the adjustment of the lengths of the shielding gas nozzles for increased CTWD.

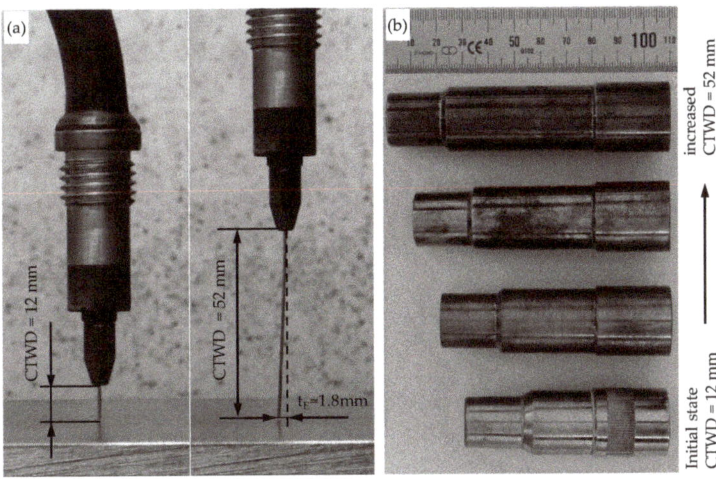

Figure 3. (**a**) Measurement of electrode spin with increased CTWD and (**b**) adaptation of shielding gas nozzles for gas metal arc welding (GMAW) with increased CTWD.

4. Results and Discussion

4.1. Analysis of Elctrical Properties in WAAM with Increased CTWD

In the first step, electrical properties were analyzed during WAAM with a short arc welding regime to characterize the stability of the arc and short arc frequency. Welding voltage and current were measured with a rate of 1000 Hz for GMAW processes with varied CTWD from 12 to 52 mm. The following Figure 4 shows welding current and voltage for a period of 350 ms (Figure 4a CTWD = 12 mm; Figure 4b CTWD = 52 mm). Furthermore, characteristic points of the electrical signals and droplet formation in short arc welding regime are exemplarily shown in Figure 4b throughout circled numbers in the diagram.

Short circuits are characterized by deflections in welding current and voltage in relation to the base values. Herein, a rapid voltage drop with increasing current could be recognized (compare Figure 4b, circle number 1). At this point, the molten droplet transferred from the end of the electrode into the melt bead. A material bridge between the fed wire and the melt pool occurred. After material transition the arc re-ignited. Therefore, high voltage was required, which was characterized by peaks in welding voltage (compare Figure 4b, circle number 2). At this point, welding current dropped slowly due to inductances in the welding circuit, leading to high welding power during arc re-ignition. The subsequent period of arc time led to the melting of the electrode tip and the formation of a new droplet (compare Figure 4b; circle number 3). Thereby, increasing arc time promoted droplet growth (compare Figure 4b; circle number 4) before the molten material transferred into the melt bead during the next short circuit phase.

Figure 4 shows differences in short circuit frequency and uniformity of the short circuits over time with increasing CTWD. Thereby, Figure 4a shows homogeneous droplet transfer with periodical short circuits of f = 42 Hz with a CTWD of 12 mm. Increasing CTWD leads to a reduction of the short circuit frequency (compare Figure 4b). This can be explained by the formation of large volume droplets due to decreased welding current respectively decreased pinch force. Thus, Figure 4b shows a short circuit frequency of f = 11 Hz with increased CTWD of 52 mm. The irregularity of the short circuit frequency arose through the formation of larger droplets (compare Figure 4b; circle number 4), which were transferred with higher kinetic energy into the melt pool causing a wave formation of the liquid melt pool. These waves partially reached the newly forming drop at the end of the wire, creating a new short circuit. This effect caused the transfer of smaller droplets with a reduced detaching frequency during short circuit. The described effect is shown in Figure 4b in the time window between 200 and 350 ms.

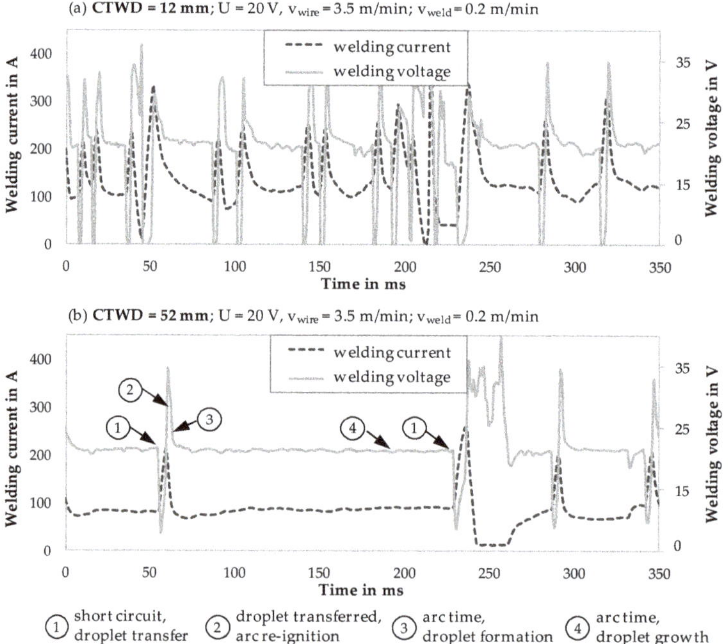

Figure 4. Analysis of welding current and voltage in GMAW with varied contact tube to work piece distance (CTWD); (**a**) CTWD = 12 mm and (**b**) CTWD = 52 mm.

It can be shown that increased CTWD led to a reduction of droplet transfers in the short arc welding regime. Concurrently, phases of arc re-ignition with high welding power decline, leading to energy reduction induced by the welding regime and respectively the arc. As a result, the energy input per unit length E_S, as a main dimension for the characterization of arc welding processes, can be reduced with a constantly set welding velocity. Equation (5) describes the correlation between welding power P_{weld} and welding velocity v_{weld} Herein, welding voltage U_{weld} and current I_{weld} can be measured directly to estimate welding power P_{weld} (compare Equation (6)).

$$E_S\left[\frac{\mathrm{kJ}}{\mathrm{cm}}\right] = \frac{P_{weld}}{v_{weld}}\left[\frac{\mathrm{kW}}{\frac{\mathrm{cm}}{\mathrm{s}}}\right] \quad (5)$$

$$P_{weld} = \frac{1}{t}\int_0^t p_{weld}(t)dt = \frac{1}{t}\int_0^t u_{weld}(t)i_{weld}(t)dt \quad (6)$$

The following Figure 5 shows the correlation of energy input per unit length E_S to the contact tube to work piece distance (CTWD) for three parameter sets. Thereby, E_S was set to values between 4.2 and 8.2 kJ/cm in the initial state of 12 mm CTWD. Increasing CTWD from 12 to 52 mm shows a steady reduction of the energy input per unit length. This relation is represented independently from the parameter set. It was evident that an increase in CTWD of 8 mm led to a reduction of energy input per unit length of 8%. A CTWD of 52 mm exhibited the maximum reduction in energy input per unit length of 40% compared to the initial state. These results correspond to the state of the art in the field of joint welding and cladding as described [25,31].

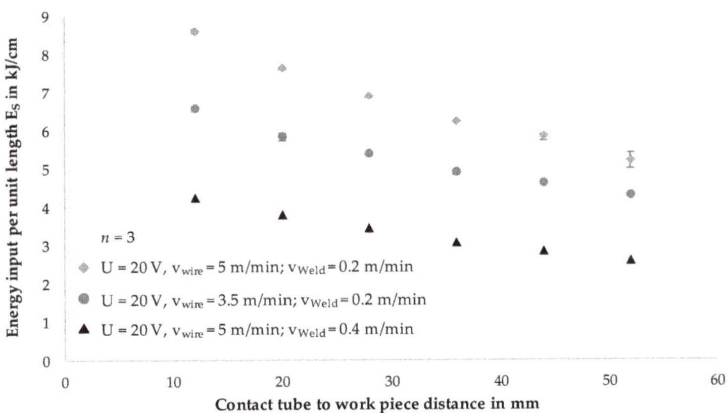

Figure 5. Analysis of energy input per unit length with increasing CTWD for differing parameter sets.

4.2. Analysis of Droplet Transfer in WAAM with Increased CTWD

Chapter 4.1 verified alterations in short circuit frequency with adapted CTWD throughout the analysis of electrical properties. Referring to the state of the art, increasing CTWD leads to the formation of large volume droplets during material transfer in GMAW [25,32]. Experimental investigations with a high velocity video recording confirmed a constant droplet growth with increasing CTWD. Figure 6 shows droplet formation in GMAW with differing CTWD of 12 mm, 28 mm and 44 mm at characteristic points during the short arc welding regime and correlated to the circled numbers in Figure 4b. The increase of CTWD is shown in uniform step size of 16 mm for equal welding parameters. Thereby, the formation of a new droplet on the electrode directly after short circuit phase is shown (compare Figure 4b, circle number 3) as well as droplet growth during arc time (compare Figure 4b, circle number 4). Furthermore, the maximum droplet size is displayed directly before the material transfer in the short circuit phase. The focus of the investigations was set to analyze droplet growth and size before material transition into the melt bead. Therefore, the short circuit phase with material transfer (compare Figure 4b, circle number 1) and arc re-ignition (compare Figure 4b, circle number 2) is not shown in Figure 6.

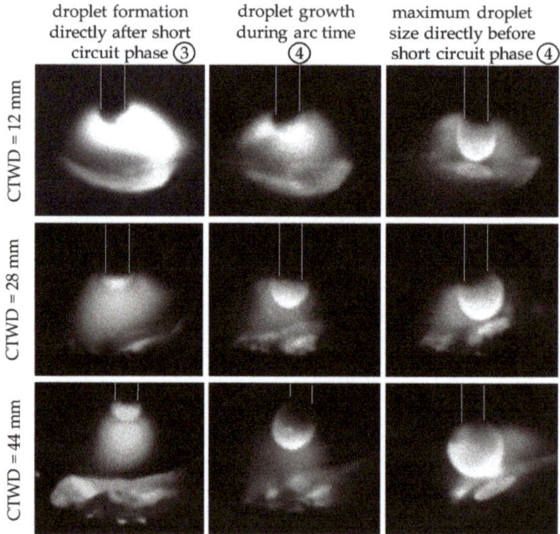

Figure 6. Droplet formation during GMAW with increasing CTWD for U = 20 V; v_{wire} = 5 m/min and v_{weld} = 0.4 m/min.

The coupled process development of increasing droplet size and reduced short circuit frequency was caused by increasing resistance heating of the free wire length with increasing CTWD. Due to high temperatures, the extended wire softened. This effect enhanced the formation of droplets with larger material volume at the end of the wire. At the same time, the pinch effect was reduced, which was induced by the Lorentz force of the magnetic field. The reason for this is the lower welding current that affects the Lorentz force squared. Figure 7a shows droplet formation for CTWD of 12 mm and the maximum investigated value of 52 mm (Figure 7b) in detail. Herein, characteristic effects in the formation of the arc and the resulting droplet can be shown with differing CTWD.

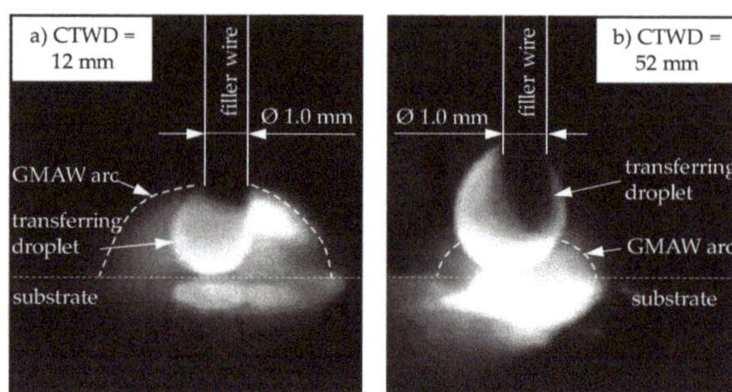

Figure 7. Correlation of droplet growth and arc behavior in GMAW with increased CTWD for U = 20 V; v_{wire} = 5 m/min and v_{weld} = 0.4 m/min.

Thereby, a contact tube to work piece distance of 12 mm indicates the arc attaching point at the filler wire, enclosing the formed droplet at the end of the electrode (compare Figure 7a). At this point, the electromagnetic pinch force was applied, constricting the molten droplet. However, increasing CTWD led to a displacement of the arc attaching point to the bottom of the formed droplet (compare Figure 7b). This resulted in a reduced pinch force to constrict the molten material, due to a decreased welding current. The formation of large volume droplets was the consequence.

The described correlation between short circuit frequency (compare Figure 4) and droplet volume (compare Figure 7) with systematically varied CTWD is shown in Figure 8. To determine the mean droplet size, high-speed camera recordings were analyzed and droplet diameters measured shortly before the transition into the melt pool on five consecutive droplets. The wire width served as a reference for the measurements in the 2-dimensional images. The short circuit frequency was determined on the basis of the number of current peaks, respectively voltage drops for a period of one second during the welding process. In the short arc welding regime the number of short circuits equaled the number of droplet transfers into the melt pool.

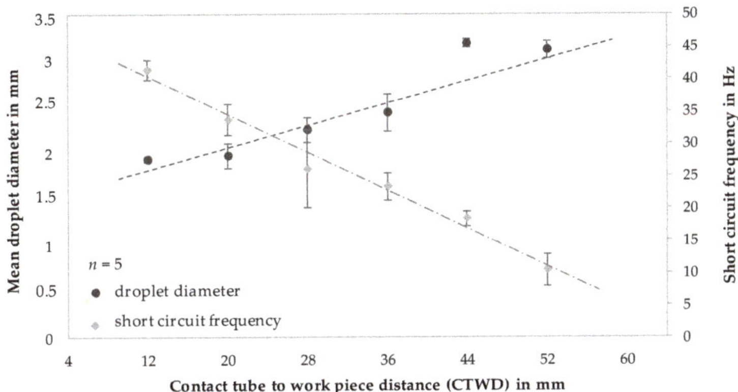

Figure 8. Correlation of mean droplet diameter and short circuit frequency with differing CTWD for U = 20 V; v_{wire} = 5 m/min and v_{weld} = 0.4 m/min.

Apparently, the short circuit frequency decreased with increasing CTWD. In the initial state of CTWD = 12 mm the short circuits occurred with a mean frequency of 42 Hz. This frequency was reduced to 11 Hz with increased distance between contact tube and substrate of 52 mm. This corresponded to 74% of the initial value. At the same time, droplet diameter increased throughout the previously described effects. The mean droplet size was determined to be 1.9 mm at 12 mm CTWD. The increase of the distance from contact tube to substrate up to 52 mm resulted in increased droplet sizes with a mean value of 3.1 mm. Thus, the average droplet size had increased by approximately 64%. During the experimental investigations, a maximum CTWD of 52 mm led to an inhomogeneous process behavior with arc deflections and excessive spatter formation. Further experimental trials were carried out with a maximum value of CTWD = 44 mm.

A side effect of decreasing short circuit frequency and respectively material transfer with large volume droplets affected the wetting behavior during GMAW. Herein, the formation of large volume droplets with low transition frequency led to a punctual wetting of the substrate. In context of WAAM, the wetting behavior was coupled to the welding velocity and was significantly relevant for the near net shape quality of the built structures. A homogeneous material transfer resulted in the formation of layers with equal geometrical properties such as width and height. Figure 9 shows the influence of CTWD adjustment and differing welding velocities on the geometrical formation of layers during WAAM. Thereby, a CTWD of 12 mm, respectively 44 mm was adjusted with differing welding velocities of 0.2–0.6 m/min. Wall structures of 100 mm length and 10 layers were manufactured. The temperature controlled WAAM process was executed with alternating welding direction in each layer and constantly set interlayer temperature of 100 °C before the next layer was applied.

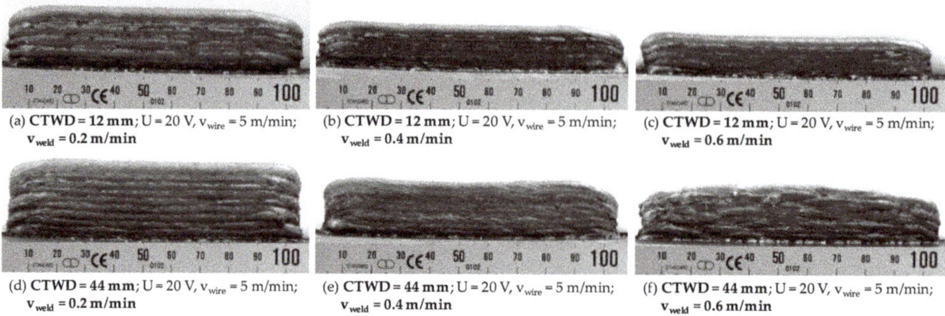

(a) **CTWD = 12 mm**; U = 20 V, v_{wire} = 5 m/min; v_{weld} = **0.2 m/min**
(b) **CTWD = 12 mm**; U = 20 V, v_{wire} = 5 m/min; v_{weld} = **0.4 m/min**
(c) **CTWD = 12 mm**; U = 20 V, v_{wire} = 5 m/min; v_{weld} = **0.6 m/min**
(d) **CTWD = 44 mm**; U = 20 V, v_{wire} = 5 m/min; v_{weld} = **0.2 m/min**
(e) **CTWD = 44 mm**; U = 20 V, v_{wire} = 5 m/min; v_{weld} = **0.4 m/min**
(f) **CTWD = 44 mm**; U = 20 V, v_{wire} = 5 m/min; v_{weld} = **0.6 m/min**

Figure 9. Correlation of CTWD adjustment and differing welding speeds on wetting behavior and near net shape quality during WAAM.

It becomes apparent, that surface quality and structure height of the work pieces were affected by CTWD adjustment as well as welding velocity. Systematical investigations on geometrical properties during WAAM with extended CTWD are shown in Chapter 4.3.

Figure 9a–c exhibited near net shaped quality for CTWD adjustment of 12 mm and welding velocities from 0.2 to 0.6 m/min. Thereby, little curvature at the structure beginning and end, an even overall height and low spatter formation were observed. In addition, high frequency of short circuits (compare Figure 8) led to homogeneous wetting behavior and melt bead solidification. As a result, the required amount of post-processing was limited due to geometrical properties close to the final contour. Increasing welding velocity results in the formation of structures with decreased height, which could be explained by reduced material transfer per unit length at a constant wire feeding speed.

Furthermore, WAAM with increased CTWD of 44 mm and adapted welding velocity is shown in Figure 9d–f. Thereby, the reduced short circuit frequency (compare Figure 8) affected the wetting behavior and near net shape properties. Thus, a stable welding process and the formation of layers with equal geometrical properties could be achieved with low welding velocities up to 0.4 m/min (compare Figure 9d,e). Thereby, surface roughness as well as the curvature at the structure beginning were slightly affected. As a result, requirements for near net shape quality in WAAM could be met with an increased CTWD of 44 mm.

Negative affection of process behavior and near net shape quality with increased CTWD is shown in Figure 9f. Herein, the combination of the reduced frequency of droplet transfers and increased welding velocity of 0.6 m/min led to partial wetting of the previously set layers. This resulted in inhomogeneous layer geometries during the manufacturing process. Throughout these dimensional deviations in layer height, the arc ignition after the short circuit phase as well as the arc length was affected constantly. The unstable welding process was insufficient for WAAM processing. As a result, the described properties for near net shape quality could not be met.

4.3. *Effects of Increased CTWD on Geometrical Layer Properties in WAAM*

In gas metal arc welding, increasing CTWD led to the reduction of energy input per unit length E_S respectively heat input as described (compare Figure 5). This had a direct effect on the formation of the weld bead geometry at a constant deposition rate. Due to lower energy input, the weld seams increased in height and decreased in width continuously. This could be described by lower temperatures in the melt pool and thus a lower viscosity of the melt, resulting in less flowable melt and faster solidification of the molten material. Figure 10 shows experimental results for the adaptation of CTWD from 12 to 44 mm in the context of resulting layer geometries. Three parameter sets with differing energy input per unit length were chosen. Though, layer height (compare Figure 10a) and layer width (compare Figure 10b) were measured on three positions of a wall structure with 10 layers. The mean values are shown in Figure 10.

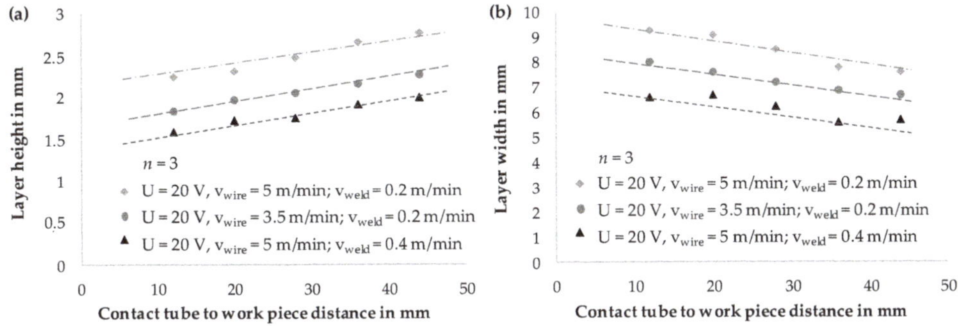

Figure 10. (**a**) Determination of layer height and (**b**) analysis of layer width for 3 parameter sets and adjusted CTWD.

The diagram in Figure 10a shows increasing values for layer height with increasing CTWD. A rise of approximately 25% was observed independently from the parameter set. The examination of the layer width (compare Figure 10b) exhibited an inverse dependence. Herein, layer width decreased to approximately 83% (CTWD = 44 mm) of the initial value (CTWD = 12 mm).

In the next step of the investigations, wall structures with a length of 160 mm and 70 layers were manufactured to analyze process behavior and near net shape quality on large volume work pieces. Figure 11 exemplarily shows the resulting work pieces for a CTWD of 12 mm (compare Figure 11a), 28 mm (compare Figure 11b) and 44 mm (compare Figure 11c). On these structures, microstructure and material hardness were tested additionally (compare Chapter 4.4). Therefore, the applied parameters were economically set to reduce processing time. The experimental trials were carried out with a welding velocity of 0.4 m/min, a voltage of 20 V and a wire feed of 5 m/min. The welding current as well as the energy input per unit length depend on the adjusted CTWD and varied from 4.23 (CTWD = 12 mm) to 2.82 kJ/cm (CTWD = 44 mm). The increase in CTWD from 12 to 44 mm led to a reduction of energy input per unit length of 34% (compare Figure 5). The interlayer temperature was set to 100 °C between the deposited layers.

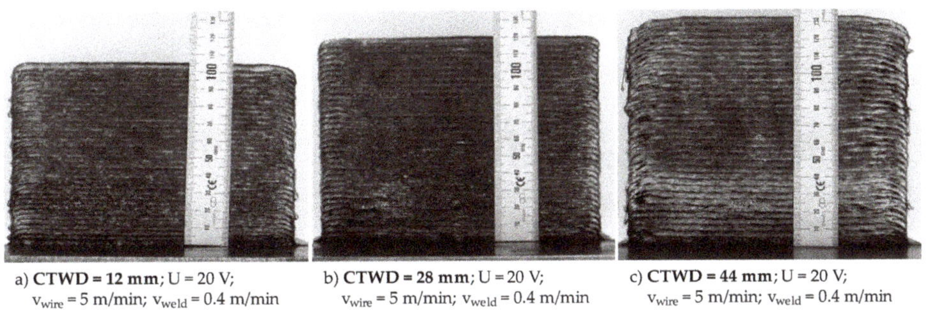

a) **CTWD = 12 mm**; U = 20 V; v_{wire} = 5 m/min; v_{weld} = 0.4 m/min

b) **CTWD = 28 mm**; U = 20 V; v_{wire} = 5 m/min; v_{weld} = 0.4 m/min

c) **CTWD = 44 mm**; U = 20 V; v_{wire} = 5 m/min; v_{weld} = 0.4 m/min

Figure 11. WAAM of wall structures with 70 layers and differing CTWD.

Figure 11 shows the resulting wall structures consisting of 70 layers. Apparently, the structures increase in height with extended CTWD as described before (compare Figure 9). At the same time, surface roughness increased due to the described droplet diameter, transferring frequency and wetting properties. The resulting mean structure height and mean layer height are shown in Table 2. Herein, the values for mean structure height were measured on three points of the wall structures and the values for mean layer height calculated by the division of mean structure height by the total number of layers. Furthermore, the results were compared to preliminary manufactured wall structures with

10 layers (compare Figure 10). The deviation is also shown in Table 2. A CTWD of 12 mm results in a structure height of approximately 107 mm, respectively a mean layer height of 1.53 mm (compare Table 2). Increased CTWD of 44 mm led to a structure height of approximately 134 mm (1.92 mm layer height), which was an increase of 25%. Little deviations < 5% compared to structures in preliminary investigations confirmed process stability and the effects of increasing CTWD on geometrical layer properties on large-volume structures.

Table 2. Structure height of large WAAM structures and comparison to previous trials (compare Figure 10).

CTWD (mm)	Mean Structure Height (mm)	Mean Layer Height (mm)	Deviation (%)
12	106.8 ± 0.3	1.53	−4.5
28	121.7 ± 0.2	1.74	−1.2
44	134.1 ± 1.4	1.92	−4.2

The side effect of adapted layer geometries was significant for the use of increased CTWD during WAAM. The change of the weld bead geometry at constant welding voltage, constant welding speed and constant wire feed has to be taken into account for path planning strategies. Especially if the value of the contact tube to work piece distance is set dynamically in the WAAM process to adapt the layer height or width.

4.4. Effects of Increased CTWD on Microstructure Development in WAAM

In wire arc additive manufacturing, the adjustment of energy input affects the cooling conditions and moreover metallurgical properties of the metallic work piece. Though, microstructure and mechanical properties can be influenced. A common approach to characterize un- and low-alloyed steel during welding is the estimation of the $t_{8/5}$-time according to DIN EN 1011-2:2001-05 [36]. This describes the time interval of cooling of the weld bead and the heat-affected zone (HAZ) from 800 to 500 °C. In this temperature range, phase transformations from γ- to α-phase take place, which are decisive for the mechanical-technological properties. Therefore, a continuous cooling transformation (CCT) diagram can be used to estimate microstructure and mechanical properties in dependence on the chemical composition and occurring cooling rates. Since no diagram of the applied filler material is available from the manufacturer, a CCT-diagram of steel with a chemical composition as similar as possible was used. Figure 12 shows the CCT-diagram of the low-alloyed steel SG36 as well as the chemical composition [37]. Furthermore, phase formation at different cooling rates and expectable hardness values are shown. Though, minor deviations in silicon and manganese content of approximately 0.3% can be seen. Due to this, slight deviations in phase proportions and hardness in comparison to the used filler wire G4Si1/SG3 may occur.

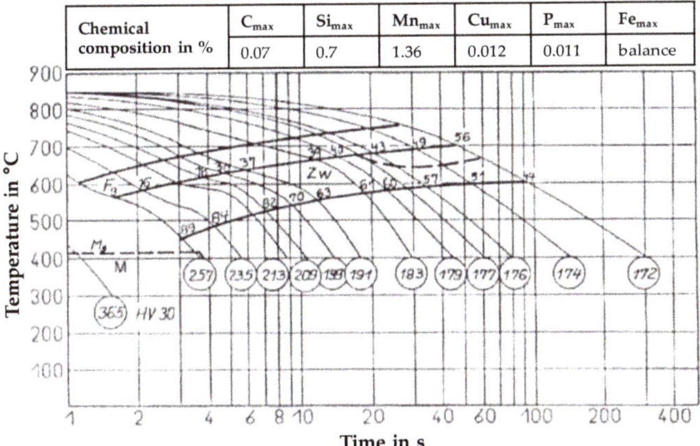

Chemical composition in %	C_{max}	Si_{max}	Mn_{max}	Cu_{max}	P_{max}	Fe_{max}
	0.07	0.7	1.36	0.012	0.011	balance

Figure 12. CCT-diagram of low-alloyed steel SG36 with comparable chemical composition to the used welding wire G4Si1/SG3 (1.5130) [37].

The filler material is a hypo-eutectoid steel with a low carbon content of C ≤ 0.07% and increased percentage of silicon and manganese. Thus, almost no pearlite was formed during cooling (compare Figure 12). The microstructure was mainly composed of different proportions of intermediate structure (bainite) and ferrite at moderate cooling rates. At very high cooling rates, e.g., in the first weld bead, martensite formed.

To correlate process adaptations and resulting metallurgical properties of the previously built structures (compare Figure 11), cooling rates of the set layers were analyzed during WAAM with adapted CTWD. Therefore, $t_{8/5}$-times were recorded layer wise by IR imaging. The analysis was carried out on each of the first 10 layers due to strongly varying cooling rates. This can be explained by heat conduction to the substrate, which featured room temperature in the initial state. As a result, an increasing number of layers led to heat accumulation and rising temperatures in the substrate until a quasi-stationary condition of heat input and heat transfer was reached. During the manufacturing of the following 10 layers, measurements were carried out in every second layer. Finally, every 10th layer was recorded with the exception of layer 35, which is the center of the structure. Figure 13 shows the resulting $t_{8/5}$-times during WAAM depending on the layer height and the set CTWD.

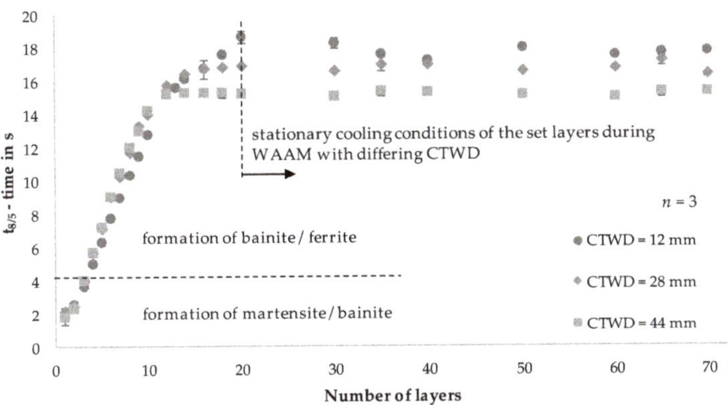

Figure 13. Analysis of the $t_{8/5}$-time in dependence on the number of layers and CTWD for U = 20 V; v_{wire} = 5 m/min and v_{weld} = 0.4 m/min (mean values of 3 measuring points per layer).

The measured $t_{8/5}$-times in the 1st weld bead of approximately 2 s show small deviations between the experimental trials with adjusted CTWD. According to the CCT-diagram, the formation of a martensitic-bainitic microstructure with high hardness values > 257 HV appeared for $t_{8/5}$-times up to 4 s. These cooling conditions were measured from the first to the third layer (compare Figure 12). With increasing structure height and thus longer $t_{8/5}$-times, the microstructure consisted of varying proportions of bainite and ferrite (compare Figure 13). The expectable hardness decreased.

Between the 4th and 10th layer, the $t_{8/5}$-times for a CTWD of 12 mm were slightly lower of approximately $t_{8/5}$ = 1.5 s in comparison to a CTWD of 28 mm, respectively 44 mm. This development is controversial to the process behavior and reduced energy input per unit length with increasing CTWD. The reason for the opposite development is the fact that the weld bead geometry and thus the conditions of heat conduction to the substrate change with increasing contact tube distance. The width of the weld bead (compare Figure 10b) decreased by approximately 0.6 mm (CTWD = 28 mm) or 0.9 mm (CTWD = 44 mm) and consequently also the effective area for heat conduction. The effect of thermal conduction is the essential driver of heat transfer close to the substrate. This observation is consistent with the research of Cunningham et al. [7].

From the 12th layer on, the resulting $t_{8/5}$-times adjusted to an increasingly stationary state between 15 and 20 s. This can be explained by the low amount of introduced thermal energy and the resulting lower temperature gradients between the structure and the substrate. As a result, stationary cooling conditions of the set layers can be observed. Herein, a difference in the $t_{8/5}$-time of approximately 2.5 s results for the different CTWD adjustments. A correlation between the energy input per unit length (compare Figure 5) and the resulting cooling rates (compare Figure 13) can be shown throughout decreasing $t_{8/5}$-times at reduced energy input. According to the CCT-diagram [37], a microstructural composition of 63% bainite and 37% ferrite occurred for $t_{8/5}$-times between 15 to 20 s with hardness values below 191 HV.

Table 3 shows the estimated phase formation and hardness values for the 1st, 35th and 70th layer. It should be noted that a CCT-diagram [37], which is designed for a chemically comparable steel was used. Due to this, slight deviations of the theoretical values to the experimentally measured values have to be taken into account.

Table 3. Estimation of phase composition and hardness depending on the $t_{8/5}$-time according to the CCT-diagram [37] (compare Figure 12).

CTWD (mm)	Number of Layer	$t_{8/5}$-Time (s)	Phase Formation (%)	Hardness (HV30)
12; 28; 44	1	1.8–1.9	Martensitic-bainitic	>257
	35	15.3–17.5	63% bainite, 37% ferrite	<191
	70	15.4–17.7	63% bainite, 37% ferrite	<191

In the next step, microstructural cross sections of the previously built wall structures with varied CTWD of 12 mm and 44 mm were taken from the middle of the 1st, 35th and 70th layer. Herein, microstructural evolution of low-alloyed steel G4Si1/SG3 is shown in dependence on differing cooling conditions during WAAM. The cross sections are shown in Figure 14.

In the cross sections of the 1st layer, a primarily bainitic structure with accumulations of ferrite grains is apparent for a CTWD of 12 mm and 44 mm. According to Seyffarth et al. [37], martensitic phase can be seen in addition to the bainitic microstructure due to $t_{8/5}$-times < 2 s in the first layer. With increasing CTWD of 44 mm, the microstructure becomes fine grained, although the measured cooling rates were equal. This phenomenon can be explained by lower energy input with increasing CTWD (compare Figure 5). Thereby, lower heat input resulted in decreased grain growth as well as decreased dilution between filler material and substrate.

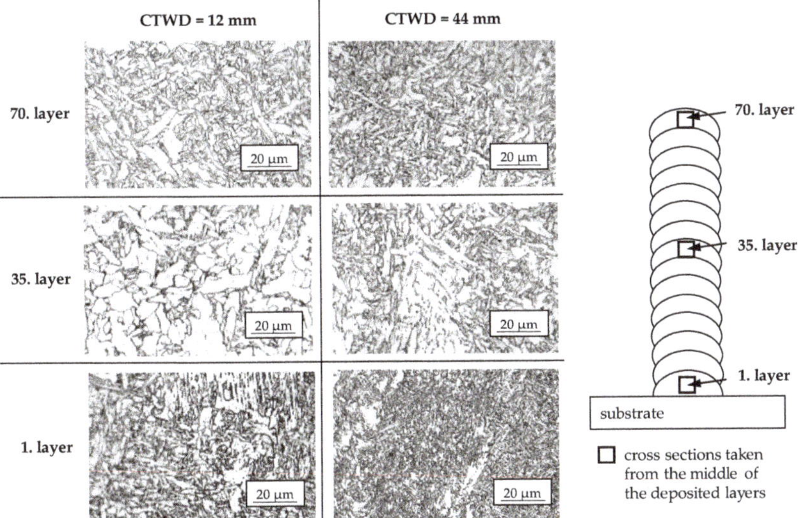

Figure 14. Microstructural cross sections of additively manufactured walls with differing CTWD for U = 20 V; v_{wire} = 5 m/min and v_{weld} = 0.4 m/min.

The cross sections of the 35th layer show a microstructural composition of coarse-grained ferrite with embedded globular bainite for a CTWD of 12 mm. An increased CTWD of 44 mm led to a reduced grain size. This correlated to the decreased energy input per unit length (compare Figure 5) and reduced $t_{8/5}$-times in the center of the structure (compare Figure 13). Furthermore, a periodical reheating of the bead throughout the following layers enhanced grain growth. In the 70th layer a bainitic-ferritic structure is visible, independently from the adjusted CTWD. This corresponds to the state of the art [37]. The fine grained microstructure can be explained by a lack of post-weld heat treatment in the last layer of the WAAM process. Further heat input throughout a following layer is not present.

Throughout microstructural testing, no structural defects such as a lack of fusion (compare Figure 11) or porosity could be found. Herein, low energy input per unit length (compare Figure 5) led to the formation of small and stable melt beads with enhanced decarburization. Furthermore, the increased silicon content of 1.0% (compare Table 1) in the filler wire provided enhanced binding of oxygen during solidification.

The verification of microstructural phases was carried out by hardness measurements on the wall structures described in Chapter 4.3. Therefore, Vickers hardness testing was conducted from the substrate to the top layer in build-up direction of the work pieces. Figure 15 shows the resulting mean values of three parallel measuring rows per structure for Vickers hardness testing with a distance of 1 mm between the measuring points.

A CTWD of 44 mm led to mean hardness values of 271 HV1 in the 1st layer, which was approximately 30 HV1 higher than the mean hardness value at 12 mm CTWD in the same position. According to the CCT-diagram of the unalloyed steel, martensitic phase exceeds hardness values of 257 HV1 [37] and can be shown for a CTWD of 44 mm. Despite equal $t_{8/5}$-times between 12 and 44 mm CTWD, this can be explained by reduced heat input and less dilution between filler material and substrate. As a result, a fine grained microstructure (CTWD = 44 mm) led to increased hardness. Moreover, the slightly different chemical composition of the filler wire to the used CCT-diagram must be considered. Increasing percentage of silicon and manganese of approximately 0.3% might affect martensite start temperature as well as phase fraction.

With increasing number of layers, the difference in hardness becomes smaller. From a distance to the substrate of approximately 20 mm a mean hardness of 180 HV1 can be observed for a CTWD of 44 mm. A decreased CTWD of 12 mm led to mean hardness values of approximately 168 HV1. These values can be measured with little deviation throughout the additively manufactured structure to a distance from the substrate of 102 mm (CTWD = 12 mm) respectively 126 mm (CTWD = 44 mm). The values correspond to the shown cross sections of the 35th layer (compare Figure 12) and verify the presence of ferritic phase with proportions of bainite according to the CCT-diagram [37].

Independently from the set CTWD an increase in hardness can be seen in the last layers of the structures up to 220 HV1 (compare Figure 15). This refers to a lack of post-heat treatment in the last layers and a fine-grained microstructure.

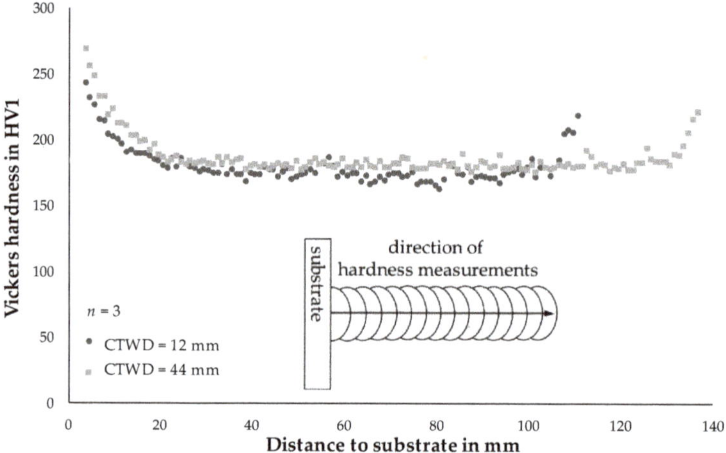

Figure 15. Hardness measurements of WAAM structures with differing CTWD for U = 20 V; v_{wire} = 5 m/min and v_{weld} = 0.4 m/min.

5. Summary and Outlook

In this study the influence of the contact tube to work piece distance (CTWD) was investigated in the context of energy reduction in wire arc additive manufacturing (WAAM). Investigations on electrical properties such as energy input per unit length E_S or short circuit frequency were analyzed. It could be shown that an increase in CTWD of 10 mm led to a reduction in E_S of 10%. A maximum reduction of 40% could be achieved. Thereby, droplet detaching frequency was reduced throughout decreasing welding current respectively reduced pinch force. As a result, the formation of increasing droplet sizes was observed. However, the adaptation of CTWD affected geometrical properties of the deposited layers as well as microstructural properties of the investigated steel. It was shown, that the grain size could be influenced by the adjustment of CTWD due to differing cooling rates. As a result, modified material hardness in the structure emerges.

In WAAM, the adaptation of CTWD is a simple and highly efficient approach, which can be applied as inline processing strategy in order to regulate energy input during the build-up process. Furthermore, the processing strategy can be used for wire arc additive manufacturing of work pieces with tailored properties within the built structures. Herein, dimensional and microstructural modulations can be implemented.

Author Contributions: Conceptualization, P.H.; Investigation, P.H., M.G., Y.A.; Methodology, P.H., M.G., Y.A. and J.R.; Project administration, J.P.B.; Supervision, J.P.B.; Writing—original draft, P.H. and J.R. All authors have read and agreed to the published version of the manuscript.

Funding: We acknowledge support for the article processing charge by the German Research Foundation (DFG) and the Open Access Publication Fund of the Technische Universität Ilmenau.

Conflicts of Interest: The authors declare no conflict of interest.

References

1. Wohlers Report 2019 Details Striking Range of Developments in AM Worldwide. Available online: https://www.3dprintingmedia.network/wohlers-report-2019-details-striking-range-of-developments-in-am-worldwide/ (accessed on 15 April 2020).
2. Karunakaran, K.P.; Bernard, A.; Suryakumar, L.D.; Taillandier, G. Rapid manufacturing of metallic objects. *Rapid Prototyp. J.* **2012**, *18*, 264–280. [CrossRef]
3. Karunakaran, K.P.; Suryakumar, S.; Pushpa, V.; Akula, S. Low cost integration of additive and subtractive processes for hybrid layered manufacturing. *Rob. Comput. Integr. Manuf.* **2010**, *26*, 490–499. [CrossRef]
4. Wu, B.; Pan, Z.; Ding, D.; Cuiuri, D.; Li, H.; Xu, J.; Norrish, J. A review of the wire arc additive manufacturing of metals: Properties, defects and quality improvement. *J. Manuf. Process.* **2018**, *35*, 127–139. [CrossRef]
5. Candel-Ruiz, A.; Kaufmann, S.; Müllerschön, O. Strategies for high deposition rate additive manufacturing by laser metal deposition. In *Proceedings of Lasers in Manufacturing (LiM)*; German Scientific Laser Society (WLT e.V.): Erlangen, Germany, 2015; pp. 584–610.
6. Bergmann, J.P.; Henckell, P.; Reimann, J.; Ali, Y.; Hildebrand, J. Grundlegende wissenschaftliche Konzepterstellung zu bestehenden Herausforderungen und Perspektiven für die Additive Fertigung mit Lichtbogen. *Weld. Cut* **2018**, *70*, 1–52.
7. Cunningham, C.R.; Flynn, J.M.; Shokrani, A.; Dhokia, V. Invited review article: Strategies and processes for high quality wire arc additive manufacturing. *Addit. Manuf.* **2018**, *22*, 672–686. [CrossRef]
8. Mehnen, J.; Ding, J.; Lockett, H.; Kazanas, P. Design study for wire and arc additive manufacture. *Int. J. Prod. Dev.* **2014**, *19*, 2–20. [CrossRef]
9. Bermingham, M.J.; Nicastro, L.; Kent, D.; Chen, Y.; Dargusch, M.S. Optimising the mechanical properties of Ti-6Al-4V components produced by wire+ arc additive manufacturing with post-process heat treatments. *J. Alloys Compd.* **2018**, *753*, 247–255. [CrossRef]
10. Asala, G.; Khan, A.K.; Andersson, J.; Ojo, O.A. Microstructural analyses of ATI 718Plus® produced by wire-ARC additive manufacturing process. *Metall. Mater. Trans. A* **2017**, *48*, 4211–4228. [CrossRef]
11. Dhinakaran, V.; Ajith, J.; Fahmidha, A.F.Y.; Jagadeesha, T.; Sathish, T.; Stalin, B. Wire Arc Additive Manufacturing (WAAM) process of nickel based superalloys—A review. *Mater. Today Proc.* **2020**, *21*, 920–925. [CrossRef]
12. Lu, X.; Zhou, Y.F.; Xing, X.L.; Shao, L.Y.; Yang, Q.X.; Gao, S.Y. Open-source wire and arc additive manufacturing system: Formability, microstructures, and mechanical properties. *Int. J. Adv. Manuf. Technol.* **2017**, *93*, 2145–2154. [CrossRef]
13. Feucht, T.; Lange, J.; Waldschmitt, B.; Schudlich, A.K.; Klein, M.; Oechsner, M. Welding Process for the Additive Manufacturing of Cantilevered Components with the WAAM. In *Advanced Joining Processes*; Springer: Singapore, 2020; pp. 67–78.
14. Ghaffari, M.; Nemani, A.V.; Rafieazad, M.; Nasiri, A. Effect of solidification defects and HAZ softening on the anisotropic mechanical properties of a wire arc additive-manufactured low-carbon low-alloy steel part. *JOM* **2019**, *71*, 4215–4224. [CrossRef]
15. Müller, J.; Grabowski, M.; Müller, C.; Hensel, J.; Unglaub, J.; Thiele, K.; Dilger, K. Design and Parameter Identification of Wire and Arc Additively Manufactured (WAAM) Steel Bars for Use in Construction. *Metals* **2019**, *9*, 725. [CrossRef]
16. Rafieazad, M.; Ghaffari, M.; Nemani, A.V.; Nasiri, A. Microstructural evolution and mechanical properties of a low-carbon low-alloy steel produced by wire arc additive manufacturing. *Int. J. Adv. Manuf. Technol.* **2019**, *105*, 2121–2134. [CrossRef]
17. Gordon, J.; Hochhalter, J.; Haden, C.; Harlow, D.G. Enhancement in fatigue performance of metastable austenitic stainless steel through directed energy deposition additive manufacturing. *Mater. Des.* **2019**, *168*, 107630. [CrossRef]

18. Wu, W.; Xue, J.; Wang, L.; Zhang, Z.; Hu, Y.; Dong, C. Forming process, microstructure, and mechanical properties of thin-walled 316L stainless steel using speed-cold-welding additive manufacturing. *Metals* **2019**, *9*, 109. [CrossRef]
19. Ali, Y.; Henckell, P.; Hildebrand, J.; Reimann, J.; Bergmann, J.P.; Barnikol-Oettler, S. Wire arc additive manufacturing of hot work tool steel with CMT process. *J. Mater. Process. Technol.* **2019**, *269*, 109–116. [CrossRef]
20. Chen, X.; Su, C.; Wang, Y.; Siddiquee, A.N.; Sergey, K.; Jayalakshmi, S.; Singh, R.A. Cold metal transfer (CMT) based wire and arc additive manufacture (WAAM) system. *J. Surf. Invest.* **2018**, *12*, 1278–1284. [CrossRef]
21. Da Silva, L.J.; Souza, D.M.; de Araújo, D.B.; Reis, R.P.; Scotti, A. Concept and validation of an active cooling technique to mitigate heat accumulation in WAAM. *Int. J. Adv. Manuf. Technol.* **2020**, 1–11. [CrossRef]
22. Shi, J.; Li, F.; Chen, S.; Zhao, Y.; Tian, H. Effect of in-process active cooling on forming quality and efficiency of tandem GMAW–based additive manufacturing. *Int. J. Adv. Manuf. Technol.* **2019**, *101*, 1349–1356. [CrossRef]
23. Fritz, A.H. neu bearbeitete und ergänzte Auflage. In *Fertigungstechnik. 12*; Springer: Berlin/Heidelberg, Germany, 2018.
24. Schwartz, M.M. Welding, Brazing and Soldering. In *ASM Handbook*; Olson, D.L., Siewert, T.A., Liu, S., Edwards, G.R., Eds.; ASM International: Cleveland, OH, USA, 1993; pp. 126–137.
25. Trube, S.; Miklos, E. *Metall-Aktivgasschweißen MAG/Metall-Inertgasschweißen MIG. Proceedings of 29*; Sondertagung: München, Germany, 2001.
26. Nadzam, J.; Armao, F.; Byall, L.; Kotecki, D.; Miller, D. *Gas Metal Arc Welding—Product and Procedure Selection*; Lincoln Global Inc.: Cleveland, OH, USA, 2014; C4.200.
27. A Green Perspective on Arc Welding Part 1: Harnessing the Power of Resistive Heating. Available online: https://ewi.org/a-green-perspective-on-arc-welding-part-1-harnessing-the-power-of-resistive-heating/ (accessed on 15 April 2020).
28. Zhan, Q.; Liang, Y.; Ding, J.; Williams, S. A Wire Deflection Detection Method Based on Image Processing in Wire + Arc Additive Manufacturing. *J. Adv. Manuf. Technol.* **2017**, *89*, 755–763. [CrossRef]
29. Dilthey, U. *Schweiß- und Schneidtechnologien, 3. bearbeitete Auflage In Schweißtechnische Fertigungsverfahren 1*; Springer: Berlin/Heidelberg, Germany, 2006.
30. Weman, K.; Lindén, G. *MIG Welding Guide*; Woodhead Publishing Limited: Cambridge, UK, 2006.
31. Barhorst, S.; Cary, H. Synergic Pulsed GMAW In Perspective. In Proceedings of the International Conference on Welding for Challenging Enviroments, Calgary, Canada, 2–5 June 1986; pp. 15–24.
32. Siewert, E. Experimentelle Analyse des Elektrodenwerkstoffübergangs beim Metallschutzgasschweißen mit gepulstem Schweißstrom. Ph.D. Thesis, Universität der Bundeswehr München, München, Germany, 2013.
33. Richter, F. The Physical Properties of Steels "The 100 Steels Programme" Part I: Tables and Figures. Ph.D. Thesis, Technische Universität Graz, Graz, Austria, 2010.
34. Matthes, K.J.; Schneider, W. *Schweißtechnik, Schweißen von metallischen Konstruktionswerkstoffen*; Carl Hanser Verlag: München, Germany, 2016.
35. *DIN EN ISO 6507-1:2018-07: Metallic Materials—Vickers Hardness Test—Part 1: Test Method (ISO 6507-1:2018)*; Beuth-Verlag: Berlin, Germany, 2018.
36. *DIN EN 1011-2:2001-05: Welding—Recommendation for Welding of Metallic Materials—Part 2: Arc Welding of Ferritic Steels*; Beuth-Verlag: Berlin, Germany, 2001.
37. Seyffarth, P.; Meyer, B.; Scharff, A. *Großer Atlas Schweiß-ZTU-Schaubilder*; DVS Media Verlag: Düsseldorf, Germany, 2018.

Article

Simulation of Ti-6Al-4V Additive Manufacturing Using Coupled Physically Based Flow Stress and Metallurgical Model

Bijish Babu [1,*], Andreas Lundbäck [2] and Lars-Erik Lindgren [2]

1 Swerim AB, Heating and Metalworking Box 812, SE-971 25 Luleå, Sweden
2 Mechanics of Sold Materials, Luleå University of Technology, SE-971 87 Luleå, Sweden; Andreas.Lundback@ltu.se (A.L.); Lars-Erik.Lindgren@ltu.se (L.-E.L.)
* Correspondence: bijish.babu@swerim.se; Tel.: +46-704197817

Received: 3 October 2019; Accepted: 18 November 2019; Published: 21 November 2019

Abstract: Simulating the additive manufacturing process of Ti-6Al-4V is very complex due to the microstructural changes and allotropic transformation occurring during its thermomechanical processing. The α-phase with a hexagonal close pack structure is present in three different forms—Widmanstatten, grain boundary and Martensite. A metallurgical model that computes the formation and dissolution of each of these phases was used here. Furthermore, a physically based flow-stress model coupled with the metallurgical model was applied in the simulation of an additive manufacturing case using the directed energy-deposition method. The result from the metallurgical model explicitly affects the mechanical properties in the flow-stress model. Validation of the thermal and mechanical model was performed by comparing the simulation results with measurements available in the literature, which showed good agreement.

Keywords: dislocation density; vacancy concentration; Ti-6Al-4V; additive manufacturing; directed energy deposition

1. Introduction

Powder Bed Fusion (PBF) is the technique of building thin layer over layer by melting the fine metal powder. Directed energy deposition (DED), on the other hand, is usually used for building features on large existing parts as well as for repairing damaged ones. PBF typically adds layers that are thinner than DED and can therefore create high-resolution structures, whereas DED produces components at a higher built rate. The primary challenge of DED is that the higher energy input from the heat source may lead to substantial distortion and higher residual stresses.

DED additive manufacturing (AM) can be considered as computer numerically controlled (CNC) multipass welding with progressive weldments made on a substrate to create free-form structures. The added metals can be in either powder or wire form and the heat source a laser or electron beam. The deposition path is generated from computer-aided design (CAD) geometry and is preprogrammed in a CNC machine, which makes the process very flexible and suitable for low volume production, eliminating the need for tooling and dies.This also enables the production of complicated geometries that are traditionally difficult to produce with conventional manufacturing processes. Additively manufactured parts of Ti-6Al-4V are traditionally found in human implants [1] and aerospace components because of the criticality of their applications. However, AM has also been used to repair aerospace components [2] that have developed defects during operation or production.

A few researchers have performed AM simulations or similar processes for Ti alloys using thermomechanical–microstructural (TMM) coupled material models. In Baykasoglu et al. [3], a thermomicrostructural model for Ti6Al-4V was presented and applied to a DED process. Salsi et al. [4]

presented a similar model and applied it on a PBF process, while Vastola et al. [5] compared the results when modelling electron-beam melting (EBM) and PBF processes. Song et al. [6] performed a welding simulation by using a TA15 alloy employing a TMM model. A similar model was utilized for performing a quenching simulation by Teixeira et al. [7] for alloy Ti17. Cao et al. [8] showed an AM simulation using electron-beam melting without including microstructural coupling. A TMM material model was employed by Ahn et al. [9] for welding simulation ignoring strain-rate dependence.

In this work, a material model combining metallurgical and flow-stress models described by Babu et al. [10] is used. This model works for arbitrary phase composition and is an improved version of that of Babu and Lindgren [11]. The AM process involves cyclic heating and cooling, resulting in nonequilibrium phase evolution, which can be addressed with this model. The metallurgical model used in this work was also utilized in the simulation of the AM case described by Charles Murgau et al. [12], which is included in the current special issue.

2. Physically Based Flow-Stress Model

An incompressible von Mises model was used here with the assumption of isotropic plasticity. Flow stress was split into two parts [11,13–15]:

$$\sigma_y = \sigma_G + \sigma^*. \tag{1}$$

Here, σ_G is a thermal stress contribution from long-range interactions of the dislocation substructure. The other term, σ^*, is the required friction stress to move dislocations within the lattice and to cross short-range barriers. Thermal vibrations can assist dislocations to overcome these barriers. Conrad [16] proposed a similar formulation after analyzing titanium systems.

2.1. Long-Range Stress Component

The long-range term from Equation (1) is derived from Seeger [13] as

$$\sigma_G = m\alpha G b\sqrt{\rho_i}. \tag{2}$$

Here, m is the Taylor factor that translates the resolved shear stress in various slip systems to effective stress, b is the magnitude of Burgers vector, $G(T)$ is the temperature-dependent shear modulus, ρ_i is the immobile dislocation density and $\alpha(T)$ is a calibrated proportionality factor.

2.2. Short-Range Stress Component:

The strain-rate-dependent part of the yield stress from Equation (1) can be derived according to the Kocks–Mecking formulation [17,18] as

$$\sigma^* = \tau_0 G\left[1 - \left[\frac{kT}{\Delta f_0 G b^3}\ln\left(\frac{\dot{\varepsilon}^{ref}}{\dot{\bar{\varepsilon}}^p}\right)\right]^{1/q}\right]^{1/p}. \tag{3}$$

Here, shear strength in the absence of thermal energy is denoted by $\tau_0 G$, and the activation energy required to overcome lattice resistance is denoted by $\Delta f_0 G b^3$. Parameters p and q define the shape of the obstacle barrier for dislocation motion. Further, k is the Boltzmann constant, T is the temperature in kelvin and ($\dot{\varepsilon}^{ref}$) and ($\dot{\bar{\varepsilon}}^p$) are the reference and plastic strain rates, respectively.

2.3. Evolution of Immobile Dislocation Density

The evolution of ρ_i in Equation (2) is modelled as having two components, hardening and restoration.

$$\dot{\rho}_i = \dot{\rho}_i^{(+)} - \dot{\rho}_i^{(-)}. \tag{4}$$

2.3.1. Hardening Process

The average distance moved by dislocations before they are annihilated or immobilized is called mean free path Λ. The Orowan equation shows that the density of dislocations and their average velocity are proportional to the plastic strain rate. Assuming that the immobile dislocation density also follows the same relation leads to

$$\dot{\rho}_i{}^{(+)} = \frac{m}{b}\frac{1}{\Lambda}\dot{\bar{\varepsilon}}^p. \tag{5}$$

The mean free path is computed from grain size (g) and dislocation subcell or subgrain diameter (s) as

$$\frac{1}{\Lambda} = \frac{1}{g} + \frac{1}{s}. \tag{6}$$

The subcell formation and evolution are modelled using a relation proposed by Holt [19].

$$s = K_c \frac{1}{\sqrt{\rho_i}}. \tag{7}$$

2.3.2. Restoration Processes

Vacancy motion is relevant to the recovery of dislocations. Restoration of the lattice commonly occurs at elevated temperatures and is therefore a thermally activated restructuring process. Creation of vacancy requires energy and increases entropy. With increasing temperature and deformation, vacancy concentration also increases. High stacking fault materials usually exhibit constant flow stress because of the balance between hardening and recovery. The current model assumes that the mechanisms of restoration are dislocation glide, dislocation climb and globularization.

$$\dot{\rho}_i{}^{(-)} = \dot{\rho}_i{}^{(glide)} + \dot{\rho}_i{}^{(climb)} + \dot{\rho}_i{}^{(globularization)}. \tag{8}$$

The model for recovery by glide can be written on the basis of the formulation by Bergström [20] as

$$\dot{\rho}_i{}^{(glide)} = \Omega \rho_i \dot{\bar{\varepsilon}}^p, \tag{9}$$

where Ω is a function dependent on temperature.

Militzer et al. [21] proposed a model for dislocation climb on the basis of Sandström and Lagneborg [22] and Mecking and Estrin [23]. With a modification of diffusivity according to Reference [11], the model can be written as

$$\dot{\rho}_i^{(climb)} = 2c_\gamma D_{app}\frac{Gb^3}{kT}\left(\rho_i^2 - \rho_{eq}^2\right), \tag{10}$$

where c_γ is a material coefficient and ρ_{eq} is the equilibrium value of the dislocation density. Here, D_{app} is the apparent diffusivity that includes the diffusivity of the $\alpha - \beta$ phases weighted by their fractions X_α and X_β, pipe diffusion D_p, as well as effects of excess vacancy concentration c_v.

Babu and Lindgren [11] proposed a model for the evolution of dislocation density during globularization where the effect of grain growth on the reduction of flow stress is only included when dislocation density is above a critical value ρ_{cr}.

if $\rho_i \geq \rho_{cr}$

$$\dot{\rho}_i{}^{(globularization)} = \psi \dot{X}_g \left(\rho_i - \rho_{eq}\right);\ \text{until } \rho_i \leq \rho_{eq} \tag{11}$$

else

$$\dot{\rho}_i{}^{(globularization)} = 0. \tag{12}$$

Here, ρ_{eq} is the equilibrium value of dislocation density, $\dot{X}_g$ is the globularization rate and ψ is a calibration constant. Thomas and Semiatin [24] modelled the two-stage process of dynamic and static recrystallization. Owing to the similarities between globularization and recrystallization, this model can be adapted.

$$X_g = X_d + (1 - X_d)\, X_s. \tag{13}$$

Here, volume fractions X_g, X_d, and X_s denote total globularization, its dynamic component and the static component, respectively.

Assuming that grain growth and static recrystallization have the same driving force, the static globularization rate can be modelled as [25,26]

$$\dot{X}_s = M\frac{\dot{g}}{g}, \tag{14}$$

where, M is a material parameter. The rate of dynamic globularization was computed on the basis of a model by Thomas and Semiatin [24] as,

$$\dot{X}_d = \frac{Bk_p\dot{\bar{\varepsilon}}_p}{\bar{\varepsilon}_p{}^{1-k_p}e^{B\bar{\varepsilon}_p^{k_p}}}, \tag{15}$$

where, B and k_p are material parameters.

2.4. Evolution of Excess Vacancy Concentration

The formation and evolution of excess vacancy concentration was modelled by Militzer et al. [21]. In the current work, Militzer's model was extended by adding the effect of temperature changes. Further, assuming that only long-range stress contributes to vacancy formation, the model can be rewritten as

$$\dot{c_v}^{ex} = \left[\chi\frac{m\alpha Gb^2\sqrt{\rho_i}}{Q_{vf}} + \zeta\frac{c_j}{4b^2}\right]\frac{\Omega_0}{b}\dot{\bar{\varepsilon}} - D_{vm}\left[\frac{1}{s^2}+\frac{1}{g^2}\right]\left(c_v - c_v^{eq}\right) + c_v^{eq}\left(\frac{Q_{vf}}{kT^2}\right)\dot{T}. \tag{16}$$

Here, $\chi = 0.1$ is the fraction of mechanical energy spent on vacancy generation, Ω_0 is the atomic volume and ζ is the neutralization effect by vacancy emitting and absorbing jogs. The concentration of jogs (c_j) and D_{vm} and the diffusivity of vacancy are given in Babu and Lindgren [11]. Additionally, Q_{vf} is the activation energy of vacancy formation.

3. Phase-Evolution Model

A simplified model [27] for the transition between the liquid and solid state was implemented to take care of temperatures above melting temperature T_{melt}. If the temperature is above T_{melt}, the volume fraction of the solid phases was set to zero. In the solid state, the Ti–6Al–4V microstructure comprises the high-temperature stable β-phase and the lower-temperature stable α-phase. Depending on temperature and heating/cooling rates, a variety of α/β morphologies can form that gives varying mechanical properties. The complex relationship between thermomechanical-processing, microstructure and mechanical properties was investigated by References [28,29]. On the basis of the literature [30–33], few microstructural features have been identified as relevant concerning mechanical properties. The three separate α-phase fractions, Widmanstatten (X_{α_w}), grain boundary ($X_{\alpha_{gb}}$), acicualr and massive Martensite (X_{α_m}) and β-phase fraction (X_β) were included in the current model. Though in the current flow-stress model individual α-phase fractions were not included separately, it is possible to incorporate them when more details about their respective strengthening mechanisms are known.

3.1. Phase Transformations

Depending on the temperature and heating/cooling rates, Ti-6Al-4V undergoes allotropic transformation. The mathematical model for the transformation is described schematically in Figure 1. Transformations denoted by F1, F2, and F3 represent the formation of α_{gb}, α_w, and α_m phases, respectively, and D3, D2 and D1 show the dissolution of the same phases. If the current volume fraction of β phase is more than β_{eq}, the excess β phase transforms into an α phase. Here, α_{gb} formation that occurs in high temperatures is the most preferred, followed by the α_w. Remaining excess β fraction is transformed to α_m if the temperature is lower than T_m, (martensite start temperature) and cooling rate is above 20°C/s. Conversely, if the current volume fraction of β is lower than β_{eq}, excess α phase is converted to β. Primarily, the α_m phase dissolves to β and α_w phases in the same proportion as the α_{eq} and β_{eq}. Remaining excess α_w and α_{gb} transform to β in that order. The equilibrium fraction of the β phase (see Figure 2) is computed by Equation (17), where T is the temperature in degrees Celsius.

$$\begin{aligned} X_\beta^{eq} &= 1 - 0.89\, e^{-\left(\frac{T^*+1.82}{1.73}\right)^2} + 0.28\, e^{-\left(\frac{T^*+0.59}{0.67}\right)^2} \\ T^* &= (T-927)/24. \end{aligned} \quad (17)$$

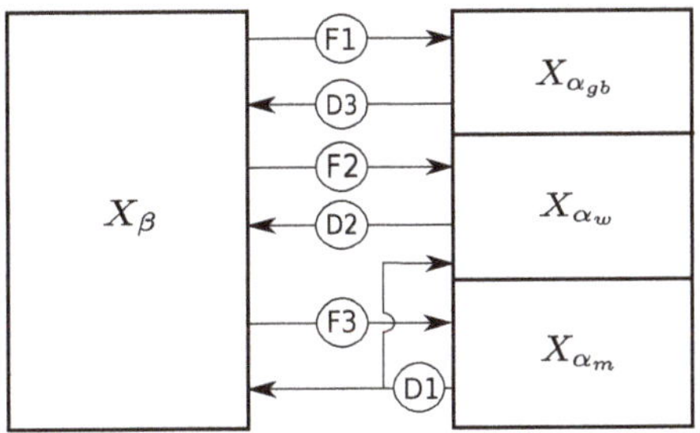

Figure 1. Phase-change mechanism.

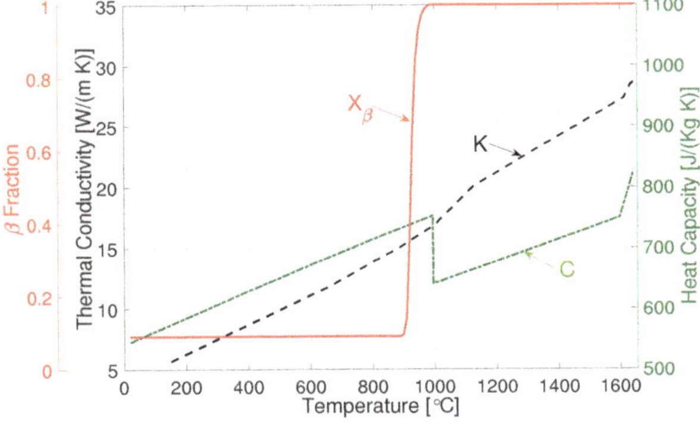

Figure 2. Thermal conductivity (K), specific heat capacity (C) and equilibrium-phase fraction (X_β).

3.2. Adaptation of Johnson–Mehl–Avrami–Kolmogrov (JMAK) Model for Diffusional Transformation

The JMAK model [34–36], originally formulated for nucleation and growth during isothermal situations, can be adapted to model any diffusional transformation. Employing the additivity principle and using sufficiently small time steps ensures that any arbitrary temperature change can be computed.

The JMAK model assumes that a single phase X_1 that is 100% in volume from the start transforms to 100% of second phase X_2 in infinite time. However, in the case of Ti-6Al-4V, this is not the case, as it is a $\alpha - \beta$ dual-phase alloy below β-transus temperature. Hence, in order to accommodate for an incomplete transformation, the product fraction is normalized with the equilibrium volume. Conversely, the starting volume of a phase can also be less than 100%, which is circumvented by assuming that the available phase volume is the total phase fraction. Another complication is the existence of the simultaneous transformation of various α phases (α_w, α_{gb}, α_m) to the β phase and back. This can be modelled by sequentially calculating each transformation within the time increment [27].

3.3. Formation of α Phase

During cooling from the β-phase, the α_{gb} and α_w phases are formed by diffusional transformation. According to the incremental formulation of the JMAK model described by Reference [27], the formation of α_{gb} and α_w can be modelled by the set of equations in rows F1 and F2, respectively, of Table 1. The Martensite phase is formed at cooling rates above 410 °C/s by diffusion-less transformation. While cooling at rates above 20 °C/s and up to 410 °C/s, massive α transformation was observed to co-occur with Martensite formation [37,38]. Owing to the similitude in crystal structures between massive-α and Martensite-α, they are not differentiated here except that, above 410 °C/s, 100% α_m was allowed to form. An incremental formulation of the Koistinen–Marburger equation described by Charles Murgau et al. [27] was used here (see equation set in row F3 of Table 1).

Table 1. Models for α-phase formation.

F1	$^{n+1}X_{\alpha_{gb}} = \left(1 - e^{-k_{gb}\left(t^*_{gb}+\Delta t\right)^{N_{gb}}}\right)\left({}^nX_\beta + {}^nX_{\alpha w} + {}^nX_{\alpha_{gb}}\right){}^{n+1}X_\alpha^{eq} - {}^nX_{\alpha w}$ $t^*_{gb} = \sqrt[N_{gb}]{-ln\left(1 - \frac{\left({}^nX_{\alpha w} + {}^nX_{\alpha_{gb}}\right)/{}^{n+1}X_\alpha^{eq}}{{}^nX_\beta + {}^nX_{\alpha w} + {}^nX_{\alpha_{gb}}}\right)\Big/k_{gb}}$
F2	$^{n+1}X_{\alpha w} = \left(1 - e^{-k_w\left(t^*_w+\Delta t\right)^{N_w}}\right)\left({}^nX_\beta + {}^nX_{\alpha w} + {}^nX_{\alpha_{gb}}\right){}^{n+1}X_\alpha^{eq} - {}^nX_{\alpha_{gb}}$ $t^*_w = \sqrt[N_w]{-ln\left(1 - \frac{\left({}^nX_{\alpha_w} + {}^nX_{\alpha_{gb}}\right)/{}^{n+1}X_\alpha^{eq}}{{}^nX_\beta + {}^nX_{\alpha_w} + {}^nX_{\alpha_{gb}}}\right)\Big/k_w}$
F3	$^{n+1}X_{\alpha_m} = \begin{cases} \left(1 - e^{-b_{km}}\left(T_{ms} - T\right)\right)\left({}^nX_\beta + {}^nX_{\alpha_m}\right); if\ (\dot{T} > 410\ ^\circ C/s) \\ \left(1 - e^{-b_{km}}\left(T_{ms} - T\right)\right)\left({}^nX_\beta + {}^nX_{\alpha_m} - {}^{n+1}X_\alpha^{eq}\right); if\ (20\ ^\circ C/s > \dot{T} > 410\ ^\circ C/s) \end{cases}$

3.4. Dissolution of α Phase

The α_m phase formed by instantaneous transformation is unstable and therefore undergoes diffusional transformation to the α_w and β phases on the basis of its current equilibrium composition. The incremental formulation of the classical JMAK model by Reference [27] and its parameters are given in row D1 of Table 2. During heating or reaching nonequilibrium phase composition, α_w and α_{gb} can transform into a β-phase controlled by the diffusion of vanadium at the $\alpha - \beta$ interface. A parabolic equation developed by Kelly et al. [39,40] derived in its incremental form by Charles Murgau et al. [27] was used here (see rows D2 and D3 of Table 2).

Table 2. Models for α-phase dissolution.

D1	${}^{n+1}X_{\alpha_m} = \left({}^{n+1}X^{eq}_{\alpha_m} - e^{-k_m (t^*_m + \Delta t)^{N_m}}\right)\left({}^{n}X_{\beta} + {}^{n}X_{\alpha_m} - {}^{n+1}X^{eq}_{\alpha_m}\right)$ $t^*_m = \sqrt[N_m]{-ln\left(\frac{\left({}^{n}X_{\alpha} - {}^{n+1}X^{eq}_{\alpha_m}\right)}{{}^{n}X_{\beta} + {}^{n}X_{\alpha_m} - {}^{n+1}X^{eq}_{\alpha_m}}\right)/k_m}$
D2	${}^{n+1}(X_{\alpha_w} + X_{\alpha_{gb}}) = \begin{cases} {}^{n+1}X^{eq}_{\alpha} f_{diss}(T)\sqrt{\Delta t + t^*}; if\ (0 < (\Delta t + t^*) < t_{crit}) \\ {}^{n+1}X^{eq}_{\alpha}; if\ (\Delta t + t^* > t_{crit}) \end{cases}$
D3	$t^* = \left(\frac{{}^{n}X_{\beta}}{{}^{n+1}X^{eq}_{\beta} f_{diss}(T)}\right)^2$

4. Coupling of Phase and Flow-Stress Models

Young's modulus and Poisson's ratio were assumed to be identical for both phases. The Wachtman model [41] for Young's modulus (E), calibrated using measurements from Babu and Lindgren [11], is written as

$$E = 107 - 0.2\,(T + 273)\,e^{-(1300/T+273)}, \tag{18}$$

where T is the temperature in degrees Celsius applied with a cut-off at T = 1050 °C (see Figure 3). A linear model for Poisson's ratio (μ) after fitting to measurements by Fukuhara and Sanpei [42] as

$$\mu \quad = \quad 0.34 + 6.34\,e^{-5}\,T, \tag{19}$$

where T is the temperature in degrees Celsius (see Figure 3).

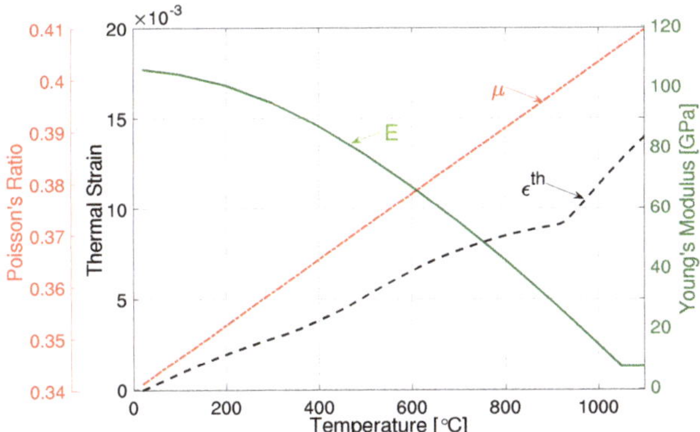

Figure 3. Poisson's ratio (μ), thermal strain (ϵ^{th}) and Youngs modulus (E).

Using X-ray diffraction, Swarnakar et al. [43] measured the volumetric expansion of unit cells of α and β phases during heating. On this basis, the average Coefficient of Thermal Expansion (CTE) of the phase mixture can be calculated using the rule of mixtures (ROM) as in Equation (20), where α_{α} and α_{β} give the CTE of α and β phases, respectively. The linear thermal strain can be computed using Equation (21), plotted in Figure 3. Here, ε^{adj} makes the ROM (Equation (20)) nonlinear.

$$\alpha_{avg} \quad = \quad X_{\alpha}\alpha_{\alpha} + X_{\beta}\alpha_{\beta} \tag{20}$$

$$\varepsilon^{th} \quad = \quad \alpha_{avg}\Delta T - \varepsilon^{adj} \tag{21}$$

$$\varepsilon^{adj} \quad = \quad 1.0e^{-8}T^2 - 8.4e^{-6}T + 3.0e^{-4}. \tag{22}$$

The thermal conductivity and specific heat capacity of the alloy taken from References [44] and [45], respectively, are given in Figure 2. The latent heat of phase transformation ($\alpha \rightarrow \beta$) and the latent heat of fusion were measured to be 64 and (290 $\pm$ 5) kJ/Kg, respectively [44].

The yield strength of the phase mixture can be written according to the linear rule of mixtures as

$$\sigma_y = X_\alpha \sigma_y^\alpha + X_\beta \sigma_y^\beta. \tag{23}$$

The plastic strain distribution can be obtained assuming an iso-work principle. According to Reference [46], this can be written as

$$\sigma_y^\alpha \dot{\bar{\varepsilon}}_\alpha = \sigma_y^\beta \dot{\bar{\varepsilon}}_\beta \tag{24}$$

$$\dot{\bar{\varepsilon}}^p = X_\alpha \dot{\bar{\varepsilon}}_\alpha + X_\beta \dot{\bar{\varepsilon}}_\beta. \tag{25}$$

The above formulation ensures that the β phase with lower yield strength has a more significant share of plastic strain as compared to the stronger α phase. For temperatures above 1100 °C, ($\sigma_y^{>1100\ ^\circ\mathrm{C}} = \sigma_y^{1100\ ^\circ\mathrm{C}}$). The stress–strain relationship predicted by the model for varying strain rates and temperatures are given in Figure 4. The rate dependence and flow-softening demonstrated by the model is visible here. A detailed comparison of model predictions and measurements along with model parameters are given in Babu et al. [10,11].

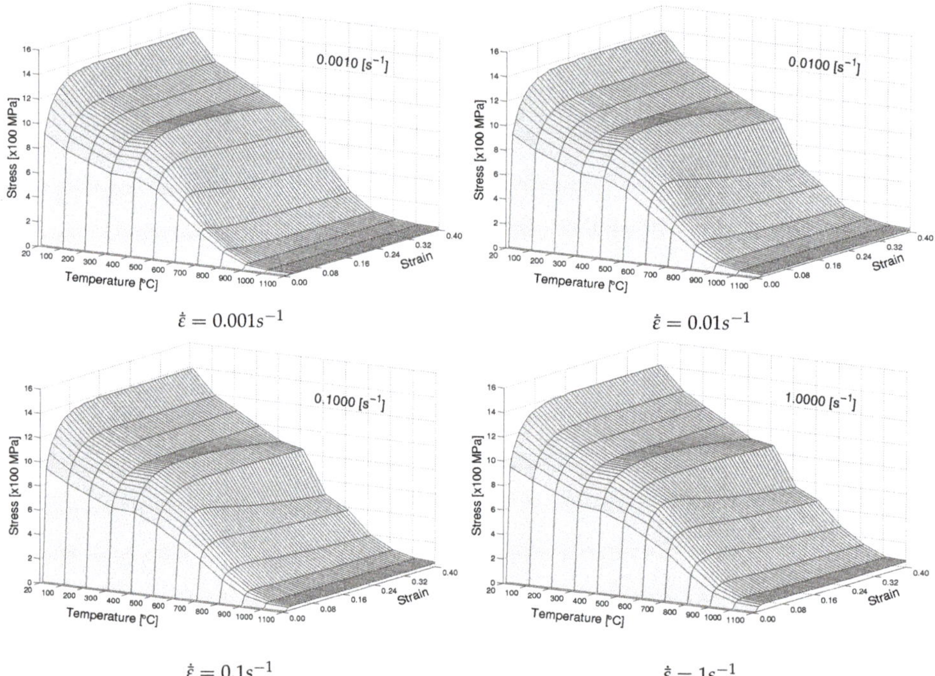

Figure 4. Stress–strain–temperature relationship.

5. Additive Manufacturing

In this article, a DED process described in Reference [47] was simulated using general-purpose Finite Element (FE) software MSC.Marc. A set of subroutines for modelling the AM process were implemented in MSC.Marc, which are explained in Lundbäck and Lindgren [48]. A coupled thermomechanical–metallurgical model described in the previous sections was also implemented as subroutines within MSC.Marc.

The dimensions of the substrate (152.4 × 38.1 × 12.7 mm) and AM component are given in Figure 5. One end of the fixture was held in position by a clamping fixture (see notations in Figure 5).

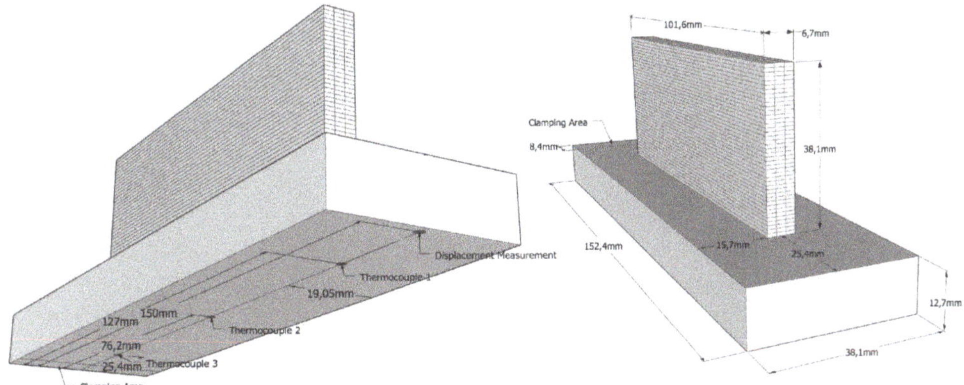

Figure 5. Dimensions of additive-manufacturing (AM) component.

Three beads were added per layer with a total width of 6.7 mm and a height of 38.1 mm; 42 discrete layers and their respective beads are also shown in Figure 5. Figure 6 shows the order of deposition starting with the middle bead, followed by one on each side. Odd layers are deposited in the direction away from the clamping, followed by even layers in the opposite direction. Figure 6 also shows the temperature contours and direction of deposition of bead three of the first layer. After each layer, a dwell time (DT) of 0, 20 and 40 s was applied for studying the effect of varying cooling rates.

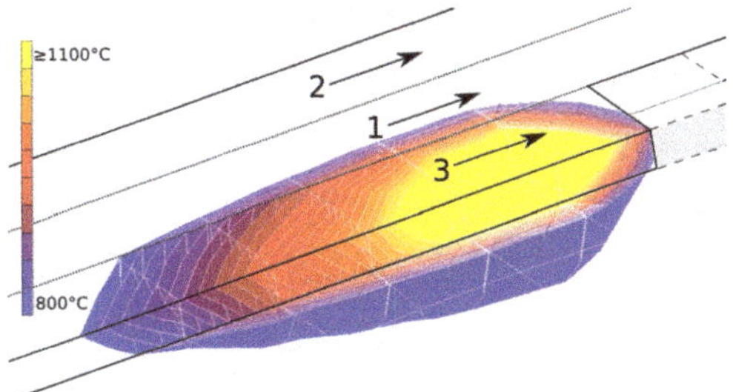

Figure 6. Temperature contours in weld pool.

6. Modelling of Additive Manufacturing

In the current work, the scope of the model was to predict microstructure evolution, the overall distortion of the component and residual stresses. This requires a solution where thermomechanical–metallurgical models are coupled by using a staggered approach. Figure 7 shows the coupling of different domains using a staggered approach. The thermal field is solved using the FE implicit iterative scheme by computing heat input and heat losses by conduction, convection and radiation. On the basis of the computed temperature in the increment, the metallurgical model computes the phase evolution for each Gauss point. The computed temperature and phase fractions are input to solve the mechanical-field equations. A large-deformation FE implicit iterative scheme was used here, where mechanical and physical properties are strongly dependent on the temperature and phase composition. Latent heat and volume changes due to phase evolution and deformation energy converted to heat are included here.

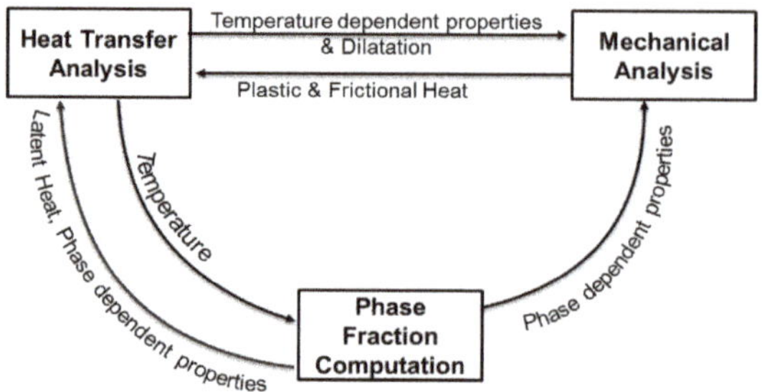

Figure 7. Coupling of thermomechanical–microstructural fields.

6.1. Heat Source

Modelling of weld-pool phenomena requires high-resolution discretization and at least one other physics domain, namely, fluid flow. This requires an impractical amount of resources for solving the problem and can be avoided considering the scope of the current work. The heat input can be modelled by using volume heat flux in a geometric region representing the weld pool, and is calibrated using measured temperatures. Goldak's [49] double ellipsoidal heat-input model was implemented in the current work, with efficiency calibrated to be 0.29. The parameters of the heat source are given in Figure 8. See Lundbäck and Lindgren [48] for details on the implementation of this model.

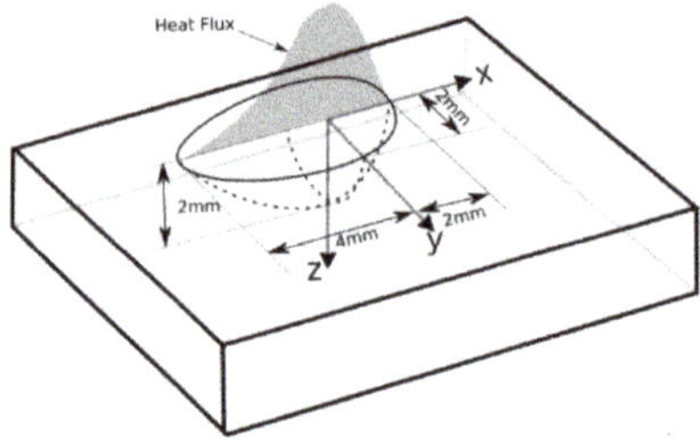

Figure 8. Gaussian distribution of density and double ellipsoid shape in xy plane.

6.2. Modelling of Material Addition

The inactive-element approach was used here, where all elements representing the added metal were deactivated before the start of the simulation, and only activated after meeting certain criteria. In each increment, the set of elements that overlapped and the geometric region represented by the current weld-pool position were thermally activated, whereas mechanical activation occurred only when the thermally active elements cooled below the solidus temperature. Before thermal activation, the elements may have had to be moved to accommodate for the distortion of the substrate and the already activated elements during the process. The moving elements maintained connectivity with the activated elements, and their volume matched the added material during that time step [48].

6.3. Boundary Conditions

In DED, much of the heat input in the initially deposited layers is absorbed by the substrate. To balance the heat input, losses by free and forced convection, and conduction to fixtures as well as radiation were included in the model. A lumped convective coefficient of 18 W/m^2/°C was applied to model the natural and forced convection from shielding gas. Both convective and radiative boundary conditions were applied on the outer surface of all thermally active elements. A surface emissivity of 0.25 was used here. In the current model, heat losses due to cooling by the fixture were achieved by using convective heat transfer with a high coefficient of 250 W/m^2/°C.

7. Comparison of Measurements and Simulations

In situ measurement of temperature and distortion was done during the AM process. Three thermocouples were attached to the bottom of the substrate at the positions shown in the left part of Figure 5. DTs allowed the component to considerably cool down during the process. In Figure 9a,c,e, dots denote the thermocouple measurements and lines, predictions from the model. The thermocouple attached to the middle of the component (TC2) registered the highest temperature, as the other two were closer to the ends that are subjected to higher convective cooling. The thermocouple attached close to the clamping (TC3) had the lowest temperature since the fixture acts as a heat sink. Raising dwell times by 20 s increased cooling, thereby reducing the peaks. As the height of the wall increased, this effect was less detectable, as this thermocouple was beneath the substrate.

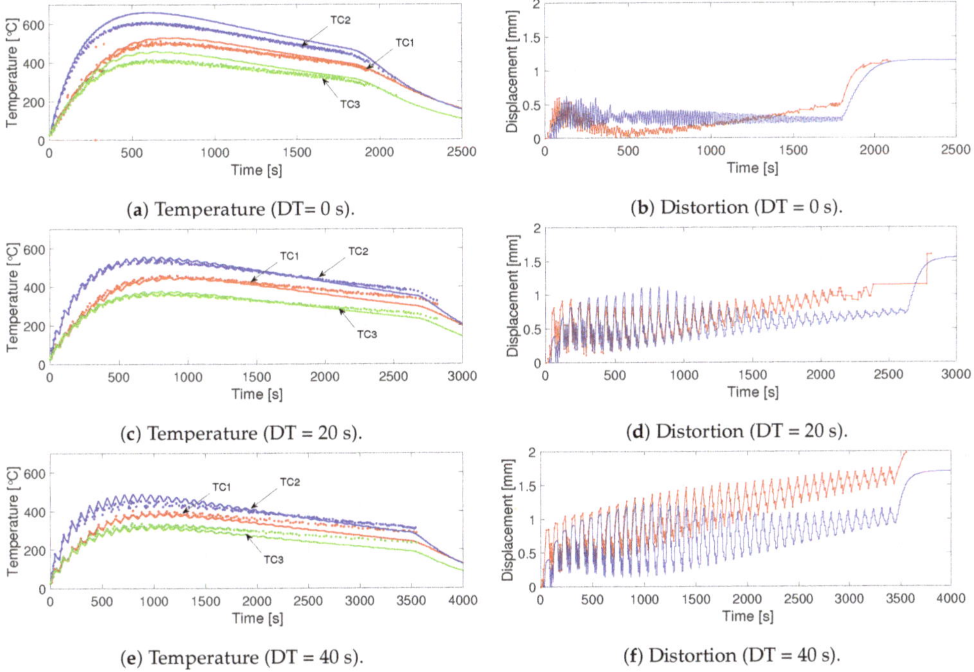

(**a**) Temperature (DT= 0 s). (**b**) Distortion (DT = 0 s).

(**c**) Temperature (DT = 20 s). (**d**) Distortion (DT = 20 s).

(**e**) Temperature (DT = 40 s). (**f**) Distortion (DT = 40 s).

Figure 9. Comparison of Measurements and Simulations.

The component distortion is measured at the free end by using a laser displacement sensor. In Figure 9b,d,f, red lines denote the measured values, and the blue lines denote predictions from the model. The addition of each layer made the component bend downwards due to the thermal gradient between the top and bottom of the substrate, which is diminished during cooling, producing oscillations. In order to compare measurement and simulation results, these oscillations were smoothed out by using the Savitzky–Golay filter [50], and are plotted in Figure 10. Here, dotted lines denote measurements, and continuous lines denote model predictions. The start and end of the linear region, its slope and the detection of the first peak can be deduced from Figure 10 and are shown in Table 3. The peak-to-peak amplitude of the oscillations at the first peak, and also at the start and end of the linear region are given in Table 3. The measured final bending of the build plate at the outer edge increased by 0.4 mm for each 20 s increase in dwell time. Simulations also gave a similar result. Simulation results are given in Table 3 within brackets.

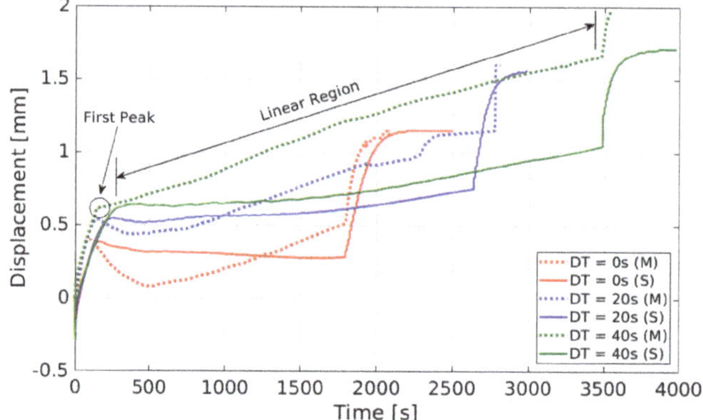

Figure 10. Comparison of distortions.

Table 3. Comparison of distortion (computed values given in brackets).

	0 s (comp)	20 s (comp)	40 s (comp)
Final displacement [mm]	1.2 (1.2)	1.6 (1.6)	2.0 (1.7)
Slope of linear region [10^{-4}mm/s]	3.4 (-0.2)	3.7 (1.1)	3.2 (1.4)
Time at 1st Peak [s]	110 (140)	170 (230)	150 (300)
Amplitude at 1st Peak [10^{-1}mm]	3.5 (3.8)	7.6 (7.0)	6.4 (6.9)
Amplitude at start of linear region [10^{-1}mm]	1.0 (0.9)	6.2 (8.4)	6.6 (6.1)
Amplitude at end of linear region [10^{-1}mm]	0.1 (0.4)	1.1 (0.8)	2.4 (2.1)

Figure 11 shows the residual stress in the welding direction along with its spread measured by Reference [47] by using hole drilling. The testing location was in the middle of the specimen at the bottom of the substrate. Results showed that, for the case with dwell times of 0 and 20 s, simulation results were close to the measurements or within the margin of error; for 40 s, it was slightly below the margin. Residual-stress distribution after cooling to room temperature in the welding direction and the von Mises effective stress for the case with dwell time of 0 s are plotted in Figure 12. The predicted temperature for the case with dwell time of 0 s at the top surface of the substrate above the location of TC2 is given in Figure 13. The computed α-phase fraction is also provided here. The addition of each bead is denoted as B1-3, and grey areas in between are the cooling time. In total, five layers are shown in Figure 13.

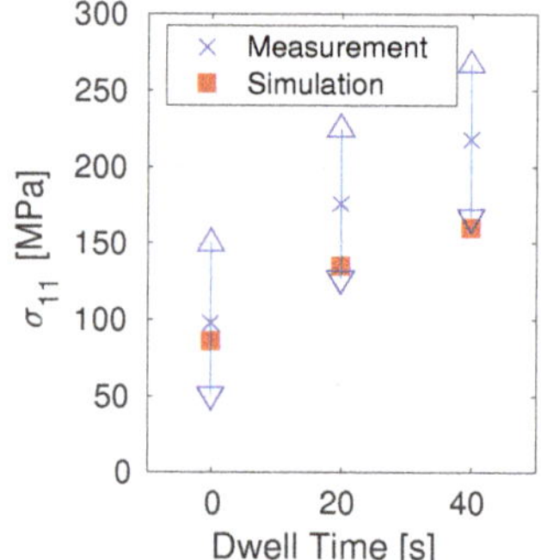

Figure 11. Residual stress.

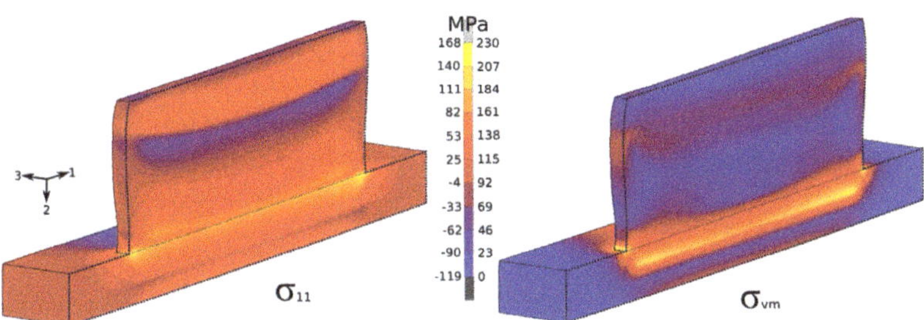

Figure 12. Residual stress for 0 s dwell time (model clipped longitudinally in midplane).

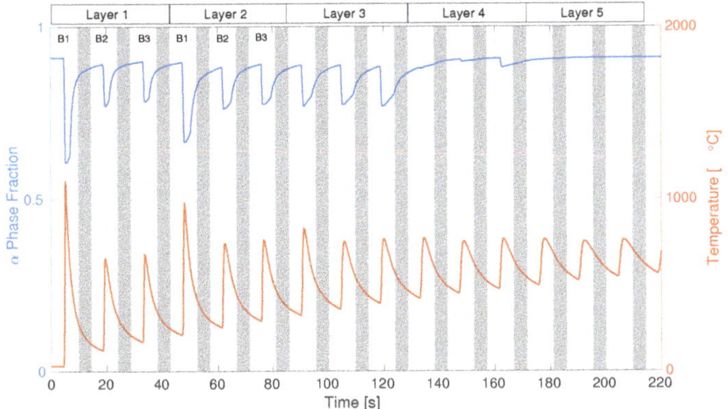

Figure 13. Prediction of phase fraction and temperature.

8. Discussion

The yield strength of the material is very much dependent on state variables of dislocation density and excess vacancy concentration. The density of these state variables changes by many orders of magnitude during heating and deformation. Deformation increases dislocation density and results in hardening, whereas an increase in vacancy concentration (due to heating and deformation) assists in the remobilization of dislocations, thereby material recovery. The advanced material model described here combining the metallurgical and flow-stress models has proven to be suitable for AM simulation. Diffusional and instantaneous transformations are included in the metallurgical model. This model was formulated in a way that it could be implemented in any standard kind of finite-element software. Temperature measurements and results from the simulations demonstrated a good overall fit. This model predicted the final distortion of the component with good accuracy except for the case with 40 s dwell time. This trend was also evident in the residual-stress measurements. The comparison of distortion before the onset of cooling showed a larger difference between model and measurements. This might be because the stress-relaxation behavior was less accurately predicted by the model. The computed phase fraction in Figure 13 showed that, after the addition of the fourth layer, the substrate underwent no significant phase evolution. Temperature peaks after 120 s had slightly less magnitude, and were therefore below the cutoff levels to introduce any phase change. Denlinger and Michaleris [51] performed the simulation of all the three cases described here. They used an approach where the plastic strain was reset to zero at a temperature of 690 °C which is a parameter calibrated for that particular AM case. This transformation-strain parameter made it possible for Denlinger and Michaleris [51] to include the effects of dwell time, whereas in the current work, mechanisms of dislocation climb and globularization resulted in the restoration of the lattice.

9. Conclusions

One of the challenges involved in the AM process is residual deformation and stresses due to the thermal dilatation of the substrate and added structure. The final properties of the AM structure are strongly influenced by the microstructure, which is dependent on the thermomechanical-processing history of the component. For the industry to fully adopt additive manufacturing and be able to qualify titanium parts for critical applications, such as in aerospace, a complete understanding of the microstructure properties and mechanical behavior is necessary. This paper showed the implementation and application of a coupled microstructural–thermal–mechanical model to an AM process. A physically based constitutive model was explicitly coupled to the microstructural model. The phase composition predicted by the microstructural model therefore affected the mechanical properties, namely, flow strength, in a direct way. Validation of the thermal and mechanical model was performed by comparing the simulation results with the available measurements in the literature. The comparison had good agreement between the results from the model and the measurements.

Author Contributions: Conceptualization, B.B. and A.L.; methodology, B.B. and A.L.; software, B.B. and A.L.; validation, B.B. and A.L.; formal analysis, B.B. and A.L.; original-draft preparation, B.B.; visualization, B.B.; supervision, L.-E.L.

Funding: This research was partly funded by the Swedish Foundation for Strategic Research (SSF), Development of Processes and Material in Additive Manufacturing, Reference number GMT14-0048.

Conflicts of Interest: The authors declare no conflict of interest.

Abbreviations

The following abbreviations are used in this manuscript:

DED	Directed Energy Deposition
TMM	Thermomechanical-Microstructural
AM	Additive Manufacturing
JMAK	Johnson–Mehl–Avrami–Kolmogrov
PBF	Powder Bed Fusion
FE	Finite Element
ROM	Rule of Mixtures
CTE	Coefficient of Thermal Expansion
CAD	Computer Aided Design
CNC	Computer Numerical Controlled
DT	Dwell Time
TC	Thermocouple

References

1. Tuomi, J.; Björkstrand, R.; Pernu, M.; Salmi, M.; Huotilainen, E.; Wolff, J.; Vallittu, P.; Mäkitie, A. In vitro cytotoxicity and surface topography evaluation of additive manufacturing titanium implant materials. *J. Mater. Sci. Mater. Med.* **2017**, *28*, 53. [CrossRef] [PubMed]
2. Dey, N.; Liou, F.; Nedic, C. Additive manufacturing laser deposition of Ti-6Al-4V for aerospace repair applications. In Proceedings of the 24th International SFF Symposium—An Additive Manufacturing Conference, Austin, TX, USA, 12–14 August 2013; pp. 853–858.
3. Baykasoglu, C.; Akyildiz, O.; Candemir, D.; Yang, Q.; To, A.C. Predicting Microstructure Evolution During Directed Energy Deposition Additive Manufacturing of Ti-6Al-4V. *J. Manuf. Sci. Eng.* **2018**, *140*, 051003. [CrossRef]
4. Salsi, E.; Chiumenti, M.; Cervera, M. modelling of Microstructure Evolution of Ti6Al4V for Additive Manufacturing. *Metals* **2018**, *8*, 633. [CrossRef]
5. Vastola, G.; Zhang, G.; Pei, Q.X.; Zhang, Y.W. Modeling the microstructure evolution during additive manufacturing of Ti6Al4V: A comparison between electron beam melting and selective laser melting. *JOM* **2016**, *68*, 1370–1375. [CrossRef]

6. Song, K.; Wei, Y.; Dong, Z.; Ma, R.; Zhan, X.; Zheng, W.; Fang, K. Constitutive model coupled with mechanical effect of volume change and transformation induced plasticity during solid phase transformation for TA15 alloy welding. *Appl. Math. Model.* **2015**, *39*, 2064–2080. [CrossRef]
7. Teixeira, J.; Denand, B.; Aeby-Gautier, E.; Denis, S. Simulation of coupled temperature, microstructure and internal stresses evolutions during quenching of a β-metastable titanium alloy. *Mater. Sci. Eng. A* **2016**, *651*, 615–625. [CrossRef]
8. Cao, J.; Gharghouri, M.A.; Nash, P. Finite-element analysis and experimental validation of thermal residual stress and distortion in electron beam additive manufactured Ti-6Al-4V build plates. *J. Mater. Process. Technol.* **2016**, *237*, 409–419. [CrossRef]
9. Ahn, J.; He, E.; Chen, L.; Wimpory, R.; Dear, J.; Davies, C. Prediction and measurement of residual stresses and distortions in fibre laser welded Ti-6Al-4V considering phase transformation. *Mater. Des.* **2017**, *115*, 441–457. [CrossRef]
10. Babu, B. Mechanism-Based Flow Stress Model for Ti-6Al-4V: Applicable for Simulation of Additive Manufacturing and Machining. Ph.D. Thesis, Luleå University of Technology, Luleå, Sweden, 2018.
11. Babu, B.; Lindgren, L.E. Dislocation density based model for plastic deformation and globularization of Ti-6Al-4V. *Int. J. Plast.* **2013**, *50*, 94–108. [CrossRef]
12. Murgau, C.; Lundbäck, A.; Åkerfeldt, P.; Pederson, R. Temperature and Microstructure Evolution in Gas Tungsten Arc Welding Wire Feed Additive Manufacturing of Ti-6Al-4V. *J. Mater.* **2019**, *12*, 3534. [CrossRef]
13. Seeger, A. The mechanism of Glide and Work Hardening in FCC and HCP Metals. In *Dislocations and Mechanical Properties of Crystals*; Fisher, J., Johnston, W.G., Thomson, R., Vreeland, T.J., Eds.; Wiley: New York, NY, USA, 1956; pp. 243–329.
14. Bergström, Y. Dislocation model for the stress-strain behaviour of polycrystalline alpha-iron with special emphasis on the variation of the densities of mobile and immobile dislocations. *Mater. Sci. Eng.* **1969**, *5*, 193–200. [CrossRef]
15. Kocks, U. Laws for Work-Hardening and Low-Temperature Creep. *J. Eng. Mater. Technol. Trans. ASME* **1976**, *98*, 76–85. [CrossRef]
16. Conrad, H. Effect of interstitial solutes on the strength and ductility of titanium. *Prog. Mater. Sci.* **1981**, *26*, 123–403. [CrossRef]
17. Kocks, U.F.; Argon, A.S.; Ashby, M.F. *Thermodynamics and Kinetics of Slip*; Volume 19, Progress in Material Science; Pergamon Press: Oxford, UK, 1975.
18. Mecking, H.; Kocks, U. Kinetics of flow and strain-hardening. *Acta Metall.* **1981**, *29*, 1865–1875. [CrossRef]
19. Holt, D.L. Dislocation cell formation in metals. *J. Appl. Phys.* **1970**, *41*, 3197. [CrossRef]
20. Bergström, Y. The plastic deformation of metals—A dislocation model and its applicability. *Rev. Powder Metall. Phys. Ceram.* **1983**, *2*, 79–265.
21. Militzer, M.; Sun, W.P.; Jonas, J.J. Modelling the effect of deformation-induced vacancies on segregation and precipitation. *Acta Metall. Et Mater.* **1994**, *42*, 133. [CrossRef]
22. Sandstrom, R.; Lagneborg, R. A model for hot working occurring by recrystallization. *Acta Metall.* **1975**, *23*, 387. [CrossRef]
23. Mecking, H.; Estrin, Y. The effect of vacancy generation on plastic deformation. *Scr. Metall.* **1980**, *14*, 815. [CrossRef]
24. Thomas, J.P.; Semiatin, S.L. Mesoscale modelling of the Recrystallization of Waspaloy and Application to the Simulation of the Ingot-Cogging Process. In *Materials Science and Technology (MS&T) 2006, Processing*; TMS: Warrendale, PA, USA, 2016; pp. 609–619.
25. Pietrzyk, M.; Jedrzejewski, J. Identification of Parameters in the History Dependent Constitutive Model for Steels. *CIRP Ann. Manuf. Technol.* **2001**, *50*, 161–164. [CrossRef]
26. Montheillet, F.; Jonas, J.J. *Fundamentals of modelling for Metals Processing*; ASM International: Materials Park, OH, USA, 2009; Volume 22A, pp. 220–231.
27. Charles Murgau, C.; Pederson, R.; Lindgren, L. A model for Ti–6Al–4V microstructure evolution for arbitrary temperature changes. *Model. Simul. Mater. Sci. Eng.* **2012**, *20*, 055006. [CrossRef]
28. Lütjering, G. Influence of processing on microstructure and mechanical properties of $(\alpha + \beta)$ titanium alloys. *Mater. Sci. Eng. A* **1998**, *243*, 32–45. [CrossRef]
29. Williams, J.C.; Lütjering, G. *Titanium*; Springer: Berlin/Heidelberg, Germany, 2003.

30. Semiatin, S.L.; Seetharaman, V.; Weiss, I. Flow behavior and globularization kinetics during hot working of Ti-6Al-4V with a colony alpha microstructure. *Mater. Sci. Eng. A* **1999**, *263*, 257. [CrossRef]
31. Semiatin, S.L.; Seetharaman, V.; Ghosh, A.K. Plastic flow, microstructure evolution, and defect formation during primary hot working of titanium and titanium aluminide alloys with lamellar colony microstructures. *Philos. Trans. Math. Phys. Eng. Sci.* **1999**, *357*, 1487–1512. [CrossRef]
32. Seetharaman, V.; Semiatin, S.L. Effect of the lamellar grain size on plastic flow behavior and microstructure evolution during hot working of a gamma titanium aluminide alloy. *Metall. Mater. Trans. A* **2002**, *33A*, 3817. [CrossRef]
33. Thomas, R.B.; Nicolaou, P.D.; Semiatin, S.L. An Experimental and Theoretical Investigation of the Effect of Local Colony Orientations and Misorientation on Cavitation during Hot Working of Ti-6Al-4V. *Metall. Mater. Trans. A* **2005**, *36A*, 129.
34. Johnson, W.; Mehl, R. Reaction Kinetics in Processes of Nucleation and Growth. *Trans. Soc. Pet. Eng.* **1939**, *135*, 416.
35. Avrami, M. Kinetics of Phase Change. I General Theory. *J. Chem. Phys.* **1939**, *7*, 1103–1112. [CrossRef]
36. Kolmogorov, A. A statistical theory for the recrystallisation of metals, Akad Nauk SSSR, Izv. *Izv. Akad. Nauk. SSSR* **1937**, *3*, 355.
37. Ahmed, T.; Rack, H.J. Phase transformations during cooling in [alpha]+[beta] titanium alloys. *Mater. Sci. Eng. A* **1998**, *243*, 206–211. [CrossRef]
38. Lu, S.; Qian, M.; Tang, H.; Yan, M.; Wang, J.; StJohn, D. Massive transformation in Ti–6Al–4V additively manufactured by selective electron beam melting. *Acta Mater.* **2016**, *104*, 303–311. [CrossRef]
39. Kelly, S.M. Thermal and Microstructure modelling of Metal Deposition Processes with Application to Ti-6Al-4V. Ph.D. Thesis, Virginia Polytechnic Institute and State University, Blacksburg, VA, USA, 2004.
40. Kelly, S.M.; Babu, S.S.; David, S.A.; Zacharia, T.; Kampe, S.L. A microstructure model for laser processing of Ti-6Al-4V. In *24th International Congress on Applications of Lasers and Electro-Optics, ICALEO 2005*; Laser Institute of America: Orlando, FL, USA; Miami, FL, USA, 2005; pp. 489–496.
41. Wachtman, J.B.; Tefft, W.E.; Lam, D.G.; Apstein, C.S. Exponential Temperature Dependence of Young's Modulus for Several Oxides. *Phys. Rev.* **1961**, *122*, 1754–1759. [CrossRef]
42. Fukuhara, M.; Sanpei, A. Elastic moduli and internal frictions of Inconel 718 and Ti-6Al-4V as a function of temperature. *J. Mater. Sci. Lett.* **1993**, *12*, 1122–1124. [CrossRef]
43. Swarnakar, A.K.; der Biest, O.V.; Baufeld, B. Thermal expansion and lattice parameters of shaped metal deposited Ti–6Al–4V. *J. Alloy. Compd.* **2011**, *509*, 2723–2728. [CrossRef]
44. Boivineau, M.; Cagran, C.; Doytier, D.; Eyraud, V.; Nadal, M.; Wilthan, B.; Pottlacher, G. Thermophysical Properties of Solid and Liquid Ti-6Al-4V (TA6V) Alloy. *Int. J. Thermophys.* **2006**, *27*, 507–529. [CrossRef]
45. Mills, K.C. *Recommended Values of Thermophysical Properties for Selected Commercial Alloys*; Woodhead Publishing: Cambridge, UK, 2002.
46. Bouaziz, O.; Buessler, P. Iso-work Increment Assumption for Heterogeneous Material Behaviour Modelling. *Adv. Eng. Mater.* **2004**, *6*, 79–83. [CrossRef]
47. Denlinger, E.R.; Heigel, J.C.; Michaleris, P.; Palmer, T. Effect of inter-layer dwell time on distortion and residual stress in additive manufacturing of titanium and nickel alloys. *J. Mater. Process. Technol.* **2015**, *215*, 123–131. [CrossRef]
48. Lundbäck, A.; Lindgren, L.E. Modelling of metal deposition. *Finite Elem. Anal. Des.* **2011**, *47*, 1169–1177. [CrossRef]
49. Goldak, J.; Chakravarti, A.; Bibby, M. A new finite element model for welding heat sources. *Metall. Trans. B* **1984**, *15*, 299–305. [CrossRef]
50. Savitzky, A.; Golay, M.J.E. Smoothing and Differentiation of Data by Simplified Least Squares Procedures. *Anal. Chem.* **1964**, *36*, 1627–1639. [CrossRef]
51. Denlinger, E.R.; Michaleris, P. Effect of stress relaxation on distortion in additive manufacturing process modelling. *Addit. Manuf.* **2016**, *12*, 51–59. [CrossRef]

Article

Fatigue Crack Growth of Electron Beam Melted Ti-6Al-4V in High-Pressure Hydrogen

M. Neikter [1,4], M. Colliander [2], C. de Andrade Schwerz [3], T. Hansson [3,4], P. Åkerfeldt [1], R. Pederson [4,*] and M.-L. Antti [1]

1. Division of Materials Science, Luleå University of Technology, 97181 Luleå, Sweden; magnus.neikter@hv.se (M.N.); pia.akerfeldt@ltu.se (P.A.); marta-lena.antti@ltu.se (M.-L.A.)
2. Department of Applied Physics, Chalmers University of Technology, 41296 Göteborg, Sweden; magnus.colliander@chalmers.se
3. GKN Aerospace Engine Systems, 461 38 Trollhättan, Sweden; claudia.schwerz@gknaerospace.com (C.d.A.S.); thomas.hansson@hv.se (T.H.)
4. Division of Subtractive and Additive Manufacturing, University West, 46132 Trollhättan, Sweden

* Correspondence: robert.pederson@hv.se

Received: 26 February 2020; Accepted: 10 March 2020; Published: 12 March 2020

Abstract: Titanium-based alloys are susceptible to hydrogen embrittlement (HE), a phenomenon that deteriorates fatigue properties. Ti-6Al-4V is the most widely used titanium alloy and the effect of hydrogen embrittlement on fatigue crack growth (FCG) was investigated by carrying out crack propagation tests in air and high-pressure H_2 environment. The FCG test in hydrogen environment resulted in a drastic increase in crack growth rate at a certain ΔK, with crack propagation rates up to 13 times higher than those observed in air. Possible reasons for such behavior were discussed in this paper. The relationship between FCG results in high-pressure H_2 environment and microstructure was investigated by comparison with already published results of cast and forged Ti-6Al-4V. Coarser microstructure was found to be more sensitive to HE. Moreover, the electron beam melting (EBM) materials experienced a crack growth acceleration in-between that of cast and wrought Ti-6Al-4V.

Keywords: fatigue crack growth (FCG); electron beam melting (EBM); Ti-6Al-4V; hydrogen embrittlement (HE)

1. Introduction

Oxygen and hydrogen are used as fuel for space rockets [1]. Hydrogen is combusted rapidly in an oxygen-rich environment and a high propelling force can be achieved, as the energy density for hydrogen is high (142 $MJkg^{-1}$) [2]. Although hydrogen is excellent for combustion it can cause hydrogen embrittlement (HE), which deteriorates the mechanical properties. HE occurs at or ahead of crack tips because at these locations tri-axial stresses are present [3]. These stresses render slightly expanded lattices, making it more energetically favorable for the hydrogen to diffuse to this location. Once at the crack tip the hydrogen causes degraded properties due to one or several HE mechanisms [4,5].

Additively manufactured Ti-6Al-4V has achieved large interest within the space industry since it can reduce weight and lead time. Ti-6Al-4V is a dual-phase alloy, consisting of both α and β phase [6]. In the α phase, which has a hexagonal close-packed crystal structure, the diffusion rate of hydrogen is lower compared to the body-centered cubic β phase [7]. The amount of β phase at room temperature is not as high compared to the α phase. The microstructure is different for Ti-6Al-4V material built with the additive manufacturing (AM) process electron beam melting (EBM) compared to conventionally wrought or cast material [8,9]. In Ti-6Al-4V manufactured with EBM, the prior β grains grow epitaxially towards the heat source, which renders a columnar structure perpendicular to the built layers [9–12], a unique morphology that is observed in neither cast nor wrought titanium. At the grain boundary of the β phase, there is nucleation of α phase when the temperature is reduced below the β transus temperature (995 °C for Ti-6Al-4V [8]). Within these columnar β grains, EBM built Ti-6Al-4V typically has a basketweave microstructure. Wrought Ti-6Al-4V microstructure consists of primary α combined with Widmanstätten microstructure [8,13], called bimodal or duplex microstructure. Cast microstructures normally consist of coarse prior β grains with large α colonies, where the α laths within the colonies are oriented in the same crystal orientation.

Diffusion is an important part of the HE mechanisms [4] where microstructure plays an important role. Tal-Gutelmacher et al. [14] investigated the effect of hydrogen solubility for Widmanstätten and bimodal microstructures. The conclusion was that fully lamellar Widmanstätten microstructure had several orders of magnitude higher hydrogen solubility than the bimodal microstructure, which was explained by the continuous β phase in the fully lamellar structure.

Texture can affect the ingress of hydrogen and the hydride formation [15,16] but the texture of EBM built Ti-6Al-4V has been shown to be weak [17]. Residual stresses, which are known to be present in various AM processes [18–21], can furthermore affect the fatigue crack growth (FCG) rate. However, Maimaitiyili et al. [22] did show that there are no or small residual stresses in as-built EBM Ti-6Al-4V.

Rozumek et al. [23] showed that the FCG rate in Ti-6Al-4V is highly depending on post processing. By performing a hardening and ageing heat treatment five times higher fatigue life was obtained.

Relevant for space applications are the cryogenic properties and temperature has been shown to have a strong effect on FCG properties. Increased temperature renders an increased diffusion rate of hydrogen [14], and two temperatures were investigated by Pittinato [24]; −129 °C and −73 °C. At −129 °C there was no difference in the FCG rate in helium and hydrogen atmospheres, whereas at −73 °C there was an increase in FCG rate in hydrogen environment.

FCG experiments have previously been performed on a wide range of Ti-6Al-4V material [24–27], but few of these studies concerned hydrogen environment and additive manufacturing. In previous studies, the FCG properties of conventionally manufactured Ti-6Al-4V in high-pressure hydrogen was performed by Gaddam et al. [26,28] showing that microstructure has an effect on the FCG properties in hydrogen. Cast Ti-6Al-4V with coarser microstructure was shown to have inferior FCG properties compared with forged material with fine microstructure.

In this work, the effect of hydrogen embrittlement on FCG properties of EBM built Ti-6Al-4V has been investigated. Samples have been exposed to either air or high-pressure (150 bar) hydrogen atmosphere at room temperature. The EBM built material has been compared to previous results [26,28] for cast and forged materials, and the differences in FCG properties have been linked to the different types of microstructures. To further investigate the hydrogen embrittlement, fractography was performed along with crack profile characterization.

2. Experimental Method

2.1. Material

EBM Ti-6Al-4V samples were manufactured with an Arcam Q20+ machine, using a layer thickness of 90 µm. Cylinders were manufactured having a length of 135 mm and a diameter of 25 mm. The layers were oriented perpendicular to the major axis of the cylinders i.e., the applied load during the FCG tests was perpendicular to the AM built layers. The powder used was Ti-6Al-4V B110 (Virgin Hoeganaes) in accordance with the aerospace material specification AMS 4992. Prior to testing, the samples received a hot isostatic pressure (HIP) treatment at 920 ± 10 °C for two hours with a pressure of 1020 bar, followed by a heat treatment at 704 ± 10 °C for two hours. Both HIP and heat treatment were conducted in argon. Out of the manufactured cylinders, Kb bars were machined (see Figure 1) using low stress grinding and polishing of the gauge section. A tensile test was performed on the post treated material and the result showed a yield strength of 890 MPa and tensile strength 990 MPa.

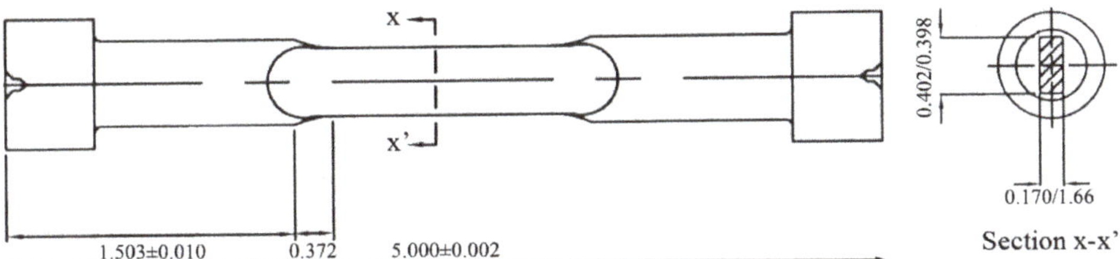

Figure 1. Sketch of Kb bar specimen. Dimensions in mm (printed with permission from Metcut Research).

2.2. FCG Experiments

The FCG test in air was conducted at Metcut Research Inc., Cincinnati, OH, USA, while the hydrogen testing was conducted at The Welding Institute (TWI) in Cambridge, UK. One Kb bar was tested in a hydrogen-rich atmosphere with a pressure of 150 bars and two Kb bars were tested in air at ambient temperature. The samples were designated H-A, Air-A, and Air-B, respectively. All tests were stress-controlled and performed at room temperature with the max loads; 645 MPa for H-A, 534 MPa for Air-A, and 528 MPa for Air-B. The FCG testing fulfills the requirements for plane strain conditions. The pre-cracking was performed using a frequency of 10 Hz. Tint temperatures between 450 °C to 350 °C were used to mark the crack propagation. The fatigue tests were performed with a uniaxial load perpendicular to the AM built layers with $R = \sigma_{min}/\sigma_{max} = 0$ and a test frequency of 0.5 Hz, using a triangular waveshape. The crack propagation was measured using potential drop. The pre-crack and final crack sizes were measured using heat tinting and these sizes were correlated to the potential drop signal. The translation from the potential drop signals to crack sizes were made using a calibration curve [29]. Corrections were made so that the measured pre-crack and final crack sizes and corresponding potential drop values were consistent with the calibration curve. The FCG rate was then computed per data point using the secant method (ASTM E647-15e1). Once the FCG crack reached a certain length the remaining material was heat tinted to reveal the final crack length and fractured in tension using a monotonically increasing load. The equation used for the experiment was:

$$K = S\sqrt{\pi\frac{a}{Q}F_s\left(\frac{a}{c}, \frac{a}{t}, \frac{a}{b}, \Theta\right)} \tag{1}$$

where S is the tensile strength, a the crack depth, Q elliptical crack shape factor, F_s boundary correction factor, t thickness of the sample, b half-width of the sample, and c half-width of the crack. See [30] for full solution.

2.3. Fractography and Microstructural Characterization

For overview images of the fracture surfaces a stereomicroscope (Nikon SMZ1270) was utilized, whilst for fractography a scanning electron microscope (SEM, Jeol IT300LV) was used. Crack profiles were made on the hydrogen and air-tested samples, first by cutting cross-sections parallel to the x-x' plane according to Figure 1. Then, by grinding and polishing carefully, the desired positions were reached in the plane perpendicular to the x-x' plane (edge of notch). The grinding was monitored using a stereomicroscope. The crack profile was characterized with light optical microscope (LOM, Nikon eclipse MA200) and SEM. For microstructural characterization, a representative cross-section was ground and polished according to the conventional sample preparation techniques for titanium, then etched using Kroll's etchant (see ASTM Standard E 407). To investigate the microstructure at low magnification a LOM was used. The software Image J version 1.52a [31] was used to measure the width of the α laths and prior β grains; 100 measurements were performed for each microstructural feature to obtain the average size.

3. Results

3.1. Microstructure

The EBM Ti-6Al-4V material consisted of columnar prior β grains. The grain boundaries are illustrated as black dotted lines in Figure 2a and they were elongated parallel to the build direction, with lengths up to approximately 2 mm. In the plane perpendicular to the build direction, the prior β grains were equiaxed, with widths of ~100 µm. The prior β grains were partially separated by discontinuous grain boundary α (GB-α) and the microstructure within the prior β grains was basketweave, with an average α lath width of 2.2 ± 0.6 µm, see Figure 2b.

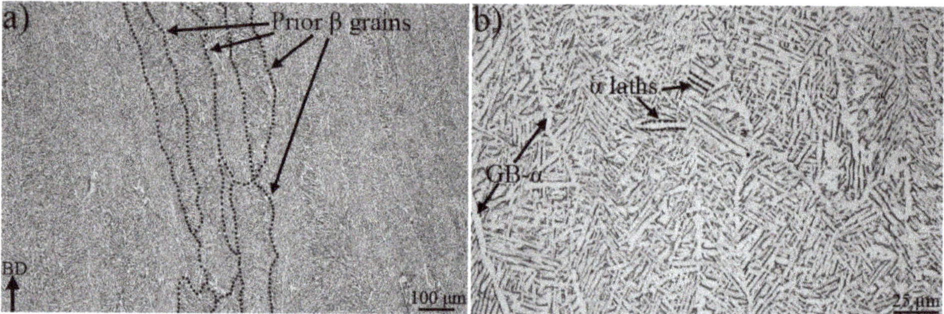

Figure 2. In (**a**) columnar prior β grains and in (**b**) the basketweave microstructure are shown. The black arrow in the bottom left corner in (**a**) points towards the build direction (BD).

3.2. Fatigue Crack Growth

Different plots of FCG test data are shown in Figures 3–5; crack length versus number of cycles, crack growth rate versus ΔK, and ratio between crack growth rate in hydrogen (specimen H-A) and air (specimen Air-B). The two air-tested samples followed the Paris law with similar inclination (see Figure 4). The hydrogen-tested sample, on the other hand, has a fluctuating crack growth rate in the first stage of the test. After the fluctuating stage, at ΔK ~23 $MPa\sqrt{m}$, a sudden increase in crack growth rate was observed. The resulting increase in crack length is shown in Figure 3: After ~5600 cycles there is a sudden offset in crack growth rate for H-A, whereas for Air-A/B no such abrupt offset is present. Figure 5 shows that below ~23 $MPa\sqrt{m}$ the relative crack propagation rate is around one, indicating no difference in air and hydrogen tests. Around this value of ΔK (23 $MPa\sqrt{m}$), the crack propagation rate ratio started to increase, and continues to do so at a constant rate up to ~32 $MPa\sqrt{m}$, where it reaches a plateau. By then, the crack propagation rate of the material tested in hydrogen is over 12 times higher than in air.

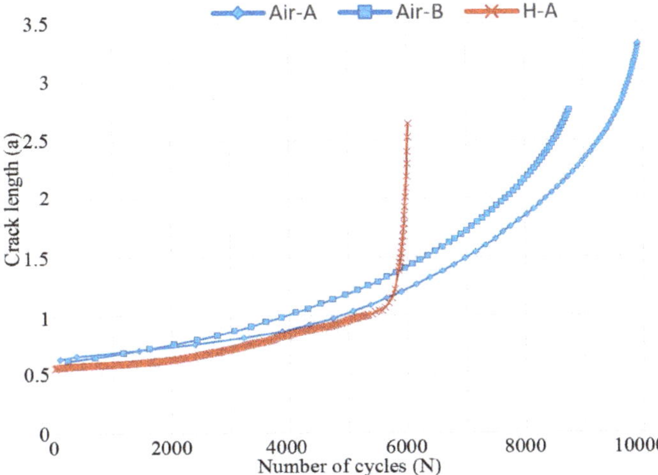

Figure 3. Crack length versus the number of cycles for all three tested electron beam melting (EBM) samples. At ~5600 cycles the crack length of the hydrogen-tested sample accelerates.

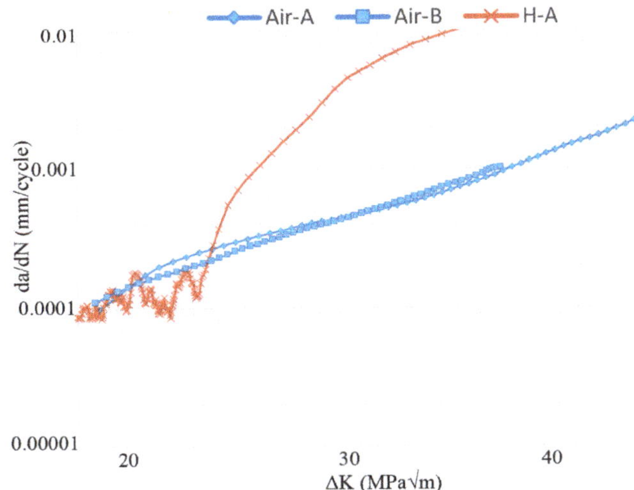

Figure 4. Crack growth rate versus ΔK. The hydrogen-tested material fluctuates below 23 $MPa\sqrt{m}$, while accelerating above. The air-tested material follows Paris law.

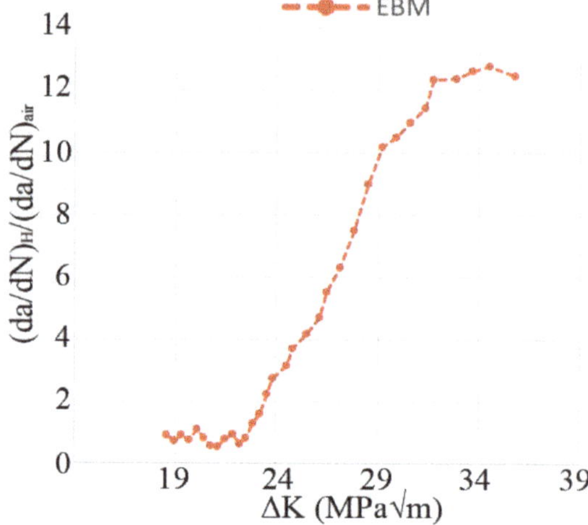

Figure 5. Relative crack propagation rate between hydrogen and air-tested material versus ΔK. At 23 $MPa\sqrt{m}$ the relative crack propagation increases steadily while reaching a plateau at 31 $MPa\sqrt{m}$.

3.3. Fractography

Figure 6 shows the fracture surface of the hydrogen-tested sample (H-A). A change in macroscopic appearance was observed at a crack length ~1 mm (this crack length is shown in Figure 3 to coincide with the accelerated crack growth), which corresponded to ΔK 23 $MPa\sqrt{m}$ and the location of this ΔK is shown by a white dashed semi-ellipse in Figure 6 (in-between X and Y line). The lines X, Y, and Z shown in Figure 6, following semi-elliptic paths, correspond to fractographic locations where the fracture surface has been characterized extra carefully. The X line surrounds the pre-crack, Y is in the middle of the crack, and Z is at the end of the crack. Table 1 presents lengths and widths of notch, pre-crack, and the fatigue crack of all samples. The lengths a and 2c of the final fatigue crack are illustrated in Figure 6.

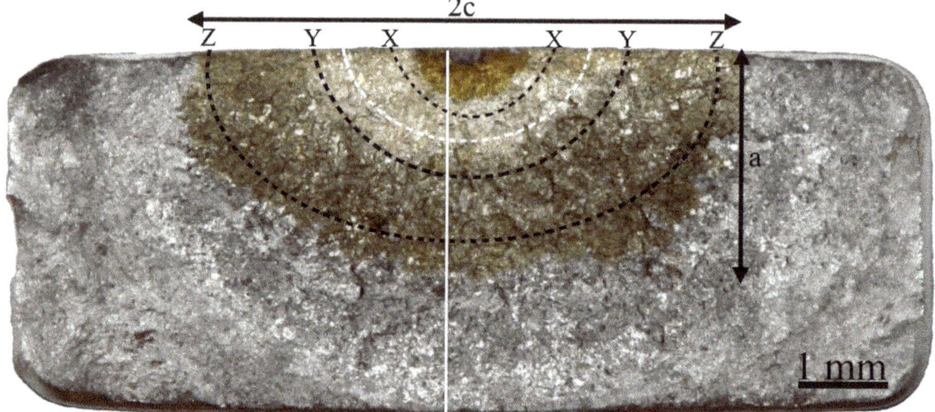

Figure 6. Overview of H-A fracture surface. The two tint temperatures, 450 and 350 °C, resulted in the two different golden color contrasts shown in the figure. "2c" represents the width and "a" the length of the crack. The three semi-elliptical dashed lines X, Y, and Z show profiles where fractography was performed in higher detail. The white dotted line in-between the X and Y lines corresponds to the ΔK 23 $MPa\sqrt{m}$. The vertical white line indicates the cross-section where crack profiles were investigated.

Table 1. Notch, pre-crack, and final fatigue crack lengths (a), widths (2c), and their ratio (a/c), for the three investigated samples Air-A/B and H-A.

Sample	Feature	a (mm)	2c (mm)	a/c
	Notch	0.18	0.35	1.03
Air-A	Pre-crack	0.64	1.25	1.02
	Fatigue crack	3.32	7.01	0.95
	Notch	0.17	0.36	0.94
Air-B	Pre-crack	0.59	1.27	0.93
	Fatigue crack	2.75	5.56	0.99
	Notch	0.18	0.34	1.06
H-A	Pre-crack	0.55	1.20	0.92
	Fatigue crack	2.64	6.02	0.88

In Figure 7a,b representative areas along Y and Z in Figure 6 of the fracture surface of the hydrogen-tested sample are shown, in Figure 7c the position of the crack profile in the ground specimen (white vertical line in Figure 6) is shown. A transition in fracture surface is shown in Figure 7a, below the white dashed semi-ellipse corresponding to crack length where the ΔK was 23 $MPa\sqrt{m}$. At this ΔK the fracture surface starts to transition from flat to rough. In Figure 7b (region Z) large cracks, exceeding lengths of 100 μm, are observed on the fracture surface.

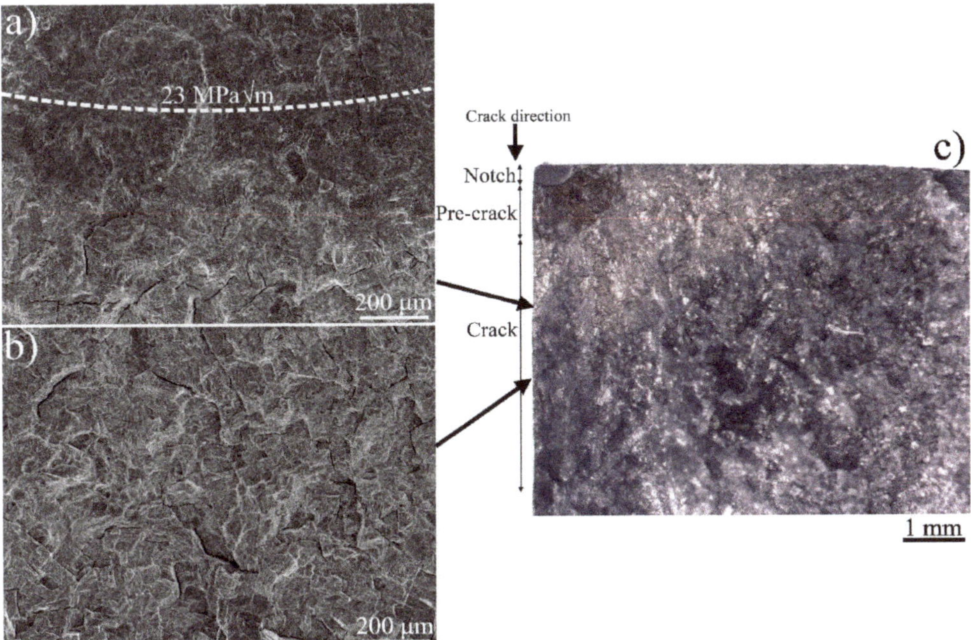

Figure 7. In (**a**) the transition zone from flat to rough fracture surface (the white dotted semi-ellipse shows ΔK 23 $MPa\sqrt{m}$), whereas (**b**) shows a rough fracture surface with large cracks. In (**c**) a section of H-A fracture surface is shown, cut as indicated by the vertical white line in Figure 6.

Figure 8 shows the characteristic fracture surface of the air-tested material, in Figure 8a higher magnification, whereas in Figure 8b lower magnification. The transition area observed in Figure 7, from flat to rough, did not exist in the air-tested material, neither do the large cracks.

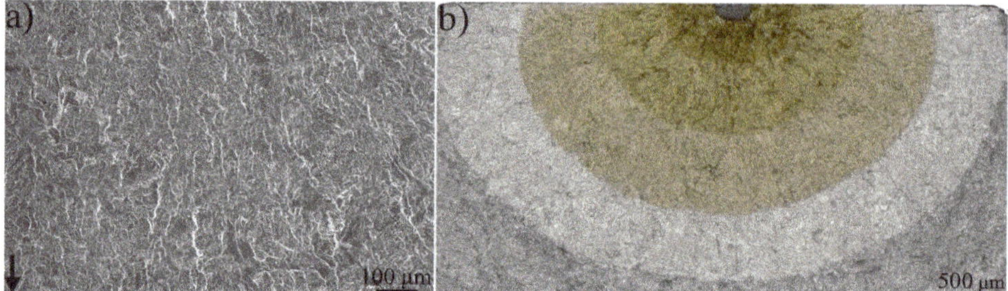

Figure 8. (**a**) 100× magnification fracture surface image of air-tested material (sample Air-B) in the region ~Y. (**b**) lower magnification image that shows an overview of the whole fracture surface. The black arrow points towards the crack propagation direction.

In Figure 9 representative images of the fracture surface are shown along the profiles X to Z (see Figure 6 for illustration of the locations on the fracture surface). In the air-tested samples, striations were observed along the whole fatigue crack, becoming increasingly larger the greater the ΔK became. In Air-B, section X, striations were only observed at higher magnification, whereas in section Z they were clearly visible for the present magnification (2000×).

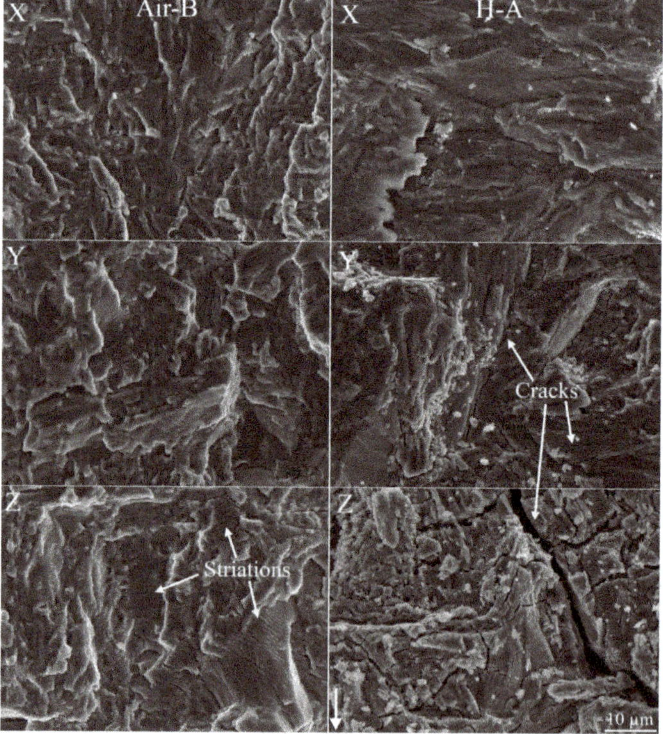

Figure 9. Images of the three sections X-Z (see Figure 6 for locations of the areas on the fracture surface) for one air (sample Air-B) and one hydrogen (sample H-A) tested sample. The images are in the same magnification and the white arrow in the bottom right indicates the crack propagation direction. In section Z of Air-B, examples of striations are indicated with the white arrows. In sections Y and Z of H-A, white arrows point at cracks; their dimensions increase from the former section to the latter.

In the hydrogen-tested sample, the fracture surface on the first stage of the fatigue crack (section X) appeared flat. With increased ΔK (above 23 $MPa\sqrt{m}$ i.e., section Y) an increase in fracture surface tortuosity aroused, along with the appearance of small cracks. Then, towards the end of the fatigue crack (section Z), larger cracks were observed with dimensions of ~100 μm. At the flat area of the H-A's fracture surface, i.e., below 23 $MPa\sqrt{m}$, features that resembled crack arrest marks (CAM) were observed. Crack propagation stops at the CAM interface, causing the formation of these marks. The origin could for example be the cleavge of a hydride. In Figure 10 features that resemble CAMs are marked by white arrows, where each plateau is the end of a CAM, which according to literature could indicate the interface between the hydride and titanium base metal [4,32]. Note that the CAM locations were not numerous.

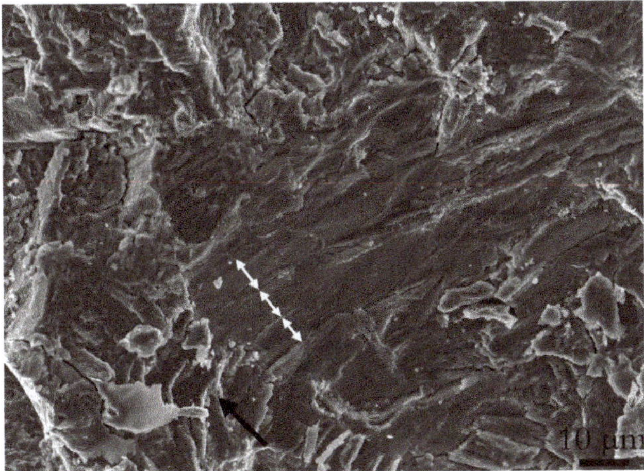

Figure 10. An area with features that resembled crack arrest marks (CAM). Each white arrow indicates a possible CAM, where each plateau could be a hydride titanium interface between cleaved hydrides. The black arrows show the crack growth direction.

Figure 11 is a high magnification image of a fracture surface cross-section in the hydrogen tested material. Secondary cracks were present across α/β interfaces. The crack path can be seen on the upper right corner, showing that the crack propagated along an α lath, seemingly in the α/β interface.

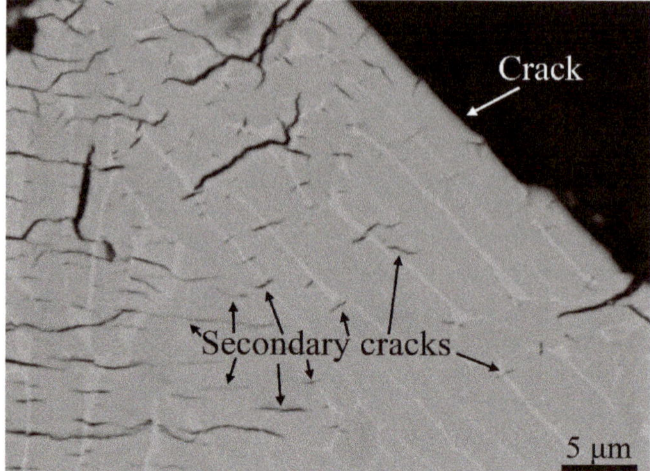

Figure 11. Image of a crack profile for sample H-A. It shows secondary cracks that grow across α/β interfaces. In the upper right corner, the crack propagated along an α lath, seemingly in the α/β interface.

In Figure 12 two crack profiles are shown (one for Air-B and one for H-A). The crack profiles correspond to the cross-section illustrated as a white vertical line in Figure 6. Crack profiles covering the whole fracture surface from notch to final tensile fracture were obtained in LOM, while the remaining images, Figure 12a–e, were obtained in an SEM using backscattered electrons.

The crack profile showed a similar pattern as was observed on the fracture surfaces. For H-A, in the region from the start of crack to ΔK 23 $MPa\sqrt{m}$, few cracks are observed i.e., image (c), from 23 $MPa\sqrt{m}$ and onwards large vertical (perpendicular to the crack direction) cracks were observed. With higher magnification also numerous small secondary cracks were observed after the ΔK 23 $MPa\sqrt{m}$, see Figure 12d. Note that these smaller secondary cracks had no connection to the main crack. For the H-A sample, three areas are shown in higher magnification; in-between pre-crack and 23 $Pa\sqrt{m}$ i.e., (c), after 23 $MPa\sqrt{m}$ i.e., (d), and before final crack i.e., (e). In (c) the crack profile was less tortuous and no vertical cracks or secondary cracks were observed. However, directly after 23 $MPa\sqrt{m}$, secondary cracks were observed as shown in Figure 12d. After area (d), deep vertical cracks appeared as shown in (e). The maximum major axis for the secondary cracks was in the dimensions of 10 to 20 μm, whereas the vertical cracks reached about 50 to 70 μm.

The hydrogen-tested sample had a tortuous crack profile, whereas the air-tested sample had a comparably straight crack path, with a homogeneous appearance throughout the various ΔK. The cracks propagated both parallel to the direction of α laths in the α/β interface and perpendicular to the laths, for both the air and hydrogen-tested material. Due to the build-orientation of the samples in regard to the load, the propagation of the fatigue crack was perpendicular to the major axis of the prior β grains i.e., perpendicular to the heat gradient, thus no propagation along grain boundary α could be observed.

For the air-tested material, no deep cracks were observed along the crack profile and features that appeared to be cracks on the fracture surface were shown to be superficial. In Figure 12b, the largest identified crack in Air-B sample is shown. The vertical cracks shown in (e) both propagated parallel to an α lath in the α/β interface, then switched to propagation across several α laths. The location of the notch, pre-crack and final crack were approximately the same for both H-A and Air-B, as was shown in Table 1.

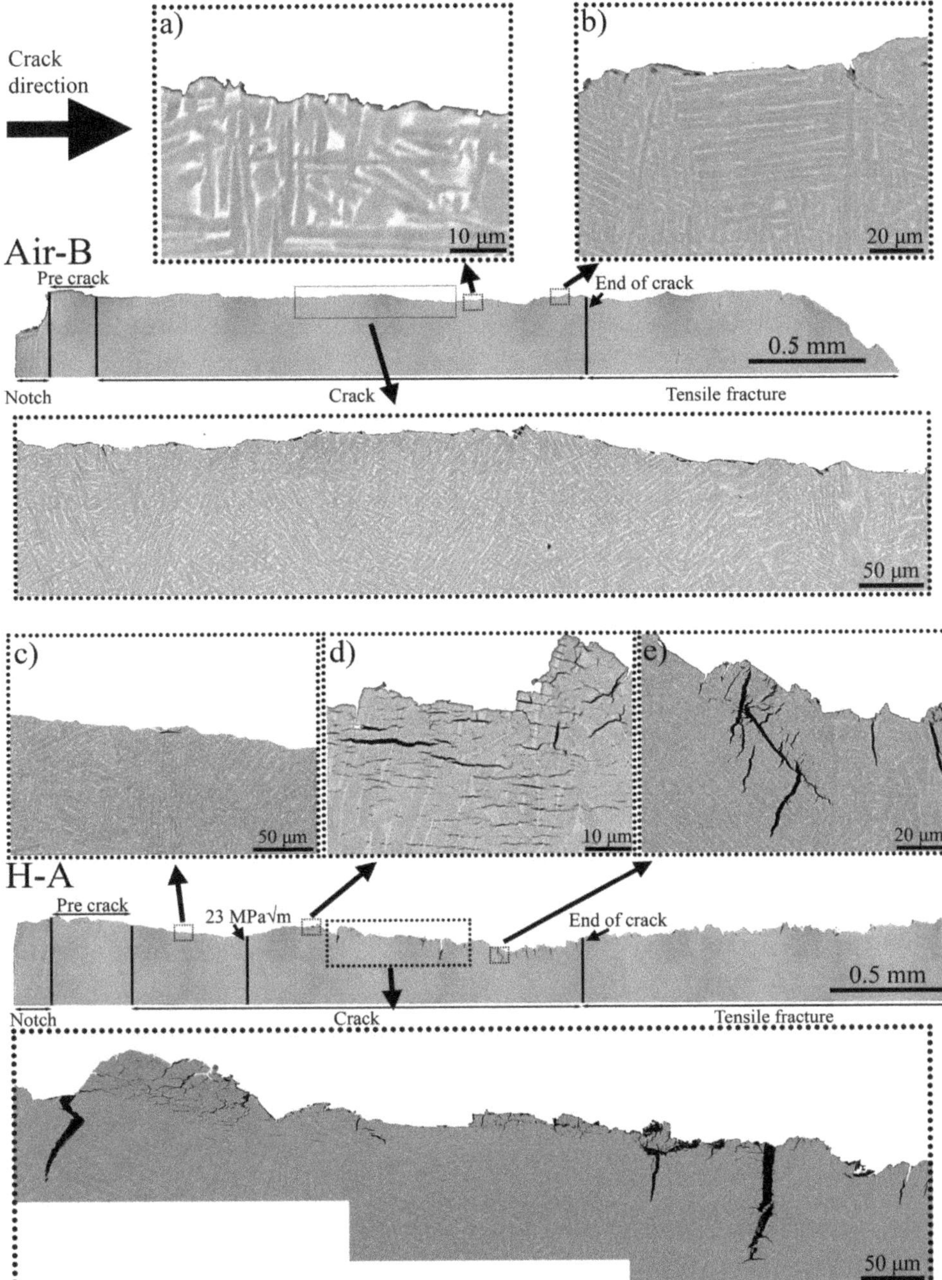

Figure 12. Crack-profiles of Air-B and H-A samples. The crack profiles correspond to a cross-section illustrated in Figure 6 as a white vertical line. In H-A, many larger cracks that grew perpendicular to the fracture surface were observed, along with numerous smaller secondary cracks that grew parallel. In H-A the crack path was tortuous, whereas in Air-B it was comparably smooth. Compared to H-A no cracks were present in Air-B. Images (**a–e**) show magnified areas.

4. Discussion

4.1. Comparison between Hydrogen and Air Atmospheres

As discussed and observed by Lynch [4,33] and others [26,34–39], the mechanical properties of titanium are expected to deteriorate when exposed to hydrogen, due to HE [3]. Such deterioration of the mechanical properties was observed in this work. When exposing the EBM built Ti-6Al-4V to hydrogen-rich atmosphere, hydrogen can be absorbed and diffused throughout the material to the crack tip. At the crack tip the hydrogen can interact with the titanium either by the nucleation of hydrides, resulting in HE through brittle fracture of titanium hydrides, or through one or several other HE mechanisms [4,34–36,40]. On the first stages of the test the FCG rate fluctuated (as shown in Figure 4 and this is believed to be due to temporarily arrests of the crack i.e., CAMs. The nucleation and cleavage of hydrides are suggested in literature to be repeated, forming a relatively flat region on the fracture surface with the presence of features such as CAMs (see Figure 10) [26,32].

At ΔK 23 $MPa\sqrt{m}$, the FCG rate abruptly increased, which was likely due to one or more of the non-hydride forming mechanisms. These mechanisms are called hydrogen enhanced local plasticity (HELP) [34], adsorption induced dislocation emission (AIDE) [4], and hydrogen enhanced de-cohesion (HEDE) [35,36], thoroughly explained in a review paper by Lynch [4]. From the fractography and the crack profiles, it could be shown that cracks appeared once the ΔK increased above 23 $MPa\sqrt{m}$. The crack profile showed that two types of cracks were present, vertical and secondary cracks. The vertical cracks were both deeper and wider than the secondary cracks. These two types of cracks were not present in the air-tested samples, thus it is evident that the hydrogen influenced the formation and subsequent propagation of these different types of cracks. The presence of secondary cracks across the interface between α laths and residual β phase is most probably related to the faster diffusion rate of hydrogen in the β phase and its larger hydrogen storage capacity. This would be coupled with the formation of hydrides since the α phase is a strong hydride forming phase [39]. Underneath the crack path, stress fields were present as well, causing further cracking of these hydrides. Thus, the presence of hydrogen influenced the crack tortuosity.

The relative crack growth rate between the hydrogen and air-tested samples, as shown in Figure 5, clearly shows the critical effect of the presence of hydrogen on titanium: At high ΔK, the FCG rates were roughly 10 times larger in a hydrogen atmosphere. As previously discussed, there was an acceleration in the FCG rate at 23 $MPa\sqrt{m}$. However, Figure 5 also shows that this acceleration did not continue until final failure. At ΔK > 31 $MPa\sqrt{m}$, the relationship between the hydrogen and air-tested samples seemed to have reached a plateau, indicating that increased ΔK in this range and increased hydrogen content ahead of the crack tip did not promote additional acceleration of the FCG rate for EBM built Ti-6Al-4V. Apart from the presence of secondary cracks, the fracture surface cross-section of the hydrogen and air-tested material differed regarding crack path. The crack path in hydrogen-tested material was tortuous in nature, whereas it was relatively flat in air-tested material. Crack paths of both specimen types crossed through α laths, as well as propagated parallel to α laths in the α/β interface.

4.2. Comparison with Cast and Wrought Ti-6Al-4V

In Figures 13 and 14, the FCG properties of the investigated Ti-6Al-4V EBM samples are compared to those of forged and cast material of the same alloy (data source: Gaddam et al. [26,28]), both in air and hydrogen atmospheres. The data in these references have been obtained through testing with similar ΔK.

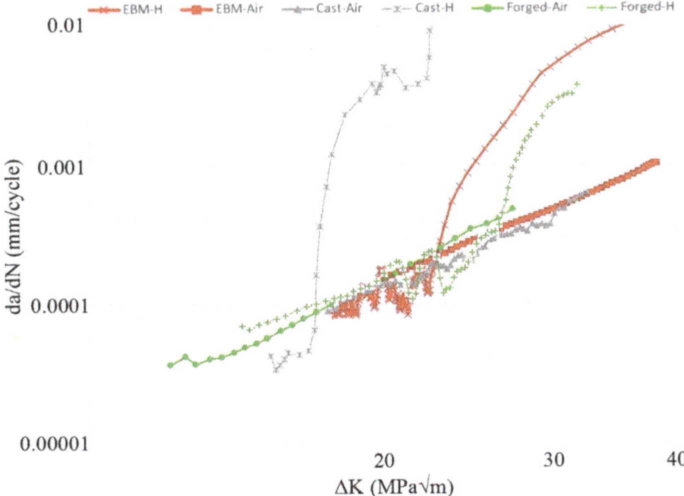

Figure 13. Fatigue crack growth (FCG) data for EBM, forged [26], and cast [28] Ti-6Al-4V. The FCG properties of material exposed to air and hydrogen atmosphere are shown, where the crack growth rate da/dN is plotted on the y-axis and the ΔK on the x-axis.

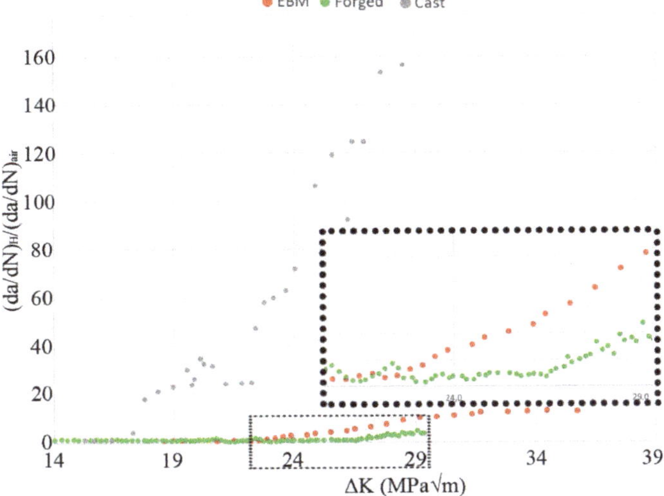

Figure 14. FCG data for EBM, forged [26], and cast [28] Ti-6Al-4V. The y-axis shows the crack propagation rate in hydrogen divided by the crack propagation rate for air, the x-axis shows ΔK. The dashed box is a magnified image to more clearly show the critical points in crack growth rate for the EBM and forged material.

In Figure 13 the crack growth rate (da/dN) is plotted against ΔK ($MPa\sqrt{m}$). In Figure 14 the crack propagation rate in hydrogen environment relative to that in air is plotted as a function of ΔK. A difference in FCG rate behavior was observed between the material exposed to the high-pressure hydrogen and air: For the air-tested material, the FCG rate followed Paris law. For the hydrogen-tested material the FCG rate fluctuated at first, then at certain ΔK, depending on the microstructure, the FCG rate increased fast. The dotted square in Figure 14 is a magnified area to make it easier to distinguish the difference between the EBM and forged materials. In cast Ti-6Al-4V, this occurrence appears at a ΔK of ~17 $MPa\sqrt{m}$; in EBM-built Ti-6Al-4V, at ~23 $MPa\sqrt{m}$; in forgings of the same alloy, at ~26 $MPa\sqrt{m}$. In Figure 14 it is furthermore shown that the maximum FCG rate difference between the hydrogen and air atmosphere was 160 times for the cast material, whereas for the EBM material it was 10 times and the forged material 5 times, all values at ΔK = 29 $MPa\sqrt{m}$.

The micrographs in the two cited papers by Gaddam et al. [26,28] were utilized as a basis for microstructural comparison. It has been observed that the microstructure of forged Ti-6Al-4V consisted of islands of primary α grains with basketweave microstructure surrounding them, i.e., bimodal microstructure. The microstructure of cast Ti-6Al-4V [28] consisted of coarse prior β grains with large α colonies with α laths with the same crystal orientation, i.e., α laths that grow parallel to each other and not perpendicular such as in the basketweave microstructure. The cast material also had prior β grains that appeared equiaxed, being surrounded with a continuous layer of grain boundary α. By comparing these results with the microstructural features of the EBM material, it can be concluded that the cast material had the coarsest microstructure of the investigated materials.

The diffusion rate has a key role in the HE mechanisms [4]. The main parameters that determine the diffusion of hydrogen in a given material are hydrogen concentration gradients, temperature, presence of hydrostatic stresses, and microstructure [14,41]. The phase distribution and grain size of the microstructure are well linked to the diffusion properties in the material [41]. As discussed by Yazdipour et al. [41] and Ichimura et al. [42] a two-fold effect exists where the number of grain boundaries affects the diffusion rate. Finer grain structure implies an increased amount of grain boundaries, which are the fastest diffusion paths and thereby enhance hydrogen diffusion rate. On the other hand, increased amounts of grains and grain boundaries render an increased grain boundary triple junction density, sites that act as hydrogen traps and decrease hydrogen diffusion rate. These effects compete, increasing or decreasing the hydrogen diffusion.

Differences in hydrogen diffusion in the materials analyzed in this work might also be due to proportions of β and α phases. Hydrogen is much more soluble in body centered cubic (BCC) β than in hexagonal close packed (HCP) α [14], due to its preferential absorption in tetrahedral sites [43], which are more abundant in BCC crystal structures than in HCP. As a result, relatively low hydrogen content generates hydrides in α titanium.

The FCG behavior of cast Ti-6Al-4V illustrated in Figure 13 can be explained as follows. Cast Ti-6Al-4V has been shown to have a higher β phase fraction than the same alloy produced by EBM [44]. In addition, the casting's coarse microstructure means a lower number of hydrogen traps. Both these reasons favor faster hydrogen diffusion. Then, due to faster diffusion rate sufficient hydrogen is diffused ahead of the crack tip, for the material to experience at least one of the HE mechanisms at 17 $MPa\sqrt{m}$, which accelerates the FCG rate even further, reaching a relative FCG rate that is ~160 times higher in hydrogen than in air. Then the same phenomena happen for the EBM and forged material at 23 $MPa\sqrt{m}$ and 26 $MPa\sqrt{m}$, respectively. The results are in well accordance to Tal-Gutelmacher et al. [14], that also found the bimodal microstructure to be less sensible to HE than the Widmanstätten microstructure with its more continuous network of residual β phase.

5. Conclusions

By performing FCG experiments of EBM built Ti-6Al-4V in hydrogen and air atmosphere and then comparing the results with already published data of cast and forged Ti-6Al-4V the following conclusions can be made:

- By exposing the EBM built Ti-6Al-4V material to a hydrogen-rich environment the FCG rate increased significantly above ΔK 23 $MPa\sqrt{m}$ compared to the air environment. Below ΔK 23 $MPa\sqrt{m}$ the hydrogen-tested material fluctuated, whereas the air-tested material followed Paris law throughout all the ΔK.
- With increased ΔK secondary cracks became numerous and large for the hydrogen-tested material. Two types of cracks were observed; smaller secondary cracks that formed across α/β interfaces,

predominantly parallel to the main crack direction and large cracks that grew perpendicular to the main crack direction, being connected to the main crack.

- The crack path of the hydrogen-tested material differed from that of the air-tested material in tortuosity, where the hydrogen-tested material was more torturous than the comparably flatter air-tested material.
- Relative to already published FCG results of wrought and cast Ti-6Al-4V, EBM built Ti-6Al-4V was found to have better FCG properties in high-pressure hydrogen compared to cast material while being slightly lower than wrought.

Author Contributions: M.N. wrote the paper and performed all the characterization work. P.Å., M.-L.A., R.P. and M.C. gave feedback on the work, helped with organization and proofread the manuscript. T.H. and C.d.A.S. were the contact persons at GKN Aerospace and helped with the fatigue crack growth experiment, they also helped with proofreading the manuscript. All authors have read and agreed to the published version of the manuscript.

Funding: The National Aviation Research Program (NRFP), the EU funded Space for innovation and growth (RIT) and the Graduate School of Space Technology at Luleå University of Technology have contributed with financial support in this project.

Acknowledgments: The author would like to acknowledge the dedicated work of Géraldine Puyoo, Staffan Brodin and Clas Andersson at GKN-Aerospace, for both the sample manufacturing that was conducted at GKN Filton but also for organizing the testing.

Conflicts of Interest: The authors declare no conflict of interest.

References

1. Mayer, W.; Tamura, H. Propellant injection in a liquid oxygen/gaseous hydrogen rocket engine. *J. Propuls. Power* **1996**, *12*, 1137–1147. [CrossRef]
2. Schlapbach, L.; Züttel, A. Hydrogen-storage materials for mobile applications. In *Anonymous Materials for Sustainable Energy: A Collection of Peer-Reviewed Research and Review Articles from Nature Publishing Group*; World Scientific: London, UK, 2011; pp. 265–270.
3. Westlake, D.G. Generalized model for hydrogen embrittlement. *Amer Soc. Met. Trans.* **1969**, *62*, 1000–1006.
4. Lynch, S. Hydrogen embrittlement (HE) phenomena and mechanisms. In *Anonymous Stress Corrosion Cracking*; Elsevier: Cambridge, UK, 2011; pp. 90–130.
5. Birnbaum, H.K. *Mechanisms of Hydrogen Related Fracture of Metals*; Office of Naval Research: Urbana, IL, USA, 1989.
6. Neikter, M.; Huang, A.; Wu, X. Microstructural characterization of binary microstructure pattern in selective laser-melted Ti-6Al-4V. *Int. J. Adv. Manuf. Technol.* **2019**, *104*, 1381–1391. [CrossRef]
7. Christ, H.-J.; Decker, M.; Zeitler, S. Hydrogen diffusion coefficients in the titanium alloys IMI 834, Ti 10-2-3, Ti 21 S, and alloy C. *Met. Mater. Trans. A* **2000**, *31*, 1507–1517. [CrossRef]
8. Lütjering, G.; Williams, J.C. *Titanium*, 2nd ed.; Springer: Berlin; Germany, 2007.
9. Neikter, M.; Åkerfeldt, P.; Pederson, R.; Antti, M.-L. Microstructure characterisation of Ti-6Al-4V from different additive manufacturing processes. *IOP Conf. Ser. Mater. Sci. Eng.* **2017**, *258*, 012007. [CrossRef]
10. Murr, L.; Gaytan, S.; Medina, F.; Martinez, E.; Martinez, J.; Hernandez, D.; Machado, B.; Ramirez, D.; Wicker, R. Characterization of Ti–6Al–4V open cellular foams fabricated by additive manufacturing using electron beam melting. *Mater. Sci. Eng. A* **2010**, *527*, 1861–1868. [CrossRef]
11. Antonysamy, A.; Meyer, J.; Prangnell, P. Effect of build geometry on the β-grain structure and texture in additive manufacture of Ti6Al4V by selective electron beam melting. *Mater. Charact.* **2013**, *84*, 153–168. [CrossRef]
12. Neikter, M.; Åkerfeldt, P.; Pederson, R.; Antti, M.-L.; Sandell, V.H. Microstructural characterization and comparison of Ti-6Al-4V manufactured with different additive manufacturing processes. *Mater. Charact.* **2018**, *143*, 68–75. [CrossRef]
13. Donachie, M.J. *Titanium: A Technical Guide*, 2nd ed.; ASM International: Dayton, OH, USA, 2000.
14. Tal-Gutelmacher, E.; Eliezer, D. The hydrogen embrittlement of titanium-based alloys. *JOM* **2005**, *57*, 46–49. [CrossRef]

15. Qin, W.; Kiran, K.; Szpunar, J.A. *Role of Crystal Misorientation in the Formation of the Interlinked Hydride Configuration in Zircaloy-4*; The Canadian Inst. of Mining, Metallurgy and Petroleum, Westmount: Quebec, QC, Canada, 2010.
16. Szpunar, J.A.; Qin, W.; Li, H.; Kumar, N.K. Roles of texture in controlling oxidation, hydrogen ingress and hydride formation in Zr alloys. *J. Nucl. Mater.* **2012**, *427*, 343–349. [CrossRef]
17. Neikter, M.; Woracek, R.; Maimaitiyili, T.; Scheffzük, C.; Strobl, M.; Antti, M.-L.; Åkerfeldt, P.; Pederson, R.; Bjerkén, C. Alpha texture variations in additive manufactured Ti-6Al-4V investigated with neutron diffraction. *Addit. Manuf.* **2018**, *23*, 225–234. [CrossRef]
18. Mishurova, T.; Cabeza, S.; Artzt, K.; Haubrich, J.; Klaus, M.; Genzel, C.; Requena, G.; Bruno, G. An Assessment of Subsurface Residual Stress Analysis in SLM Ti-6Al-4V. *Materials* **2017**, *10*, 348. [CrossRef] [PubMed]
19. Yadroitsev, I.; Yadroitsava, I. Evaluation of residual stress in stainless steel 316L and Ti6Al4V samples produced by selective laser melting. *Virtual Phys. Prototyp.* **2015**, *10*, 67–76. [CrossRef]
20. Szost, B.A.; Terzi, S.; Martina, F.; Boisselier, D.; Prytuliak, A.; Pirling, T.; Hofmann, M.; Jarvis, D.J. A comparative study of additive manufacturing techniques: Residual stress and microstructural analysis of CLAD and WAAM printed Ti–6Al–4V components. *Mater. Des.* **2016**, *89*, 559–567. [CrossRef]
21. Shi, X.; Ma, S.; Liu, C.; Wu, Q.; Lu, J.; Liu, Y.; Shi, W. Selective laser melting-wire arc additive manufacturing hybrid fabrication of Ti-6Al-4V alloy: Microstructure and mechanical properties. *Mater. Sci. Eng. A* **2017**, *684*, 196–204. [CrossRef]
22. Maimaitiyili, T.; Woracek, R.; Neikter, M.; Boin, M.; Wimpory, R.; Pederson, R.; Strobl, M.; Drakopoulos, M.; Schäfer, N.; Bjerkén, C. Residual Lattice Strain and Phase Distribution in Ti-6Al-4V Produced by Electron Beam Melting. *Materials* **2019**, *12*, 667. [CrossRef]
23. Rozumek, D.; Hepner, M. Influence of microstructure on fatigue crack propagation under bending in the alloy Ti–6Al–4V after heat treatment: Einfluss des Gefüges der wärmebehandelten Titanlegierung Ti–6Al–4V auf die Ermüdungsrisswahrscheinlichkeit unter Biegebeanspruchung. *Materialwissenschaft Und Werkstofftechnik* **2015**, *46*, 1088–1095. [CrossRef]
24. Pittinato, G.F. Hydrogen-enhanced fatigue crack growth in Ti-6Al-4V ELI weldments. *Met. Mater. Trans. A* **1972**, *3*, 235–243. [CrossRef]
25. Riemer, A.; Richard, H.A.; Brüggemann, J.J.; Wesendahl, J.N.; Riemer, A. Fatigue crack growth in additive manufactured products. *Frat. Integr. Strutt.* **2015**, *9*, 34.
26. Gaddam, R.; Pederson, R.; Hörnqvist, M.; Antti, M.-L. Fatigue crack growth behaviour of forged Ti–6Al–4V in gaseous hydrogen. *Corros. Sci.* **2014**, *78*, 378–383. [CrossRef]
27. Zhai, Y.; Lados, D.; Brown, E.J.; Vigilante, G.N. Fatigue crack growth behavior and microstructural mechanisms in Ti-6Al-4V manufactured by laser engineered net shaping. *Int. J. Fatigue* **2016**, *93*, 51–63. [CrossRef]
28. Gaddam, R.; Hörnqvist, M.; Antti, M.-L.; Pederson, R. Influence of high-pressure gaseous hydrogen on the low-cycle fatigue and fatigue crack growth properties of a cast titanium alloy. *Mater. Sci. Eng. A* **2014**, *612*, 354–362. [CrossRef]
29. Axelsson, J. Baric, Improved Data Regression Methods for Crack Growth Characterization. Master's Thesis, Chalmers University, Gothenburg, Swedem, 2016.
30. Newman, J.; Raju, I. *Stress-Intensity Factor Equations for Cracks in Three-Dimensional Finite Bodies*; NASA TM 83200; Langley Research Center: Hampton, VA, USA, 2009; pp. 1–54.
31. Abràmoff, M.D.; Magalhães, P.J.; Ram, S.J. Image processing with ImageJ. *Biophoton. Int.* **2004**, *11*, 36–42.
32. Nelson, H.G. A film-rupture model of hydrogen-induced, slow crack growth in acicular alpha-beta titanium. *Metall. Trans. A* **1976**, *7*, 621–627. [CrossRef]
33. Lynch, S. Metallographic and fractographic techniques for characterising and understanding hydrogen-assisted cracking of metals. In *Gaseous Hydrogen Embrittlement of Materials in Energy Technologies*; Elsevier: Cambridge, UK, 2012; pp. 274–346.
34. Beachem, C.D. A new model for hydrogen-assisted cracking (hydrogen "embrittlement"). *Met. Mater. Trans. A* **1972**, *3*, 441–455. [CrossRef]
35. Oriani, R.A. A mechanistic theory of hydrogen embrittlement of steels. *Ber. Bunsenges. Phys. Chem.* **1972**, *76*, 848–857.
36. Troiano, A.R. The Role of Hydrogen and Other Interstitials in the Mechanical Behavior of Metals. *Met. Microstruct. Anal.* **2016**, *5*, 557–569. [CrossRef]

37. Birnbaum, H.; Sofronis, P. Hydrogen-enhanced localized plasticity—A mechanism for hydrogen-related fracture. *Mater. Sci. Eng. A* **1994**, *176*, 191–202. [CrossRef]
38. Nelson, H.G.; Williams, D.P.; Stein, J.E. Environmental hydrogen embrittlement of an α-β titanium alloy: Effect of microstructure. *Met. Mater. Trans. A* **1972**, *3*, 473–479. [CrossRef]
39. Tal-Gutelmacher, E.; Eliezer, D. Hydrogen-Assisted Degradation of Titanium Based Alloys. *Mater. Trans.* **2004**, *45*, 1594–1600. [CrossRef]
40. Neikter, M. Microstructure and Hydrogen Embrittlement of Additively Manufactured Ti-6Al-4V. Ph.D. Thesis, Luleå University of Technology, Luleå, Sweden, 2019.
41. Yazdipour, N.; Dunne, D.; Pereloma, E.V. Effect of Grain Size on the Hydrogen Diffusion Process in Steel Using Cellular Automaton Approach. *Mater. Sci. Forum* **2012**, *706*, 1568–1573. [CrossRef]
42. Ichimura, M.; Sasajima, Y.; Imabayashi, M. Grain Boundary Effect on Diffusion of Hydrogen in Pure Aluminum. *Mater. Trans. JIM* **1991**, *32*, 1109–1114. [CrossRef]
43. Wipf, H.; Kappesser, B.; Werner, R. Hydrogen diffusion in titanium and zirconium hydrides. *J. Alloys Compd.* **2000**, *310*, 190–195. [CrossRef]
44. Neikter, M.; Woracek, R.; Durniak, C.; Persson, M.; Antti, M.-L.; Åkerfeldt, P.; Pederson, R.; Zhang, J.; Vogel, S.C.; Strobl, M. Texture of electron beam melted Ti-6Al-4V measured with neutron diffraction. In Proceedings of the 14th World Conference on Titanium, Nantes, France, 10–14 June 2019.

Article

Residual Lattice Strain and Phase Distribution in Ti-6Al-4V Produced by Electron Beam Melting

Tuerdi Maimaitiyili [1,2,*], Robin Woracek [3,4], Magnus Neikter [5], Mirko Boin [6], Robert C. Wimpory [6], Robert Pederson [7], Markus Strobl [3,4,8], Michael Drakopoulos [9], Norbert Schäfer [10] and Christina Bjerkén [2]

1 Photons for Engineering and Manufacturing Group, Paul Scherrer Institute, 5232 Villigen, Switzerland
2 Department of Materials Science and Applied Mathematics, Malmö universitet, 20506 Malmö, Sweden; christina.bjerken@mau.se
3 European Spallation Source ERIC, 22100 Lund, Sweden; Robin.Woracek@esss.se (R.W.); markus.strobl@psi.ch (M.S.)
4 Nuclear Physics Institute of the CAS, 250 68 Husinec—Řež, Czech Republic
5 Division of Materials Science, Luleå University of Technology, 971 81 Luleå, Sweden; magnus.neikter@ltu.se
6 Department of Microstructure and Residual Stress Analysis, Helmholtz-Zentrum Berlin für Materialien und Energie, 14109 Berlin, Germany; boin@helmholtz-berlin.de (M.B.); robert.wimpory@helmholtz-berlin.de (R.C.W.)
7 Department of Engineering Science, University West, 46132 Trollhättan, Sweden; robert.pederson@hv.se
8 Neutron Imaging and Applied Materials Group, Paul Scherrer Institute, 5232 Villigen, Switzerland
9 Imaging and Microscopy Group, Diamond Light Source Ltd., Oxfordshire OX11 0DE, UK; michael.drakopoulos@diamond.ac.uk
10 Department of Nanoscale Structures and Microscopic Analysis, Helmholtz-Zentrum Berlin für Materialien und Energie, 14109 Berlin, Germany; norbert.schaefer@daimler.com
* Correspondence: tuerdi.maimaitiyili@psi.ch; Tel.: +46(0)76-2945402

Received: 29 January 2019; Accepted: 20 February 2019; Published: 23 February 2019

Abstract: Residual stress/strain and microstructure used in additively manufactured material are strongly dependent on process parameter combination. With the aim to better understand and correlate process parameters used in electron beam melting (EBM) of Ti-6Al-4V with resulting phase distributions and residual stress/strains, extensive experimental work has been performed. A large number of polycrystalline Ti-6Al-4V specimens were produced with different optimized EBM process parameter combinations. These specimens were post-sequentially studied by using high-energy X-ray and neutron diffraction. In addition, visible light microscopy, scanning electron microscopy (SEM) and electron backscattered diffraction (EBSD) studies were performed and linked to the other findings. Results show that the influence of scan speed and offset focus on resulting residual strain in a fully dense sample was not significant. In contrast to some previous literature, a uniform α- and β-Ti phase distribution was found in all investigated specimens. Furthermore, no strong strain variations along the build direction with respect to the deposition were found. The magnitude of strain in α and β phase show some variations both in the build plane and along the build direction, which seemed to correlate with the size of the primary β grains. However, no relation was found between measured residual strains in α and β phase. Large primary β grains and texture appear to have a strong effect on X-ray based stress results with relatively small beam size, therefore it is suggested to use a large beam for representative bulk measurements and also to consider the prior β grain size in experimental planning, as well as for mathematical modelling.

Keywords: residual stress/strain; electron beam melting; diffraction; Ti-6Al-4V; electron backscattered diffraction; X-ray diffraction

1. Introduction

Titanium-based alloys have been widely used as engineering materials in many industries because of their excellent combination of a high strength/weight ratio and good corrosion resistance [1]. Application of the titanium and its alloys can be found in many different industries ranging from the aerospace to consumer goods, e.g., heat exchangers, automotive, offshore petroleum, gas exploration and medical implants [1]. Among various titanium alloys, Ti-6Al-4V is by far the most commonly used and accounting for some 50% of the total titanium output in the world [1]. However, extracting high purity titanium and the production of usable raw Ti-6Al-4V is a difficult and expensive process.

Many additive manufacturing (AM) techniques have been developed to fabricate geometrically complex and fully dense metal parts for a variety of applications. The most common AM methods use a laser or electron beams as the power source. In comparison with conventional methods, the electron beam melting (EBM) method has the advantage of increased component complexity with limited manufacturing expertise, shorter lead times, reduced material waste and minimum or almost zero tooling cost [2–4]. In the EBM, electromagnetic lenses are used for focusing and guiding of high-energy electron beams to selectively melting the ingredient powder to form a three-dimensional solid object point by point and layer-by-layer bases. The process is a fully computer controlled automatic system. Electron beam melted Ti-6Al-4V and its microstructure have previously been described by [3–6]. Despite the young age of the EBM process, it has already been demonstrated that it can create defect-free complex products with good mechanical properties, such as functionally graded cellular structures [7–10]. However, the EBM process is complex, and the results depend upon the variables of the system, such as beam power (beam current), beam size, scan speed, and scan direction/scan strategy. These are collectively referred to as process parameters. Thus, each set of process parameter settings produces a different built environment and cooling conditions, and as a consequence, different microstructures are observed [2,11] as well as possible different states of residual stress/strain (RS) and texture [12–15].

RS is internal, self-balanced stress that exists in alloy systems without any external applied forces and may appear during mechanical, thermal or thermochemical processing [16–18]. Depending on the compressive or tensile nature and magnitude of the RS, it significantly affects the mechanical properties of the materials. Lack of understanding RS distribution and the effect of different processing on the stress may create serious consequences [19]. Therefore, it is crucial to know the nature and magnitude of RS in the material for safe and economical operation. AM processes, such as EBM have already proven to be a potential manufacturing process in various fields with many possibilities. However, during AM processes, because of the repetitive selective point by point melting and solidification process, strong heat gradients are generated between different parts of the built material, which potentially can lead to unfavourable residual stresses. Despite several decades of development of AM processes, metal AM is still in its infancy and the relation between different AM process combinations with RS, microstructure, texture and material properties still lack significant understanding. Quality and performance of additively manufactured material can only be reliably controlled and optimized by understanding the effects of different process combinations on the RS and microstructure of the built material. There is limited research concerning RS on titanium alloys produced with conventional methods [17], wire/arc based additive manufacturing [20,21] and selective laser melting (SLM) [22–24]. However, there are only a few concerning RS in EBM Ti-6Al-4V [25–27]. Therefore, this study aims to explore the effect of a subset of EBM process parameters on residual strain/stress in Ti-6Al-4V. RS can be investigated by e.g., hole drilling techniques, ultrasonic and magnetic methods and diffraction-based methods [16,18,28]. Each RS measurement technique has its own advantages and disadvantages. Among them, diffraction (e.g., X-ray and neutron diffraction) is one of the most accurate and well-developed methods of quantifying local and global residual stresses in the material [16,18]. Compared with conventional RS measurement techniques, the diffraction method offers many advantages. The residual stresses in the material are calculated from the strain measured in the crystal lattice. Generally, RS diffraction measurements are not significantly influenced by material properties such as hardness, the degree of cold work, or preferred orientation. In fact, all can be quantified by diffraction techniques. In addition, they are non-destructive, have a high spatial resolution (ranging in the order if μm to mm depending on the radiation source)), almost no specimen geometry restrictions, and can be applied to all types of crystalline material. Furthermore, with the high-energy synchrotron X-ray diffraction setup used in this work, it is possible to characterize and conduct quantitative studies on multi-component systems containing phase quantities as low as 0.1–0.7% [29–31] with a high temporal resolution (0.2 s) [31]. This means that phase specific RS in the target system can be determined dynamically with high accuracy. Last but not the least, diffraction-based RS techniques are already well established—there is good level of expertise, and standards have been developed [16,18,32].

To investigate the effect of EBM process parameters on residual strain and stress development, various specimens produced with different beam sizes, scan speeds and build thicknesses were investigated. In addition, original powder specimens that were used to produce these solid specimens have also been tested. Diffraction measurements were carried out using the high energy (50–150 KeV) beamline I12-JEEP (joint engineering, environment and processing) at the Diamond Light Source (UK) with an energy dispersive X-ray diffraction (EDXRD) setup [33], as well as the dedicated RS neutron diffractometer (ND) E3 [34] at Helmholtz Zentrum Berlin (Germany). After data collection, the Pawley pattern fitting method was performed for the EDXRD data and single peak fitting for the ND data using the structure analysis software package TOPAS-Academic [35].

2. Experimental Setup and Data Analysis

2.1. EBM Process and Material

The EBM process is described in greater detail e.g., in [36]. However, the EBM process is quite complicated owing to numerous parameters (e.g., feed material, beam current, feed rate, build layer thickness and scan speed), which potentially change the build environment and cause a different micro-structure in the final product that may not be wanted. The solidification rate, surface smoothness, and microstructural homogeneity of EBM processed parts are strongly influenced by the process parameters.

The electron beam parameters, such as beam speed, beam current and scan length can be varied in a controlled sequence throughout the build according to algorithms developed by the manufacturer. According to the literature, the resolution of the build in EBM is influenced by layer thickness [37,38] powder size [39] and spot size [2]. It has been reported that the size of the powder and layer thickness has a direct influence on build part quality [38,39]. Commonly, a smaller powder gives finer surface finish and it has been demonstrated that the EBM can process powder with size 25–45 μm [39]. The benefit with smaller particles is finer surface finish and more compact layer i.e., no/little shrinkage. Nonetheless, due to the charging from the electron beam, smaller powder particles are not possible for EBM i.e., a phenomenon called smoking can otherwise occur. A benefit with larger particle size and layer thickness, however, is an increased build rate. To achieve the highest quality, the current standard EBM layer thickness has been reduced from 100 to 50–70 μm and powder with size 45–100 μm was used [40]. The powder used in this study was gas atomized and size within the range of 45–100 μm. The chemical composition is Ti-6Al-4V-0.1Fe-0.15O-0.01N-0.003H (in wt %).

The mechanical properties of the EBM built Ti-6Al-4V are affected by scan speed [41]. It has been reported that the rapid heating and cooling, and repetitive thermal cycles of AM processes in laser-based additively manufactured material create RS in as-built material [42,43]. According to Klingbeil et al. [43], preheating of the build platform and powder layer can reduce RS and negative effects, such as part warping. In the EBM process, for Ti-6Al-4V the whole build chamber is heated to approximately 700 °C prior to building, and each layer of powder is preheated before melting. The minimum temperature of the build chamber is maintained $\geq$ 600 °C throughout the build process [40]. Therefore, the EBM built component is naturally annealed immediately after building, and thus remaining residual stresses may be limited in magnitude. However, the relation between scan speed, offset focus and RS has not yet been verified experimentally to the authors' knowledge.

Ti-6Al-4V blocks of 50 mm × 50 mm × 5 mm were built and measured at the centre of each specimen both along the build direction (vertical, "V") and in the build plane (horizontal, "H") by neutron diffraction. To investigate the finger prints of the phase specific residual stresses, specimens were cut from the as-build blocks in varied thicknesses (see Table 1) and investigated with synchrotron XRD in two directions (Figure 1). For the analysis of the RS state and microstructure, three different standard controlling factors were chosen: build thickness, scan speed and offset focus. The details of specimen production parameters are tabulated in Table 1. Three specimens were made for each set of processing parameters. Because of limited beam time, only one specimen was measured during neutron diffraction experiment.

Table 1. Specimen process parameters.

Specimen ID	Contour Scan Speed (mm/s)	Scan Speed (mm/s)	Current (mA)	Contour Current (mA)	Offset Focus (mA)	Specimen Thickness (mm)
S4	180	575	8	9	15	2.47/5 *
S5	180	650	8	9	15	2.68
S6	180	651	8	9	15	2.70/5 *
S7	180	652	8	9	15	1.99/5 *
S8	180	575	8	9	30	3.02
S9	180	652	8	9	30	2.93
S10	180	653	8	9	30	2.77
S11	180	650	8	9	30	2.97
S12	180	700	8	9	12	2.50/5 *

* is the as-built specimen thickness that was used in the neutron diffraction studies.

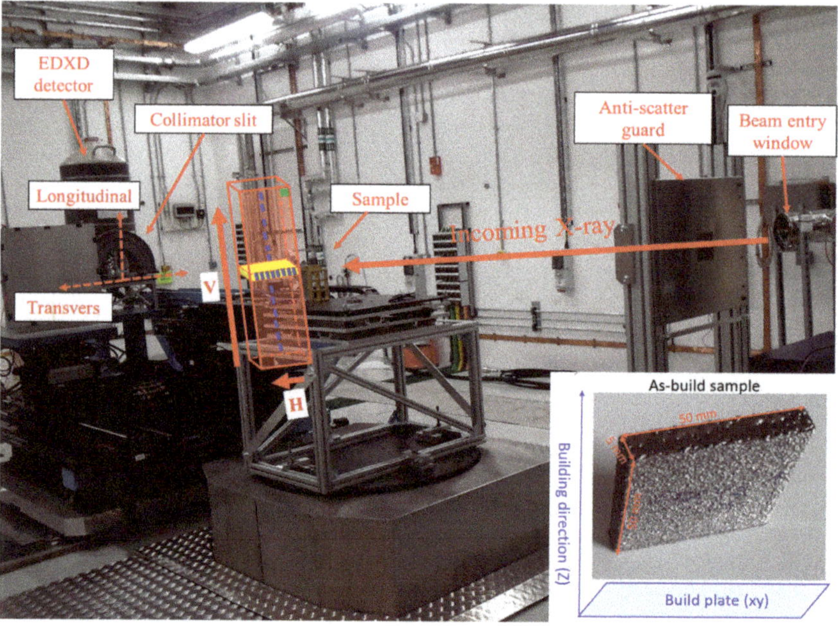

Figure 1. Beam line setup and illustration of measurements. "V" is the vertical direction and "H" is horizontal direction. Each blue and green rectangle represents the measurement locations. Embedded image is a photo of one of the as-built specimens.

2.2. Synchrotron X-ray Measurements (EDXRD)

Synchrotron X-ray diffraction is a well-known and powerful tool for structural analysis, quality characterization [18,44] and conducting quantitative studies on multi-component systems [30,31]. Herein reported experiments were performed at the high energy (50–150 KeV) beamline I12-JEEP at Diamond Light Source in the UK with EDXRD setup [33]. All specimens were irradiated with a white/continuous X-ray beam with a photon flux range from 1.8×10^{11} to 9.4×10^{10} photons s^{-1} depending on the energy of the X-ray beam. The EDXRD data were recorded with a "horseshoe" detector, consisting of 23 liquid-nitrogen-cooled germanium (Ge) energy-sensitive detector elements, positioned 2 m behind the specimen position with a take-off angle of 5° from the incident beam. The energy resolution of Ge-detector ranges from 7×10^{-3} at 50 keV to 4×10^{-3} at 150 keV. As the 23 detector elements are equally spaced in steps of 8.2°, they have full azimuthal coverage from 0° to 180°. During the measurements, each detector element independently records diffraction patterns under a different Bragg angle. As the sequence of energy values of each detector element is approximately the same for all, the 23 diffraction patterns can be summed to form one pattern after normalization of each data set for detailed phase analysis. The actual EDXRD setup with schematics of measurements and specimen can be seen in Figure 1.

In the experiment, a 2-mm thick copper filter and slit sizes of 0.1×0.1 mm^2 and acquisition time of 100 s were used. To map strain variations in the material along build direction and in the build plane as complete as possible in the limited synchrotron beam time, measurements were made at an average of 55 points in the centre of the specimen along the vertical, and 10 in the horizontal direction. The volume of the material contributing to the diffraction pattern corresponds to the intersection of the incident and diffracted beams, typically defined by slits and collimators, respectively. To ensure the gauge volume is fully inside the specimen, the centre of the measurements points close to the edges was positioned about 300 μm from the edge during horizontal and vertical line scans in the centre of the specimen. The clearance of corner scan, however, is about 0.5 mm from the corner edge. Each detector element in the setup measures the lattice spacing in a specific direction, which means that strains in 23 different directions in 0–180° range can be measured in a single measurement without specimen rotation. The measured strain direction can be defined by the X-ray scattering vector, which bisects the incident and diffracted beams. With this EDXRD setup, the three-dimensional strain/stress field can be calculated by including additional information or by adopting assumptions, such as the plane strain criterion. For that reason, only strains along the longitudinal and transversal directions are presented in this work. Detector elements 1 and 23 were used to measure transverse strain, and elements #11 or #12 were used to measure longitudinal strain. Compared with the angle dispersive diffraction setup used in the lab X-ray source, the JEEP (I12) at Diamond has many orders of magnitude higher X-ray flux with high energy. At JEEP it is possible to probe metal specimens in millimeter to centimeter thickness in relatively short times. This gives the possibility to determine the RS state of the various phases in one experiment. This was important for the present study as thicker specimens allowed to obtain better statistical average (larger diffracting volume, i.e., more grains) for proper RS state probing.

2.3. Neutron Diffraction Studies

To avoid setup or instrument dependent error and justify the observations, the same specimens were investigated with the neutron diffractometer E3 [34] and the ESS test beam line V20 [45,46] at the research reactor BER-II in the Helmholtz-Zentrum Berlin (HZB). E3 is a constant wavelength and angular dispersive instrument (detector is moved around 2θ), whereas V20 is a time of flight beamline, where a wavelength dispersive setup with a fixed detector position was used.

2.4. Microstructure Studies

All specimens used for microscopic studies were prepared by using standard metallographic preparation routines. To investigate the microstructure of the specimens, visible light microscope

(Nikon ECLIPSE L150, Amsterdam, The Netherlands) and scanning electron microscope (SEM, ZEISS EVO LS10, Zeiss, Oberkochen, Germany) were used. The SEM was operated using an acceleration voltage of 25 keV with magnifications ranging between 1000 and 5000×. Electron backscatter diffraction (EBSD) scans were performed in an SEM equipped with a field emission source and an automated EBSD acquisition system. Here, the SEM was operated at an accelerating voltage of 20 kV and a beam current of 5 nA. The EBSD scans were acquired with the specimen tilted to 70° at a working distance of 10 mm. Two EBSD maps have been obtained from two different locations at the centre of the specimen along the build direction.

2.5. Diffraction Data Analysis

After measurement, the software MATLAB was used to extract the data in xy binary format, which is friendly to many structure refinement programs. Then, the structure analysis software packages TOPAS-Academic was used to carry out least-square refinement (Pawley method) and standard single peak fitting routines [35] for stress/strain analysis. Phase quantities of the sample were determined by the Rietveld analysis [35]. The basic crystal structure information of the various phases needed for the Pawley and the Rietveld refinement were obtained from the Inorganic Crystal Structure Database (ICSD) [47]. All structural data used, such as reference and refined parameters, are tabulated in Table 2.

Table 2. Some known and calculated crystallographic information of different Ti phases.

Phase	Structure	Space Group	*a*(Å)	*c*(Å)	(°C)	Atomic Site	Reference (ICSD)
α(Ti)	HCP	P63/mmc	2.9511	4.6843	25	0.3333 0.6667 0.25	43416
			2.9508	4.6855	-		52522
			2.9064	4.6667	20		99778
			2.951	4.682	-		43614
			2.916	4.631	-		168830
			2.9232(7)	4.6700(4)	25		Polycrystal
			2.9230(6)	4.6697(6)	25		Polycrystal corner
			2.9210(6)	4.6644(8)	25		Powder
β(Ti)	BCC	$Im\bar{3}m$	3.2765	3.2765	-	000	653278
			3.2068(7)	3.2068(7)	25		Polycrystal
			3.2082(7)	3.2082(7)	25		Polycrystal corner
			3.2023(8)	3.2023(8)	25		Powder
Omega Ti	Loose HCP	P6/mmm	4.6	2.82	High pressure	Ti1 000; Ti2 0.3333 0.6667 0.5	52521

As overall diffraction patterns (or peak shapes) are a convolution of background, specimen and instrument, a standard cerium oxide specimen with lattice parameters 5.41165(1) Å was measured under the same configurations for calibration and instrument profile function extraction purpose. Then, the Pawley [35] refinements were performed to determine the profile function. After a good fit was obtained from cerium oxide data, the instrument profile function was set fixed together with zero error correction, and then a Pawley batch fitting was carried out to refine the structure of various phases to define best possible unit cell parameters of each measurement. The Pawley fitting was performed on the full diffractogram, which spans from 0 to 145 degrees in 2θ. The peak profiles were modelled with a modified Thompson–Cox–Hastings pseudo-Voigt (pV-TCHZ) profile function [35]. The background was fitted with a Chebyshev function with six coefficients and the zero-shift error together with axial divergence calibrated with a standard cerium oxide reference specimen. During batch refinement all instrument related parameters were kept fixed and only unit cell parameters together with background were refined. To avoid any complication introduced by oxides, elemental composition variation, and phase distribution to the measured strain, a single peak fitting was also performed together with the Pawley fitting in limited diffraction ranges, i.e., 1.19639–1.79480 Å.

2.6. Strain Calculations

Diffraction patterns are routinely used as a fingerprint of a material's crystalline structure. Any external stimulus, such as applied load or heat gradient can alter the interplanar lattice (d) spacing of the material, causing a change of the diffraction pattern. Therefore, the atomic lattice spacing in crystalline materials can be used as a natural strain gauge [16,48]. For instance, tensile stress will cause an increase of the lattice spacing in a given direction and compressive stress leads to the opposite. The average elastic lattice macrostrain (ε_{hkl}) in the sampled volume can be calculated by comparing the measured lattice spacing d_{hkl} with that of the unstrained (stress-free) lattice spacing ${d^0}_{hkl}$ as [16,48]:

$$\varepsilon_{hkl} = \frac{d_{hkl} - d^0_{hkl}}{d^0_{hkl}} \tag{1}$$

If the target material is isotropic and all grains/domains in the material have the same responses to external stimulus, then the Equation (1) is sufficient to determine the stress/strain state in the material for a given direction of the stress/strain tensor. However, most crystalline materials have anisotropic properties, including the stiffness of the unit cell, and hence accurate stress/strain analysis requires consideration of multiple reflections. While individual peak fitting can be used to investigate the elastic and plastic anisotropy of individual lattice plane families, a fit of a complete diffraction profile—as used for results reported herein—can be used to determine the average lattice parameter from which the strain can be calculated that is closely representative of the bulk macroscopic strain (for the same direction).

The strain in a cubic material/phase (e.g., β-Ti) can be calculated by [48,49]:

$$\varepsilon_{cubic} = \frac{a - a_0}{a_0} \tag{2}$$

Similarly, the strain in a hexagonal material can be calculated by [48,49]:

$$\varepsilon_{hexagonal} = \frac{2\left(\frac{a}{a_0} - 1\right) + \left(\frac{c}{c_0} - 1\right)}{3} \tag{3}$$

where a and c are the lattice parameter of the given phase; a_0 and c_0 are the strain-free lattice parameters. Unstrained lattice parameter can be extracted by several methods, e.g., summarized by Withers et al. [32]. In this work, the d^0_{hkl}, a_0 and c_0 are obtained by (1) measuring the raw powder used for building these specimens used in the study; (2) measuring from the corner of each specimen; (3) taking the mean values of all scans of each sample.

3. Results and Discussion

3.1. Microstructure

Depending on processing and alloying elements, the Ti-6Al-4V microstructure may include α-Ti (hcp), β-Ti (bcc), α′ (hcp martensite), as well as different amount of dislocations, substructures and twinning [1,4]. The EBM built Ti-6Al-4V, however, commonly consists of two phases, namely α-Ti (hcp) and β-Ti (bcc) twinning [2,4,5,50]. Typical visible light micrographs of one of the specimens from build-direction and build plane are shown in Figure 2a,b, respectively. All materials studied show a microstructure which consists of intertwined α lath colonies where individual α-laths are separated by a thin layer of retained β-phase. This type of microstructure is formed when fast cooling rates from the β-phase field are achieved [51], the presence of grain boundary α does, however, indicate that the cooling rate was not fast enough to suppress the grain boundary α nucleation, which can occur for AM processes with even faster cooling rates. High-resolution SEM micrographs corresponding to Figure 2a,b are shown in Figure 2c,d, respectively. In these SEM micrographs, the dark colour corresponds to the alpha phase and white colour corresponds to the β phase. The microstructure shown in Figure 2 is similar to the microstructure of EBM built Ti-6Al-4V reported by others [2–4,11,15]. The columnar nature of prior β grains that is shown in Figure 2 is a direct consequence of the thermal gradient that exists in the build direction [15].

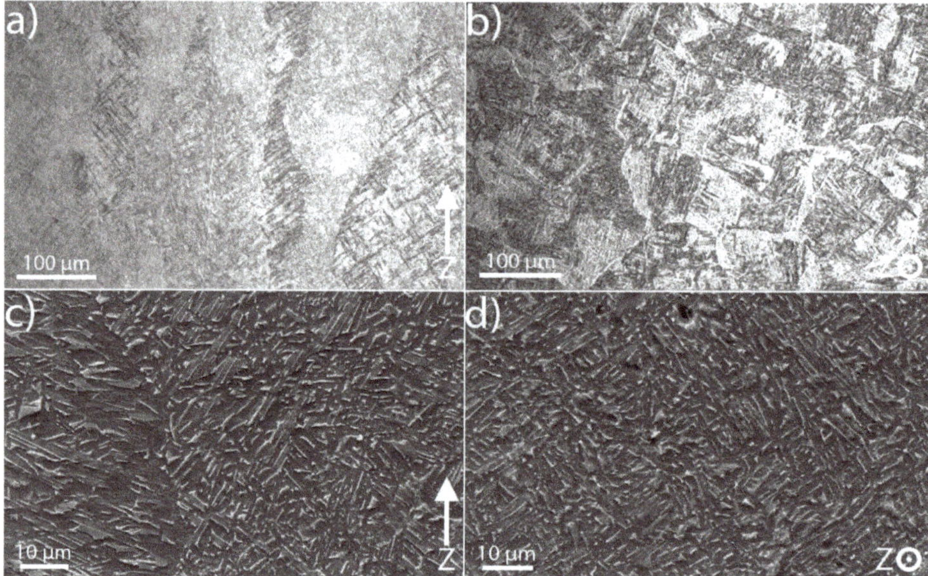

Figure 2. Light optical microscopy image of S7V (**a**), and S7H (**b**). In (**c**) and (**d**), the images are taken with a SEM where the specimens used are S8V and S8H respectively. Z indicates the build direction and etching procedure employed to obtain the alloy microstructures was by immersion in Kroll's etchant for 10 s.

Synchrotron X-ray and neutron diffraction work confirmed the α+β microstructure observed by microscopy. Figure 3 shows the raw XRD diffraction patterns of as-received Ti-6Al-4V powders used for building all the specimens investigated in this study together with a representative diffraction pattern of EBM built material collected at the centre and one of the top corners (close to the final layer) of the same specimen. The colour coded, small vertical tick marks in Figure 3 (and all other figures throughout) represent the hkl peak positions of labelled phases in the same colour. As expected, the diffraction patterns from the corner and centre of the built material showed close similarity, but they both differ significantly from the powder pattern. According to Figure 3, it is evident that the as-received Ti-6Al-4V powders and the EBM specimens (independent of build location) contain predominantly α-phase with the HCP structure, consistent with previous studies [52]. In addition, the system also contained detectable amount of BCC β-phase (1.5–10 wt %) together with limited quantity of Ti-oxide (<1 wt %). The peak position and the intensity of the oxide diffraction peaks measured at different parts of the specimen did not show observable variation. Therefore, the effect of the oxide is excluded from the discussion. The presence of the β-phase can be clearly seen by the corresponding high angle diffraction peaks at d-spacing = 2.26 Å and d-spacing = 1.60 Å for the (011) and (002) reflections as reported in [53,54].

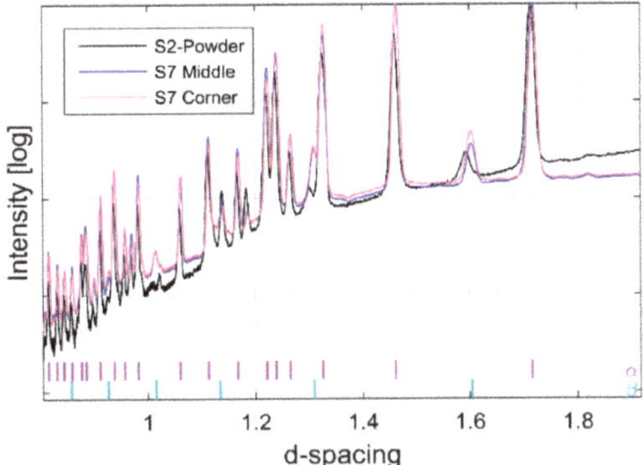

Figure 3. Diffraction pattern obtained by EDXRD of Ti-6Al-4V powder and EBM built material. The colour coded, small vertical tick marks represent the hkl peak positions of labelled phases in the same colour (α = purple, β = magenta).

It is known that the mechanical properties of the Ti-alloys strongly depend on the distribution of alloying elements and phases present in the system [1]. Generally, Ti-alloys show higher strength and higher density but lower creep strength with an increase in β-Ti phase content [1]. To ensure and control the properties of the material for a specific application, it is important to understand how different material processing affect the material phase composition of the alloy system. The microstructure and mechanical properties of Ti-6Al-4V for specific applications can be tailored by post heat treatment after manufacturing. Commonly, the post thermal process operations will take place in the α+β phase and/or single β-phase field. The volume fraction of β-phase increases with temperature and therefore governs the mechanical properties [55]. Therefore, it is important to understand how different AM processes will influence the formation/distribution of the β-phase in the built material. However, because of the limited phase quantity at room temperature of the β-phase in Ti-6Al-4V and its low diffraction strength, the β-phase has not been observed [56] or has been excluded from the analysis. While studying Ti-6Al-4V fabricated by the selective laser melting process (SLM), Chen et al. [56] did not observe a clear β-phase from either the as-received powder or the built solid specimens from their XRD work. According to Chen et al. [56], the reason behind weak or absent β-phase in their system is the high cooling rate used during the build. They argued that [56], SLM processed specimens maintain the original composition and crystal structure of the powders; all possible existing β-phase at high temperatures might all transform into the α or martensitic α′ phase via rapid solidification during the SLM process. However, it is interesting that even the as received Ti-6Al-4V powder used in their work did not show detectable β-phase. We think that such phase absence might be explained by the sensitivity of the conventional XRD method and the textures present in the system in addition to the rapid cooling of SLM processing and powder manufacturing.

Ti-6Al-4V has been shown to have texture [1,12], and its development strongly depends on processing type and temperature [55]. Because of localized melting and solidification together with repetitive heating processes, the microstructure and texture formed in additively manufactured material are different from that in cast or wrought materials. In order to emphasize the complexity and variation in textures in EBM built Ti-6Al-4V, two grain orientation maps together with corresponding inverse pole figures obtained from the same specimen measured along the build direction at two different areas in the specimen are shown in Figure 4. Figure 4a is close to the last melted layer and Figure 4b is a few mm below. As can be seen in the figure, the material shows preferred grain orientation (texture), and it varies from one location to another despite that these two measurements were done in the same vertical plane along the same line and only a few mm apart. Microstructure and textures presented in Figures 2 and 4 indicate the complexity of the Ti-6Al-4V microstructure produced by EBM. Because of this highly varying microstructure and its strong correlation with the AM process parameters and build geometry, the mechanical properties of the material reported in the literature shows a large scatter [5,50,52].

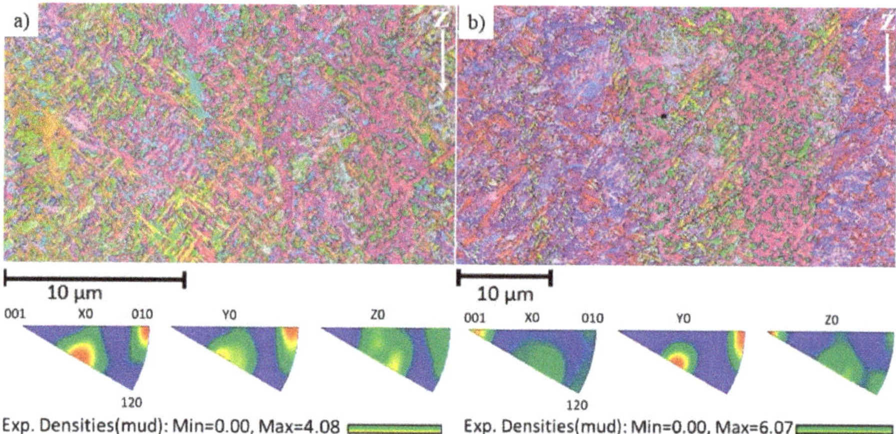

Figure 4. Inverse pole figure maps of Ti-6Al-4V (S7V). The build direction is indicated by the white arrows. (**a**) is close to the last layer; (**b**) few mm below from (**a**); Images under (a,b) is the corresponding inverse pole figures.

3.2. Peak Intensities and Phase Compositions

All observed diffraction patterns from one diffraction angle (from detector element 15) collected during a vertical scan of one of the specimens, and the corresponding specimen patterns obtained from all 23 directions from the same specimen are presented in Figure 5a,b, respectively. As shown in Figure 5a, some of the α- and β-phase diffraction peaks are absent in certain parts of the build material from the diffraction patterns collected from only one scattering direction. From Figure 5a, one can see that there is no correlation between the absences of α- and β-phase diffraction peaks. In addition, such an absence does not coincide with the additive build layer thickness. Similar observations have also been observed for all other specimens studied in this work. The summed diffraction patterns of all measurements in all 23 scattering directions, however, showed all corresponding α- and β-phase diffraction peaks with similar peak intensity and width with a sign of α- and β-phases homogeneity as shown in Figure 5b. This difference is thought to be a result from the preferred orientation of the different grains (i.e., the α-phase is also present in probed volumes that do not show any α-peak for a single detector, but the peaks are observed in other scattering angles from the same volume). In Figure 5, HKL indices of α- and β-phases with significant changes are given at corresponding peak Bragg peak positions and interesting regions where β-phase peaks are absent in detector element 15 also marked with (1), (2) and (3) with green curly brackets.

Similarly, diffraction patterns from one of the EBM Ti-6Al-4V specimens collected along the build direction and in the build plane are shown in Figures 6a and 7a as a two-dimensional plot, viewed down the intensity axis, with the diffraction pattern number along the ordinate, and peak positions along the abscissa. For clarity and emphasizing the existence of β-Ti in the system, the peak intensity of well separated (002) β-Ti peak from the measurement presented in Figures 6a and 7a is plotted in Figures 6b and 7b, respectively. Peak intensity is colour coded to maintain consistency with previous figures. Note that the first and last data acquisition location (gauge volume) in Figure 7 does not exactly correspond to the edges of the specimen. There is about 300-μm clearance from each side. The peak intensity drop is shown in Figure 7b might be caused by the scan strategies used in this work. The outermost boundaries between the build and powder bed were melted first to form a contour. Thereafter powders inside the contour were melted in the manner of serpents' path (from one side to another, then one step downwards followed by return, then one step downwards etc.), which is also known as hatching. As shown in Figures 5b and 6, none of the diffraction patterns showed any sign of α′ martensite in the top layer of the build material unlike that reported in [15]. According to the Galarraga et al. [13], the observation of a martensitic structure is indicative of a high cooling rate imposed during solidification and subsequent cooling in the solid state of the last layer, and a cooling rate of >410 K/s may result in such a structure in last layer. Al-Bermani's [15] and Galarraga et al. [13] have also reported that faster cooling rates after solution heat treatment produce a greater amount of α′ martensitic phase, with water-cooling at a rate of 650 K/s resulting in a fully α′ martensitic microstructure. However, apart from the last layer, the first layer can also vary in microstructure compared to the bulk, due to the faster cooling rate close to the thermal conductive substrate. And these pronounced effects of cooling rates on microstructure also render microstructural differences. Nonetheless, the volume of these microstructural extremes is relatively small. Interestingly, Galarraga et al. [13] did not observe α′ martensite in their unprocessed EBM built Ti-6Al-4V ELI (Extra Low Interstitial) specimens. However, based on the weakening of the (200) β phase diffraction peak at d-spacing = 1.60 Å and change in microstructure, tensile test results and microhardness values after post build heat treatment, they concluded that even air-cooled specimens may contain α′ martensite. The difference between the observation reported by Al-Bermani's [15] and this study might be because of differences in specimen build geometry, process parameters and post-build heat treatments used in the two different studies.

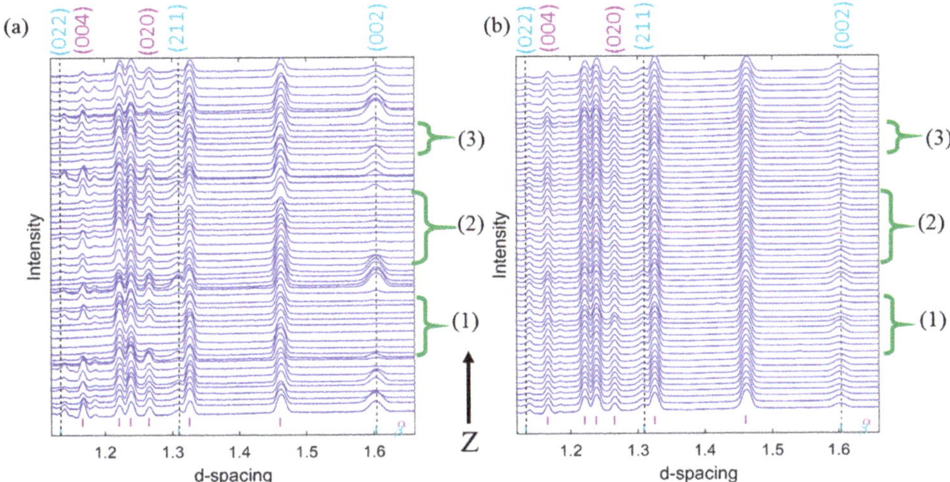

Figure 5. Assembled diffraction patterns of a vertical scan (along the build length) of one of the EBM Ti-6Al-4V samples (S7V). (**a**) is from the detector element 15 alone and (**b**) is the sum of all 23 elements of the same scan. The black arrow indicates the build direction from bottom to top.

During EBSD and SEM studies of EBM built Ti-6Al-4V, Al-Bermani et al. [15] observed 100% β-phase within the build height 0–300 μm and mostly α-phase at other parts. According to Al-Bermani et al. [15], the β phase occurs due to the co-melting and diffusion of the alloying elements in the austenitic stainless-steel base plate and the initial Ti-6Al-4V layers. The β phase stabilizing elements, such as Cr, Fe, and Ni were provided by melting of stainless-steel plate. Since all measurements presented in this work have been performed from 13 mm upwards, we are unable to confirm that observation. Nonetheless, based on observations presented in Figures 4–7 it is safe to conclude that globally there is no significant phase difference in different parts of the EBM built Ti-6Al-4V parts at some distance away from the base plate despite the varied microstructure. This conclusion can be further supported by neutron diffraction studies. The deep penetration power into Ti and the larger sampling volume of neutrons provide the possibility to evaluate the bulk average phase compositions and residual stress/strain of all present phases in the system.

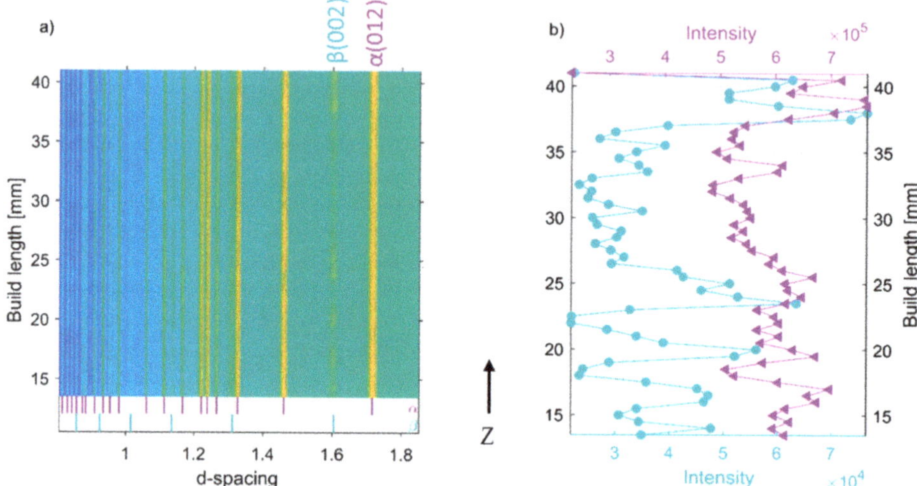

Figure 6. (**a**) Accumulated diffraction patterns of vertical line scan along the build direction of specimen S8, and (**b**) corresponding peak intensity of (002) β-Ti (magenta) and (012) α-Ti. (purple).

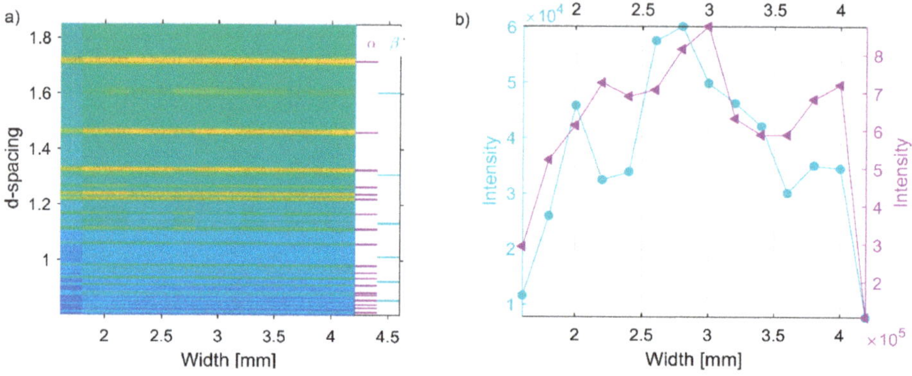

Figure 7. (**a**) Accumulated diffraction patterns of horizontal line scan along the build plane of specimen S8, and (**b**) corresponding peak intensity of (002) β-Ti (magenta) and (012) α-Ti (purple).

To cross-check for potential variations among larger sampling volumes, data sets collected at the ESS test beamline at HZB on the same specimen (S8) are presented in Figure 8a. The beamline operates in time of flight mode, where neutrons start to travel from a source at time t = 0 and travel tens of meters to the specimen at which point they have separated by their different velocities and hence wavelengths. In this measurement, an incident beam size of 10 mm × 10 mm was used, with the detector positioned at a scattering angle of 2θ = 90°. The results in Figure 8a present a relative direct comparison (the setup remained unchanged besides translating the specimen) between the upper half (red) and lower half (blue) of the specimen. Both diffraction patterns are almost identical to each other with the indication that there is no significant phase variation between the upper and lower part of the specimen. These results agree well with those in Figure 5b and supports the conclusion that phase distributions in the material are uniform.

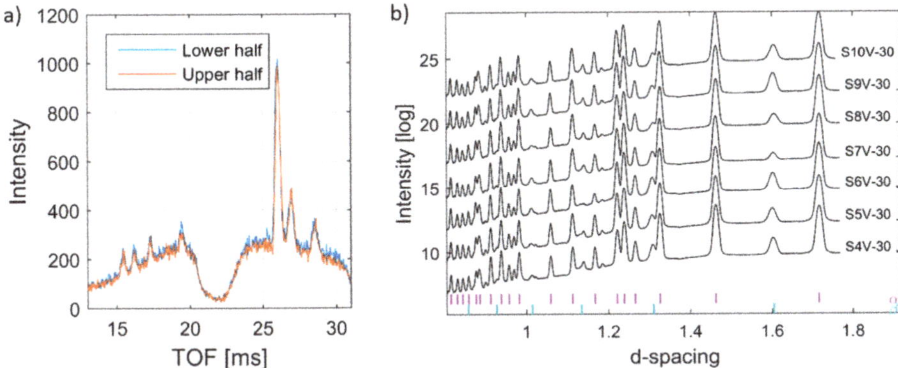

Figure 8. Time-of-flight neutron data for specimen S8 (**a**). Raw diffraction patterns depicting a relative comparison between the upper and lower half of specimen S7, supporting the observations in Figure 6. (**b**) Collection of diffraction patterns for seven different EBM Ti-6Al-4V specimens.

To investigate whether different EBM process parameter combinations may lead to phase differences, an EDXRD pattern from the middle of different specimens is presented in Figure 8b. As shown in the figure, there is no phase difference between specimens other than intensity. As expected all specimens contain both α- and β-phases. Figure 9 represents a typical Pawley refinement result of data presented in Figures 6 and 7. In that figure, the blue dots represent observed intensities, the red line represents the calculated, and the green line is the difference curve on the same scale. The color-coded tick marks indicate the calculated positions of Bragg peaks. As seen from the difference curve, the fit is highly satisfactory. All diffraction patterns contained some of the major α-Ti and β-Ti peaks, which are commonly reported in the literature [57] together with many other minor and high angle peaks which have not been included in any other reported studies. The matches of the interplanar spacings (d-spacings) obtained from these peaks and unit cell parameters of α-Ti and β-Ti were not in perfect agreement with the data of the ICSD (see, Table 2). This difference might be caused by that different EBM process parameters were used to manufacture the materials investigated in these studies, which may influence the lattice strains and hence the overall lattice parameter.

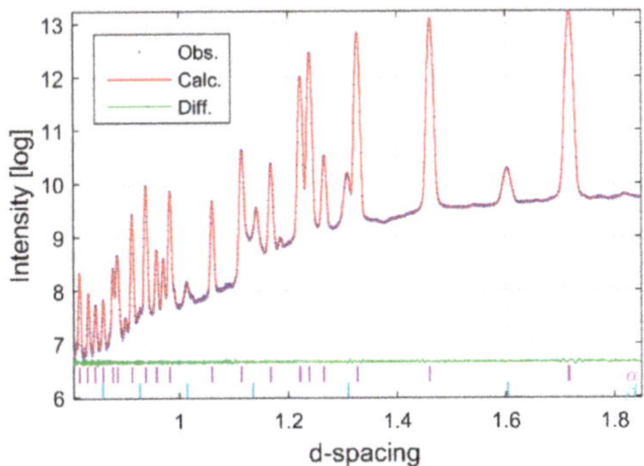

Figure 9. One typical Pawley refinement result: R_{wp} 1.73, Goodness of fit (GOF) 1.52.

3.3. Residual Strain

One of the biggest challenges in diffraction-based residual strain measurements of additively manufactured material is obtaining the strain-free reference lattice parameters. Commonly, the reference can be determined from a stress-free specimen of identical material. Because of the localized heating effect, microstructure variation and possible texture, specimens built by AM are expected to show some differences and even at different locations in the same specimen. Considering the microstructure difference between the ingredient powder, the corner of the build material and the center of the build material, only mean value-based strain calculations have been selected as the suitable strain reference extraction method. The lattice strain of α- and β-Ti calculated by using unit cell parameters (a and c) obtained from Pawley fitting of diffraction data from vertical and horizontal scans are shown in Figures 10 and 11, respectively. Each coloured line in these figures corresponds to one specific scan belonging to a different specimen. The colour code is the same for both figures. The error bars are equal to or smaller than the marker size.

Recently, Tiferet et al. [27] have reported that none of SLM or EBM specimens contain residual macro- nor micro-strains, instead the ingredient powder contains residual strain. Interestingly, lattice parameters of α-Ti obtained from ingredient powder from our studies show almost identical result as their bulk specimen. In addition, in contrast to Tiferet et al. [27], lattice parameters obtained from specimen corners, ingredient powder and bulk specimens from our measurements are similar. Such small lattice parameter differences indicate that the residual strain in EBM built material is low. The difference between our results and Tiferet et al. [27] may be related to the different processing approaches that were used.

In addition to lattice strains calculated from unit cell parameters of α- and β-phases obtained from Pawley pattern fitting, a lattice specific strain was also calculated for selected α- and β-phase hkl reflections after single peak fitting. Despite the availability of other hkl reflections, only selected α-phase ((010), (012), (110), (004)), and β-phase ((011), (002), (211)) hkl reflections, which do not overlap or are well-separated from other phases, were used for calculation. The general trends of the strains from single peak-based calculations are similar to whole pattern fittings shown in Figures 10 and 11. However, each lattice plane families yielded some disparity from one another. For simplicity and considering the superiority of strain calculation by using lattice parameters obtained by whole pattern fitting over single peak fitting, the results of single peak fitting are not shown.

From Figure 10, it is shown that the measured strain in both longitudinal and transverse directions in α- and β-Ti are fluctuating and they do not show any obvious trend with respect to deposition layers. However, closer inspection of the strain result of the α-Ti phase shows a slight compressive to tensile transition trend in both strain measurement directions (Figure 10) along the build. Even though the α-Ti strain result fluctuation does not match with the deposition layer, the frequency of the fluctuation appears to correlate with the length of the columnar grains. Therefore, to obtain reliable RS results from the experiment and modelling, we recommend considering the size of the primary β-grains in the measurement strategy, data interpretation and in theoretical simulations. Unlike α-Ti, the β-Ti did not show any observable transition and the strain appears to fluctuate more than the α-Ti (Figure 11). The reason for such a strong fluctuation might be that the β-Ti phase is the minority phase with high crystallographic symmetry and strong texture. Hence, the β-Ti phase in the Ti-6Al-4V system could be working as buffer phase, which causes local equilibrium of the stress/strain distribution in the system.

The material is built layer by layer via selectively melting the raw material. Therefore, each deposited layer in an additive manufactured structure undergoes multiple heating and cooling cycles with repetitive stress relaxation and accumulation processes. Hence, large thermal RS differences are to be expected at different heights in the build. However, at same build height or in same deposition layer, the heat treatment is expected to be the same. For that reason, the RS in the same layer should be the same. As predicted, the RS in the centre of the specimen measured along the scan direction showed linearity in both longitudinal and transverse directions (Figure 11). The magnitude and nature of the strain measured near the edge of most of the measurements appear similar to the result obtained from the centre or has a compressive nature.

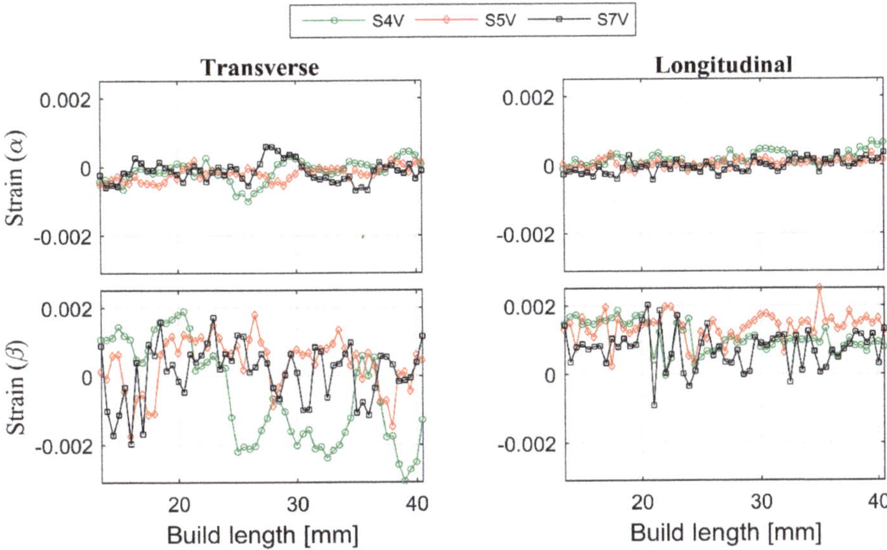

Figure 10. The lattice strain from the central vertical line of specimens perpendicular to the EBM build plane. Transverse direction at left and longitudinal direction at right. Color code of each specimen can be seen from the legend at top of the figure.

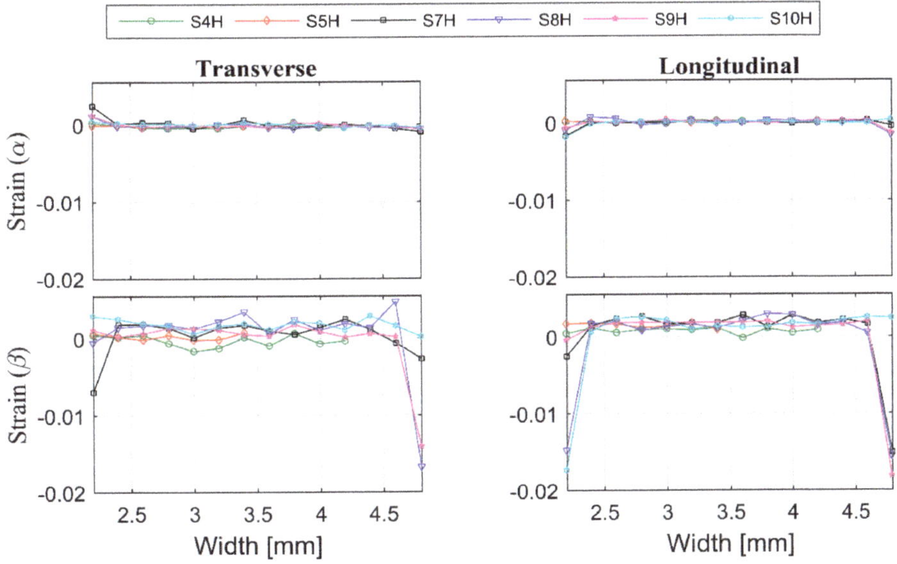

Figure 11. The lattice strain from the central horizontal line of a specimen that is parallel to the EBM build plane. Transverse direction at left, and longitudinal direction at right.

The transverse strain in α-Ti, however, appears compressive on one side and tensile on the other. The EBM scan strategy used for building is expected to induce a sharp strain gradient from the surface/edge to the centre/inside of the build. During the EBM process, each layer of the target object is built in two steps. Firstly, the outer boundary is melted (contour). Then, in the second step, the actual part is built within the contours. The contour provides an interface between the actual build and the surrounding powder. Since molten metal is deposited on a colder contour wall, thermal contraction of the solidified material occurring during solidification creates a tensile stress in the deposit and compressive stress in the contour wall. The origin of this is not currently known. Whether it is build thickness or beam power requires further investigation. These observed differences within and between specimens may be attributed to the differences in their structure at the temperatures experienced during EBM processes. As shown in Figure 2a, the material microstructure consists of large prior β-grains with α-lamellas. Therefore, the scatter in the strain results presented could be understood to be a result of the grain statistics as the synchrotron X-ray experiment was conducted with a relatively small gauge volume. In order to justify whether this is the case, more measurements have been carried out on selected specimens with the E3 instrument at HZB. Similar to the synchrotron result, neutron diffraction measurements also did not show any significant difference from one another at different locations within the specimens (Figure 12). From Figure 12, one can see that there is neither any directional peak shift of the (112) peak of the α-Ti phase along the build direction nor peak position differences between specimens. This observation agrees with the results presented in Figures 10 and 11. The main difference between the neutron diffraction results in Figure 12 and the X-ray diffraction results in Figures 10 and 11 is that the neutron diffraction results seemed more linear than the X-ray diffraction results. This could indicate that there is indeed some grain size effect on the X-ray results. Therefore, to overcome similar problems, a larger beam size should be used in X-ray based measurements.

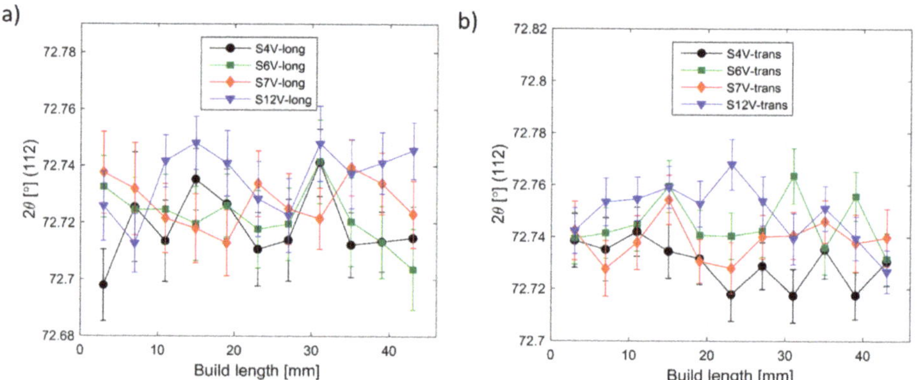

Figure 12. Variation of Ti(112) peak position with respect to build height by neutron diffraction. (**a**) Longitudinal direction, (**b**) transverse direction. It can be seen that variations are within the error bars.

Residual stresses are important to control with regard to quality and integrity of parts built by AM. Therefore, it is crucial to know the status and magnitude of any residual stresses to be able to reduce any detrimental effects. Since residual stresses in additive manufactured materials arise mainly from temperature gradients, the stress levels formed in the parts could possibly be modified by changing the heat gradients during manufacturing. The scan speed, beam current and offset focus of the EBM process influences the heat gradients and are likely to cause different states of residual stresses. However, according to the present observations, the influence of these process parameters is not that obvious (at least not for the parameters selected in this work). The reason for not observing any strain/stress trends or the effects of different process parameters is most likely because of the high preheating temperature with slow cooling used in the EBM process. In SLM it was reported that heating the build platform ≥ 200 °C can reduce the RS levels [58]. Therefore, in the future RS studies, the role of the preheating temperature should be considered.

There are some limitations in the current work that could be addressed by future work, concerning process parameter diversity. In AM there are many processes parameters that affect the cooling rate of the build and consequently the microstructure and RS. Therefore, the authors propose that further investigations be performed with other processes parameters in mind. In Table 1 the process parameters that have been used in this work are shown. An example is that in this work, only a current of 8 mA has been used, this process parameter greatly affects the cooling rate and would undoubtedly render different RS results.

4. Conclusions

Residual strains in Ti-6Al-4V specimens built by AM from nominal 45–100 μm diameter gas atomised powder using an electron beam melting technique have been evaluated by using state-of-the-art neutron and synchrotron X-ray diffraction techniques. Based on the present studies of the EBM built Ti-6Al-4V materials, the following conclusions can be made:

1. Despite strong texture and a columnar microstructure in the build direction, a fairly uniform phase distribution has been observed. The microstructure is mostly Widmanstätten type with majority are α-Ti phase with some amount of β phase (1.5–10 wt %).
2. Based on synchrotron X-ray studies, longitudinal strain in α-Ti changes slightly from compression to tensile along the build direction, but no such trend is observed in β-Ti. Neutron studies showed no clear trend of RS in neither α- nor β-Ti.
3. Neither transverse nor longitudinal strains in α- and β-Ti in the build plane change significantly. The longitudinal strains in both α- and β-Ti, however, showed a compressive nature close to the edges of the built materials.
4. No clear residual strain differences between deposition layers was found.
5. The strain variation along the build direction may be related to the elongated prior β-grains and should be considered in future experimentation and analysis.

Author Contributions: Conceptualization, T.M.; Data curation, T.M.; Formal analysis, T.M.; Funding acquisition, R.P., M.S. and C.B.; Investigation, T.M., R.W., M.B., R.C.W., M.D. and N.S.; Methodology, T.M. and R.W.; Project administration, T.M.; Resources, T.M. and C.B.; Supervision, R.P., M.S. and C.B.; Visualization, T.M.; Writing—original draft, T.M.; Writing—review & editing, T.M, R.W., M.N., M.B., R.C.W., R.P., M.S. and C.B.

Funding: InterReg ESS & MaxIV: Cross Border Science and Society financially supported the project (MAH-003) along with "Nationellt rymdtekniskt forskningsprogram" (NRFP), the EU funded "Space for innovation and growth" (RIT), the "Graduate School of Space Technology" at Luleå University of Technology and the OP RDE, MEYS, under the project "European Spallation Source–participation of the Czech Republic—OP", Reg. No. CZ.02.1.01/0.0/0.0/16 013/0001794. Diamond light source (EE7858) and HZB provided beam time.

Acknowledgments: We acknowledge in-depth discussions with Adnan Safdar, J. Blomqvist, M.S. Blackmur, Axel Steuwer and their help during the synchrotron experiment. EBSD measurements were performed at the HZB Corelab Correlative Microscopy and Spectroscopy (CCMS) and authors thank Christiane Förster for EBSD sample preparation.

Conflicts of Interest: The authors declare no conflicts of interest.

References

1. Gerd, L.; James, C.W. *Titanium*, 2nd ed.; Springer: Heidelberg, Germany, 2007.
2. Safdar, A.; He, H.Z.; Wei, L.; Snis, A.; Chavez, d.P. Effect of process parameters settings and thickness on surface roughness of EBM produced Ti-6Al-4V. *Rapid Prototyp. J.* **2012**, *18*, 401–408. [CrossRef]
3. Murr, L.E.; Esquivel, E.V.; Quinones, S.A.; Gaytan, S.M.; Lopez, M.I.; Martinez, E.Y.; Medina, F.; Hernandez, D.H.; Martinez, E.; Martinez, J.L.; et al. Microstructures and mechanical properties of electron beam-rapid manufactured Ti-6Al-4V biomedical prototypes compared to wrought Ti-6Al-4V. *Mater. Charact.* **2009**, *60*, 96–105. [CrossRef]

4. Murr, L.E.; Quinones, S.A.; Gaytan, S.M.; Lopez, M.I.; Rodela, A.; Martinez, E.Y.; Hernandez, D.H.; Martinez, E.; Medina, F.; Wicker, R.B. Microstructure and mechanical behavior of Ti-6Al-4V produced by rapid-layer manufacturing, for biomedical applications. *J. Mech. Behav. Biomed. Mater.* **2009**, *2*, 20–32. [CrossRef] [PubMed]
5. Tan, X.; Kok, Y.; Tan, Y.J.; Descoins, M.; Mangelinck, D.; Tor, S.B.; Leong, K.F.; Chua, C.K. Graded microstructure and mechanical properties of additive manufactured Ti-6Al-4V via electron beam melting. *Acta Mater.* **2015**, *97*, 1–16. [CrossRef]
6. Neikter, M.; Åkerfeldt, P.; Pederson, R.; Antti, M.-L.; Sandell, V. Microstructural characterization and comparison of Ti-6Al-4V manufactured with different additive manufacturing processes. *Mater. Charact.* **2018**, *143*, 68–75. [CrossRef]
7. Zhang, L.; Liu, Y.; Li, S.; Hao, Y. Additive Manufacturing of Titanium Alloys by Electron Beam Melting: A Review. *Adv. Eng. Mater.* **2018**, *20*, 1700842. [CrossRef]
8. Liu, Y.J.; Wang, H.L.; Li, S.J.; Wang, S.G.; Wang, W.J.; Hou, W.T.; Hao, Y.L.; Yang, R.; Zhang, L.C. Compressive and fatigue behavior of beta-type titanium porous structures fabricated by electron beam melting. *Acta Mater.* **2017**, *126*, 58–66. [CrossRef]
9. Liu, Y.J.; Li, S.J.; Wang, H.L.; Hou, W.T.; Hao, Y.L.; Yang, R.; Sercombe, T.B.; Zhang, L.C. Microstructure, defects and mechanical behavior of beta-type titanium porous structures manufactured by electron beam melting and selective laser melting. *Acta Mater.* **2016**, *113*, 56–67. [CrossRef]
10. Zhao, S.; Li, S.J.; Wang, S.G.; Hou, W.T.; Li, Y.; Zhang, L.C.; Hao, Y.L.; Yang, R.; Misra, R.D.K.; Murr, L.E. Compressive and fatigue behavior of functionally graded Ti-6Al-4V meshes fabricated by electron beam melting. *Acta Mater.* **2018**, *150*, 1–15. [CrossRef]
11. Safdar, A.; Wei, L.-Y.; Snis, A.; Lai, Z. Evaluation of microstructural development in electron beam melted Ti-6Al-4V. *Mater. Charact.* **2012**, *65*, 8–15. [CrossRef]
12. Neikter, M.; Woracek, R.; Maimaitiyili, T.; Scheffzük, C.; Strobl, M.; Antti, M.-L.; Åkerfeldt, P.; Pederson, R.; Bjerkén, C. Alpha texture variations in additive manufactured Ti-6Al-4V investigated with neutron diffraction. *Addit. Manuf.* **2018**, *23*, 225–234. [CrossRef]
13. Galarraga, H.; Warren, R.J.; Lados, D.A.; Dehoff, R.R.; Kirka, M.M.; Nandwana, P. Effects of heat treatments on microstructure and properties of Ti-6Al-4V ELI alloy fabricated by electron beam melting (EBM). *Mater. Sci. Eng. A* **2017**, *685*, 417–428. [CrossRef]
14. de Formanoir, C.; Michotte, S.; Rigo, O.; Germain, L.; Godet, S. Electron beam melted Ti-6Al-4V: Microstructure, texture and mechanical behavior of the as-built and heat-treated material. *Mater. Sci. Eng. A* **2016**, *652*, 105–119. [CrossRef]
15. Al-Bermani, S.S.; Blackmore, M.L.; Zhang, W.; Todd, I. The Origin of Microstructural Diversity, Texture, and Mechanical Properties in Electron Beam Melted Ti-6Al-4V. *Metall. Mater. Trans. A* **2010**, *41*, 3422–3434. [CrossRef]
16. Fitzpatrick, M.E.; Lodini, A. *Analysis of Residual Stress by Diffraction using Neutron and Synchrotron Radiation*; CRC Press: Boca Raton, FL, USA, 2003.
17. Yang, X.; Richard Liu, C. Machining titanium and its alloys. *Mach. Sci. Technol.* **1999**, *3*, 107–139. [CrossRef]
18. Hauk, V. Structural and Residual Stress Analysis by Nondestructive Methods. In *Structural and Residual Stress Analysis by Nondestructive Methods*; Hauk, V., Ed.; Elsevier Science B.V.: Amsterdam, The Netherlands, 1997.
19. Withers, P.J.; Bhadeshia, H.K.D.H. Residual stress. Part 2—Nature and origins. *Mater. Sci. Technol.* **2001**, *17*, 366–375. [CrossRef]
20. Szost, B.A.; Terzi, S.; Martina, F.; Boisselier, D.; Prytuliak, A.; Pirling, T.; Hofmann, M.; Jarvis, D.J. A comparative study of additive manufacturing techniques: Residual stress and microstructural analysis of CLAD and WAAM printed Ti–6Al–4V components. *Mater. Des.* **2016**, *89*, 559–567. [CrossRef]
21. Shi, X.; Ma, S.; Liu, C.; Wu, Q.; Lu, J.; Liu, Y.; Shi, W. Selective laser melting-wire arc additive manufacturing hybrid fabrication of Ti-6Al-4V alloy: Microstructure and mechanical properties. *Mater. Sci. Eng. A* **2017**, *684*, 196–204. [CrossRef]
22. Ali, H.; Ma, L.; Ghadbeigi, H.; Mumtaz, K. In-situ residual stress reduction, martensitic decomposition and mechanical properties enhancement through high temperature powder bed pre-heating of Selective Laser Melted Ti6Al4V. *Mater. Sci. Eng. A* **2017**, *695*, 211–220. [CrossRef]

23. Mishurova, T.; Cabeza, S.; Artzt, K.; Haubrich, J.; Klaus, M.; Genzel, C.; Requena, G.; Bruno, G. An Assessment of Subsurface Residual Stress Analysis in SLM Ti-6Al-4V. *Materials* **2017**, *10*, 348. [CrossRef] [PubMed]
24. Yadroitsev, I.; Yadroitsava, I. Evaluation of residual stress in stainless steel 316L and Ti6Al4V samples produced by selective laser melting. *Virtual Phys. Prototyp.* **2015**, *10*, 67–76. [CrossRef]
25. Hrabe, N.; Gnäupel-Herold, T.; Quinn, T. Fatigue properties of a titanium alloy (Ti–6Al–4V) fabricated via electron beam melting (EBM): Effects of internal defects and residual stress. *Inter. J. Fatigue* **2017**, *94*, 202–210. [CrossRef]
26. Vastola, G.; Zhang, G.; Pei, Q.X.; Zhang, Y.-W. Controlling of residual stress in additive manufacturing of Ti6Al4V by finite element modeling. *Addit. Manuf.* **2016**, *12*, 231–239. [CrossRef]
27. Tiferet, E.; Rivin, O.; Ganor, M.; Ettedgui, H.; Ozeri, O.; Caspi, E.N.; Yeheskel, O. Structural investigation of selective laser melting and electron beam melting of Ti-6Al-4V using neutron diffraction. *Addit. Manuf.* **2016**, *10*, 43–46. [CrossRef]
28. Withers, P.J.; Bhadeshia, H.K.D.H. Residual stress. Part 1—Measurement techniques. *Mater. Sci. Technol.* **2001**, *17*, 355–365. [CrossRef]
29. Maimaitiyili, T.; Bjerken, C.; Steuwer, A.; Wang, Z.; Daniels, J.; Andrieux, J.; Blomqvist, J.; Zanellato, O. In situ observation of γ-ZrH formation by X-ray diffraction. *J. Alloy. Compd.* **2017**, *695*, 3124–3130. [CrossRef]
30. Maimaitiyili, T.; Steuwer, A.; Blomqvist, J.; Bjerkén, C.; Blackmur, M.S.; Zanellato, O.; Andrieux, J.; Ribeiro, F. Observation of the δ to ε Zr-hydride transition by in-situ synchrotron X-ray diffraction. *Cryst. Res. Technol.* **2016**, *51*, 663–670. [CrossRef]
31. Maimaitiyili, T.; Blomqvist, J.; Steuwer, A.; Bjerkén, C.; Zanellato, O.; Blackmur, M.S.; Andrieux, J.; Ribeiro, F. In situ hydrogen loading on zirconium powder. *J. Synchrotron Radiat.* **2015**, *22*, 995–1000. [CrossRef] [PubMed]
32. Withers, P.J.; Preuss, M.; Steuwer, A.; Pang, J.W.L. Methods for obtaining the strain-free lattice parameter when using diffraction to determine residual stress. *J. Appl. Crystallogr.* **2007**, *40*, 891–904. [CrossRef]
33. Korsunsky, A.M.; Song, X.; Hofmann, F.; Abbey, B.; Xie, M.; Connolley, T.; Reinhard, C.; Atwood, R.C.; Connor, L.; Drakopoulos, M. Polycrystal deformation analysis by high energy synchrotron X-ray diffraction on the I12 JEEP beamline at Diamond Light Source. *Mater. Lett.* **2010**, *64*, 1724–1727. [CrossRef]
34. Wimpory, R.C.; Mikula, P.; Šaroun, J.; Poeste, T.; Li, J.; Hofmann, M.; Schneider, R. Efficiency Boost of the Materials Science Diffractometer E3 at BENSC: One Order of Magnitude Due to a Horizontally and Vertically Focusing Monochromator. *Neutron News* **2008**, *19*, 16–19. [CrossRef]
35. Coelho, A. TOPAS Academic: Technical Reference. TOPAS Academic: Technical Reference 2004, V4.1. Available online: http://www.TOPAS-academic.net/ (accessed on 1 January 2017).
36. Neikter, M. Microstructure and Texture of Additive Manufactured Ti-6Al-4V. Licentiate Thesis, Lulea University of Technology, Luleå, Sweden, 31 January 2018.
37. Cansizoglu, O.; Harrysson, O.L.A.; West, H.A.; Cormier, D.R.; Mahale, T. Applications of structural optimization in direct metal fabrication. *Rapid Prototyp. J.* **2008**, *14*, 114–122. [CrossRef]
38. Algardh, J.K.; Horn, T.; West, H.; Aman, R.; Snis, A.; Engqvist, H.; Lausmaa, J.; Harrysson, O. Thickness dependency of mechanical properties for thin-walled titanium parts manufactured by Electron Beam Melting (EBM)®. *Addit. Manuf.* **2016**, *12*, 45–50. [CrossRef]
39. Karlsson, J.; Snis, A.; Engqvist, H.; Lausmaa, J. Characterization and comparison of materials produced by Electron Beam Melting (EBM) of two different Ti–6Al–4V powder fractions. *J. Mater. Process. Technol.* **2013**, *213*, 2109–2118. [CrossRef]
40. Electron Beam Melting. Available online: http://www.arcam.com/technology/electron-beam-melting/2016 (accessed on 1 October 2016).
41. Wang, X.; Gong, X.; Chou, K. Scanning Speed Effect on Mechanical Properties of Ti-6Al-4V Alloy Processed by Electron Beam Additive Manufacturing. *Procedia Manuf.* **2015**, *1*, 287–295. [CrossRef]
42. Qian, L.; Mei, J.; Liang, J.; Wu, X. Influence of position and laser power on thermal history and microstructure of direct laser fabricated Ti–6Al–4V samples. *Mater. Sci. Technol.* **2005**, *21*, 597–605. [CrossRef]
43. Klingbeil, N.W.; Beuth, J.L.; Chin, R.K.; Amon, C.H. Residual stress-induced warping in direct metal solid freeform fabrication. *Inter. J. Mech. Sci.* **2002**, *44*, 57–77. [CrossRef]

44. Cullity, B.D.; Stock, S.R. *Elements of X-ray Diffraction*, 3rd ed.; Pearson Education Limited: London, UK, 2013.
45. Strobl, M.; Bulat, M.; Habicht, K. The wavelength frame multiplication chopper system for the ESS test beamline at the BER II reactor—A concept study of a fundamental ESS instrument principle. *Nucl. Instrum. Methods Phys. Res. A* **2013**, *705*, 74–84. [CrossRef]
46. Woracek, R.; Hofmann, T.; Bulat, M.; Sales, M.; Habicht, K.; Andersen, K.; Strobl, M. The test beamline of the European Spallation Source—Instrumentation development and wavelength frame multiplication. *Nucl. Instrum. Methods Phys. Res. A* **2016**, *839*, 102–116. [CrossRef]
47. Belsky, A.; Hellenbrandt, M.; Karen, V.L.; Luksch, P. New developments in the Inorganic Crystal Structure Database (ICSD): Accessibility in support of materials research and design. *Acta Crystallogr. B Struct. Sci.* **2002**, *58*, 364–369. [CrossRef]
48. Daymond, M.R.; Bourke, M.A.M.; Dreele, R.B.V. Use of Rietveld refinement to fit a hexagonal crystal structure in the presence of elastic and plastic anisotropy. *J. Appl. Phys.* **1999**, *85*, 739–747. [CrossRef]
49. Lundbäck, A.; Pederson, R.; Colliander, M.H.; Brice, C.; Steuwer, A.; Heralic, A.; Buslaps, T.; Lindgren, L. Modeling And Experimental Measurement with Synchrotron Radiation of Residual Stresses in Laser Metal Deposited Ti-6Al-4V. In *Proceedings of the 13th World Conference on Titanium;* Venkatesh, V., Pilchak, A.L., Allison, J.E., Ankem, S., Boyer, R., Christodoulou, J., Fraser, H.L., Imam, M.A., Kosaka, Y., Rack, H.J., et al., Eds.; Wiley: Hoboken, NJ, USA, 2016; pp. 1279–1282. [CrossRef]
50. Liu, S.; Shin, Y.C. Additive manufacturing of Ti6Al4V alloy: A review. *Mater. Des.* **2019**, *164*, 107552. [CrossRef]
51. Kelly, S.M.; Kampe, S.L. Microstructural evolution in laser-deposited multilayer Ti-6Al-4V builds: Part I. Microstructural characterization. *Metall. Mater. Trans. A* **2004**, *35*, 1861–1867. [CrossRef]
52. Zhao, X.; Li, S.; Zhang, M.; Liu, Y.; Sercombe, T.B.; Wang, S.; Hao, Y.; Yang, R.; Murr, L.E. Comparison of the microstructures and mechanical properties of Ti–6Al–4V fabricated by selective laser melting and electron beam melting. *Mater. Des.* **2016**, *95*, 21–31. [CrossRef]
53. Da Silva, S.L.R.; Kerber, L.O.; Amaral, L.; dos Santos, C.A. X-ray diffraction measurements of plasma-nitrided Ti–6Al–4V. *Surf. Coat. Technol.* **1999**, *116–119*, 342–346. [CrossRef]
54. Xu, W.; Lui, E.W.; Pateras, A.; Qian, M.; Brandt, M. In situ tailoring microstructure in additively manufactured Ti-6Al-4V for superior mechanical performance. *Acta Mater.* **2017**, *125*, 390–400. [CrossRef]
55. Murty, S.V.S.N.; Nayan, N.; Kumar, P.; Narayanan, P.R.; Sharma, S.C.; George, K.M. Microstructure–texture–mechanical properties relationship in multi-pass warm rolled Ti–6Al–4V Alloy. *Mater. Sci. Eng. A* **2014**, *589*, 174–181. [CrossRef]
56. Chen, L.Y.; Huang, J.C.; Lin, C.H.; Pan, C.T.; Chen, S.Y.; Yang, T.L.; Lin, D.Y.; Lin, H.K.; Jang, J.S.C. Anisotropic response of Ti-6Al-4V alloy fabricated by 3D printing selective laser melting. *Mater. Sci. Eng. A* **2017**, *682*, 389–395. [CrossRef]
57. Muguruma, T.; Iijima, M.; Brantley, W.A.; Yuasa, T.; Ohno, H.; Mizoguchi, I. Relationship between the metallurgical structure of experimental titanium miniscrew implants and their torsional properties. *Eur. J. Orthod.* **2011**, *33*, 293–297. [CrossRef] [PubMed]
58. Mercelis, P.; Kruth, J.-P. Residual stresses in selective laser sintering and selective laser melting. *Rapid Prototyp. J.* **2006**, *12*, 254–265. [CrossRef]

Article

Effect of Process Parameters and High-Temperature Preheating on Residual Stress and Relative Density of Ti6Al4V Processed by Selective Laser Melting

Martin Malý [1,*], Christian Höller [2], Mateusz Skalon [3], Benjamin Meier [4], Daniel Koutný [1], Rudolf Pichler [2], Christof Sommitsch [3] and David Paloušek [1]

1 Faculty of Mechanical Engineering, Institute of Machine and Industrial Design, Brno University of Technology, Technická 2896/2, 616 69 Brno, Czech Republic; Daniel.Koutny@vut.cz (D.K.); David.Palousek@vut.cz (D.P.)

2 Institute of Production Engineering, Graz University of Technology, Inffeldgasse 25F, 8010 Graz, Austria; christian.hoeller@tugraz.at (C.H.); rudolf.pichler@tugraz.at (R.P.)

3 Institute of Materials Science, Joining and Forming, Graz University of Technology, Kopernikusgasse 24/I, 8010 Graz, Austria; mateusz.skalon@tugraz.at (M.S.); christof.sommitsch@tugraz.at (C.S.)

4 Materials—Institute for Laser and Plasma Technology, Joanneum Research, Leobner Straße 94, 8712 Niklasdorf, Austria; benjamin.meier@joanneum.at

* Correspondence: Martin.Maly2@vut.cz; Tel.: +420-541-144-927

Received: 26 February 2019; Accepted: 18 March 2019; Published: 20 March 2019

Abstract: The aim of this study is to observe the effect of process parameters on residual stresses and relative density of Ti6Al4V samples produced by Selective Laser Melting. The investigated parameters were hatch laser power, hatch laser velocity, border laser velocity, high-temperature preheating and time delay. Residual stresses were evaluated by the bridge curvature method and relative density by the optical method. The effect of the observed process parameters was estimated by the design of experiment and surface response methods. It was found that for an effective residual stress reduction, the high preheating temperature was the most significant parameter. High preheating temperature also increased the relative density but caused changes in the chemical composition of Ti6Al4V unmelted powder. Chemical analysis proved that after one build job with high preheating temperature, oxygen and hydrogen content exceeded the ASTM B348 limits for Grade 5 titanium.

Keywords: Selective Laser Melting; Ti6Al4V; residual stress; deformation; preheating; relative density; powder degradation

---

1. Introduction

One of the most popular additive manufacturing technologies is Selective Laser Melting (SLM). SLM technology allows the production of nearly full density metallic parts with mechanical properties comparable to the ones produced by conventional methods. Components are made layer-by-layer directly from powdered material, where each layer is selectively melted in an inert atmosphere by a laser beam [1–3].

Due to non-uniform spot heating and fast cooling, thermal gradients are formed in materials, which lead to the development of residual stresses [1]. Residual stress (RS) is described as stress which remains in the material when the equilibrium with surrounding environment is reached [4]. Unwanted RS in SLM can cause part failure due to distortions, delamination or cracking.

The RS measurement is possible by mechanical and diffraction methods; magnetic and electric techniques; and by the ultrasonic and piezoelectric effect. Mechanical measurements are usually based on material removal and its relaxation or measuring part distortion. Typical mechanical measurement is a drilling method with an accuracy of ±50 MPa [4]. The main advantage of this method is its ability to measure RS to the depth of 1.2 times the diameter of the drilled hole. The Bridge Curvature Method (BCM) is often used in case of the SLM for the fast comparison of process parameters and their influence on the RS [5]. The principle is based on measuring the distortion angle, after the sample is cut off from the base plate, using the bridge-like samples. Measuring accuracy could be affected by imprecise cutting or by angle evaluating method, thus measurement of the top surface inclination was proposed as a better technique [6,7]. Value of the RS can be determined by the simulation of measured distortion in Finite Element Method (FEM) analysis [8].

Ali et al. [9] proved that higher exposure time and lower laser power with preserved energy density lowered the RS due to lower cooling rate and temperature gradient. The variation of laser power and exposure time did not cause a change in yield strength of Ti6Al4V, but elongation increased with lower laser power and higher exposure time. The cooling rate and also the RS can be affected by layer thickness. Higher layer thickness prolonged the cooling rate and RS was lower, but with higher layer thickness the relative density was lowered [9,10].

Scanning strategy has a significant effect on the RS. Ali et al. [11] observed that with longer scanning vectors the RS increased. Due to prolonged time between scanning adjacent scan tracks, higher thermal gradients were induced. The lowest RS had a stripe strategy with the ninety-degree rotation. This conclusion was also confirmed by Robinson et al. [12]. Ali et al. [11] did not observe the positive or negative effect of the scanning strategy on mechanical properties nor relative density.

Powder bed preheating can significantly reduce the amount of RS [5,13,14]. Ali et al. [15] demonstrated that for Ti6Al4V-ELI material preheating of the build platform to the temperature of 570 °C effectively eliminated the RS. A positive influence of preheating on microstructure and mechanical properties of H13 tool steel was observed by Mertens et al. [16]. The preheating up to 400 °C in his study improved mechanical properties and the parts had more homogeneous microstructure. Formation of cracks can be also affected by preheating, which was observed during printing aluminium of 2618, 7075 [17,18] and tool steel [19].

In this study, the BCM samples made of Ti6Al4V were used to evaluate the effect of process parameters on the relative density and the RS. Investigated parameters were hatch laser speed, hatch laser power border laser velocity, waiting time between adjacent layers and powder bed preheating up to 550 °C. The design of experiment and the surface response method were used for a comprehensive evaluation of the effects of observed process parameters. Furthermore, the influence of high-temperature preheating on powder degradation was evaluated.

2. Materials and Methods

2.1. Powder Characterization

In this study, Ti6Al4V gas atomized powder (SLM Solutions Group AG, Lübeck, Germany) was used. The chemical composition of virgin powder delivered by the manufacturer is in Table 1. The powder shape was checked by scanning electron microscopy (SEM) LEO 1450VP (Carl Zeiss AG, Oberkochen, Germany). Figure 1a shows that the powder particles have a spherical shape with a low amount of satellites. The particles size distribution was analysed by laser diffraction analyser LA-960 (Horiba, Kioto, Japan). Measured particle mean size was 43 μm and median size 40.9 μm. The particles up to 29.97 μm represented 10% of particle distribution while particles up to 58.61 μm represented 90% (Figure 1b).

Table 1. Chemical composition of virgin Ti6Al4V powder.

Al (wt %)	C (wt %)	Fe (wt %)	V (wt %)	O (wt %)	N (wt %)	H (wt %)	Ti (wt %)
6.38	0.006	0.161	3.96	0.087	0.008	0.002	Bal.

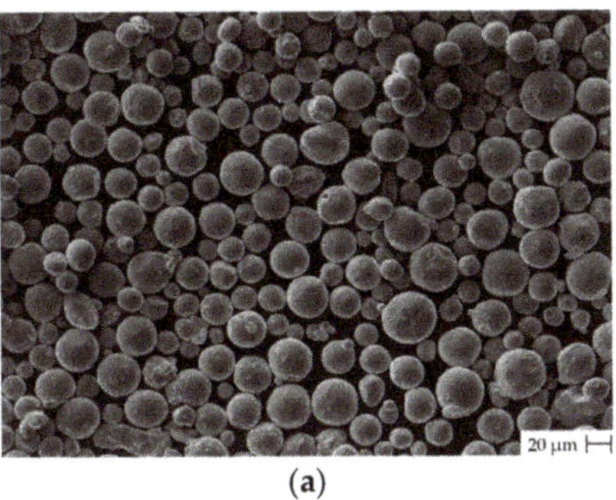

(a)

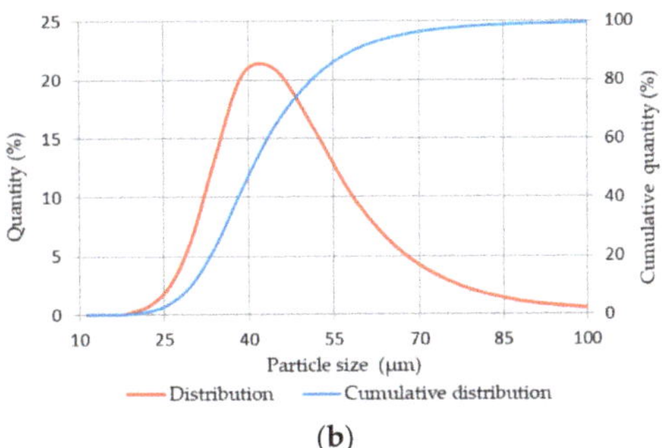

(b)

Figure 1. Ti6Al4V gas atomized powder characterization (**a**) shape evaluation by SEM; (**b**) particles size distribution.

The chemical composition of used powder was evaluated by the following methods. The aluminium content was checked by the inductively coupled plasma atomic emission spectroscopy. Oxygen and nitrogen contents were evaluated by hot extraction in helium by LECO TCH 600 (LECO Corporation, Saint Joseph, MO, USA). The hydrogen concentration was verified by the inert gas fusion thermal conductivity method JUWE H-Mat 2500 (JUWE Laborgeraete GmbH, Viersen, Germany). The accuracy of all methods is Al ± 0.327 wt %, O ± 0.008 wt % and N ± 0.0025 wt %.

2.2. Sample Fabrication

The samples were manufactured on the SLM 280^{HL} (SLM Solutions Group AG, Lübeck, Germany) 3D printer. The machine is equipped with 400 W ytterbium fibre laser YLR-400-WC-Y11 (IPG Photonics, Oxford, MS, USA) with a focus diameter of 82 µm and a Gaussian shape power distribution. Argon was used as a protective atmosphere during the process and the O_2 content was kept below 0.05 %. Before each experiment, the humidity of the powder was measured by the hydro thermometer Hytelog (B + B Thermo-Technik GmbH, Donaueschingen, Germany) with an accuracy of ±2%. The powder humidity was kept under 10%. The heating platform (SLM Solutions Group AG, Lübeck, Germany) was used to preheat the powder. This device is able to preheat the build platform up to 550 °C, but the build area is reduced to a cylindrical shape with 90 mm in diameter and 100 mm in height. For the preheating a resistive heating element is used and the temperature is controlled by a thermocouple placed below the base plate. The temperature of a printed component may be slightly lower than the measured temperature by the thermocouple. However, the maximum height of parts printed in this study is 12 mm, thus the temperature field should be relatively homogeneous. Build data were prepared in Materialise Magics 22.03 (Materialise NV, Leuven, Belgium).

2.3. Sample Geometry

The geometry of samples was designed according to the BCM shape (Figure 2a) [5], therefore the effect of chosen process parameters on distortion and RS can be evaluated. Support structures were used for all samples to simulate the condition during the printing of real components. To restrict distortion during the SLM process (before cutting off) the 4 mm high block supports were reinforced with 1 mm block spacing, while fragmentation was switched off. Teeth top length was set to 1 mm. Support structures were added just under the pillars. Samples were rotated to 20° from recoating direction to ensure consistent powder spreading. Samples were cut in the middle of support structures and the evaluated parameter was top surface angle distortion α, which is the sum of α_1 and α_2 (Figure 2b).

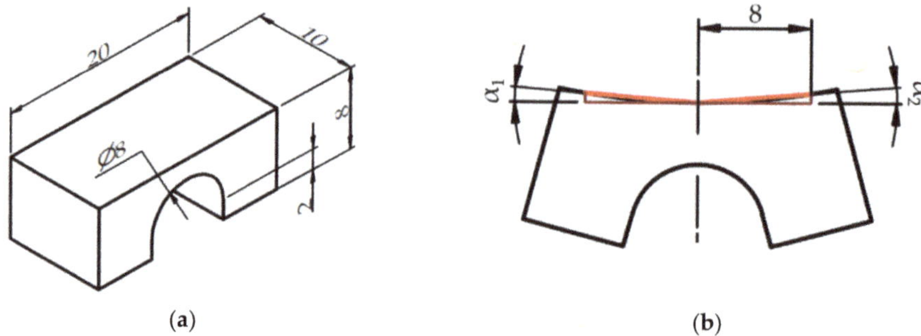

Figure 2. Samples geometry: (**a**) Dimensions of the BCM sample; (**b**) Measured bridge top surface angle distortion α is the sum of α_1 and α_2. Dimensions presented in mm.

2.4. Design of Experiment

For data evaluation Design of Experiment (DoE) and Surface Response Design (SRD) were used. Hatch laser power (H LP), hatch laser velocity (H LV), border laser velocity (B LV), delay time (DT) and preheating temperature (T) were chosen as the variable factors. The range of parameters with central points is summarized in Table 2. The DT value is waiting delay between two adjacent layers, which was set in the printing machine. The real delay (RD) value which was used for result evaluation is composed of set DT between two layers and 13 s recoating time. If the DT is zero, then the RD value is composed of 13 s recoating time and scanning time. The temperature range was set from the common preheating temperature 200 °C to the maximum temperature of 550 °C that our equipment is capable to evolve.

Table 2. Table of used process parameters for Design of Experiment (DoE) and Surface Response Design (SRD).

Values/Parameters	H LP (W)	H LV ($m \cdot s^{-1}$)	B LV ($m \cdot s^{-1}$)	DT (s)	T (°C)
Minimum value	100	700	350	0	200
Middle point	187.5	900	575	30	375
Maximum value	275	1100	800	60	550

The half fraction of the SRD was built with the five continuous variable parameters. This means twenty-six samples plus four repetition central points. To minimize the number of global parameters, which has an influence on the whole build job, the face-centered design was used.

For evaluation, Minitab 17 (Minitab Inc., State College, PA, USA) was used. Data of the top surface angle distortion α were evaluated with full quadratic terms with 95% confidence level for all intervals and with backward elimination of 0.1. Relative density data were evaluated on samples 1–16 as the half fraction of factorial design. Then the data were assessed with 95% confidence level and insignificant term combinations were manually deleted.

Border laser power was set to 100 W and hatch spacing to 0.12 mm. The layer thickness of 50 μm and stripe strategy with a maximum stripe length of 10 mm and a rotation of 67° was used. Fill contour was turned off. Other parameters were set as standard.

2.5. Distortion Evaluation

The 3D optical scanner Atos TripleScan 8M (GOM GmbH, Braunschweig, Germany) was used for assessing distortions of the bridges. Each sample was scanned after cut-off from the base plate and after coating by TiO_2 mating spray with the thickness of around 3 μm [20]. The 3D scanned surface data were evaluated in GOM Inspect 2018 (GOM GmbH, Braunschweig, Germany).

The top surface angle distortion α was measured on the top surface of the bridge as is shown in Figure 2b. First, the Computer-aided Design (CAD) data of the undeformed bridge was fitted by Gaussian best fit function on the scanned data. Then three cross sections were created parallel to the YZ plane in distance 0, 8 and −8 mm. Lines using Gaussian best fit function were fitted on the top surface in each cross section (Figure 3a). Next, points in distance 0, 3 and −3 mm in Y direction were created on those three lines (Figure 3b). Then the distance was measured between middle and side points. Finally, X and Z components from each measured distance were used for calculating angle distortions by tangent function. Left and right sides were calculated separately. Therefore, the α value was calculated as the sum of angles on both sides. The result of the top surface angle distortion α value is the mean value of three measurements of one sample.

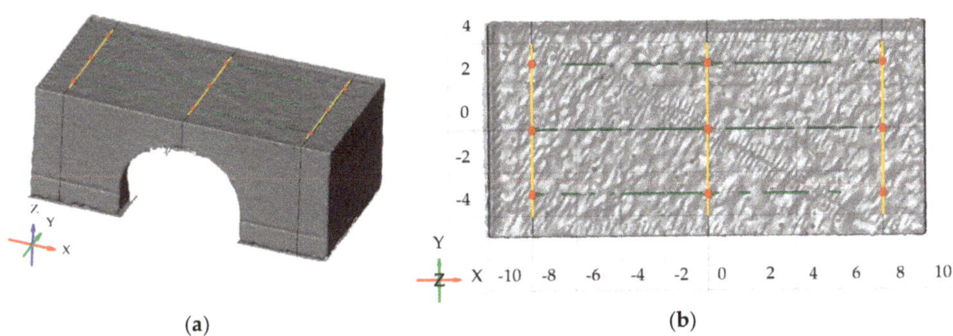

Figure 3. Distortion evaluation, fitted lines are yellow, line points are red and measured distances are green: (**a**) Isometric view on scanned data; (**b**) Top view of scanned data.

2.6. Relative Density Measurement

Relative density was determined using an optical method and was calculated as the mean value of parallel to build cross sections (Figure 4a). Value of relative density was evaluated in ImageJ v. 1.52k (National Institutes of Health, Bethesda, MD, USA). First, the picture of the cross section was converted to 8-bit type. Next, an automatic threshold was applied and relative density was evaluated in the areas defined by red rectangles (Figure 4b).

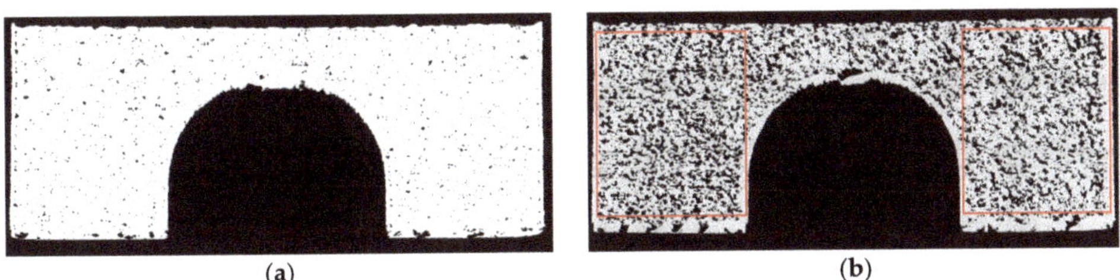

Figure 4. Cross sections of the BCM samples: (**a**) Cross section of sample 1; (**b**) Cross section of the sample made with the lowest energy density (Sample 3), red rectangles show area for relative density evaluation.

3. Results

3.1. Top Surface Distortion and Relative Density

The experimental design matrix and results of top surface angle distortion α and measured relative density are summarized in Table 3. Samples were sorted in printing order and horizontal lines represent a group of samples which were printed together in one build job.

Table 3. DoE and SRD test matrix with process parameters, the value of top surface angle distortion α and relative density.

Sample Number	H LP (W)	H LV (mm/s)	B LV (mm/s)	TD (s)	RD (s)	T (°C)	α (°)	Relative Density (%)
1	100	700	800	0	22	200	1.499	97.59
2	275	700	350	0	22	200	0.294	98.68
3	100	1100	350	0	22	200	1.201	74.04
4	275	1100	800	0	22	200	1.110	99.97
5	275	1100	350	60	73	200	1.127	99.60
6	100	700	350	60	73	200	1.413	97.06
7	100	1100	800	60	73	200	1.416	81.33
8	275	700	800	60	73	200	0.859	99.37
9	275	1100	350	0	22	550	0.437	99.69
10	100	700	350	0	22	550	0.389	98.68
11	275	700	800	0	22	550	0.406	99.43
12	100	1100	800	0	22	550	0.456	82.33
13	100	1100	350	60	73	550	0.917	91.17
14	100	700	800	60	73	550	0.520	98.93
15	275	1100	800	60	73	550	0.377	99.35
16	275	700	350	60	73	550	0.244	99.51
17	100	900	575	30	43	375	0.905	-
18	187.5	1100	575	30	43	375	0.943	-
19	187.5	900	575	30	43	375	1.000	-
20	187.5	900	800	30	43	375	0.771	-
21	275	900	575	30	43	375	0.454	-
22	187.5	900	350	30	43	375	0.764	-
23	187.5	700	575	30	43	375	0.795	-
24	187.5	900	575	30	43	375	0.740	-
25	187.5	900	575	0	17	375	0.174	-
26	187.5	900	575	60	73	375	0.876	-
27	187.5	900	575	30	43	550	0.392	-
28	187.5	900	575	30	43	200	0.809	-
29	187.5	900	575	30	43	375	0.716	-
30	187.5	900	575	30	43	375	0.668	-

3.2. *Surface Response Model for Top Surface Angle Distortion* α

Minitab 17 was used to establish a regression model for prediction of the top surface angle distortion α responses to the H LP, H LV, B LV, RD and T. Equation (1) represents the SRM-based mathematical model of significant parameters, which represent the relation between observed parameters. Table 4 and Figure 5 show results from an analysis of variance (ANOVA). The correlation of the regression model for α value is confirmed by determination coefficients $R^2 = 91.82\%$ and adjusted $R^2 = 86.04\%$.

$$\begin{aligned} \alpha = {} & 4.01 - 0.00766\,\text{H LP} - 0.00854\,\text{H LV} + 0.003232\,\text{B LV} + 0.0382\,\text{RD} - 0.001802\,\text{T} \\ & + 0.000005\,\text{H LV}\cdot\text{H LV} - 0.000252\,\text{RD}\cdot\text{RD} + 0.000004\,\text{H LP}\cdot\text{H LV} + 0.000005\,\text{H LP}\cdot\text{T} \\ & - 0.000002\,\text{H LV}\cdot\text{B LV} - 0.000018\,\text{B LV}\cdot\text{RD} - 0.000002\,\text{B LV}\cdot\text{T} \end{aligned} \quad (1)$$

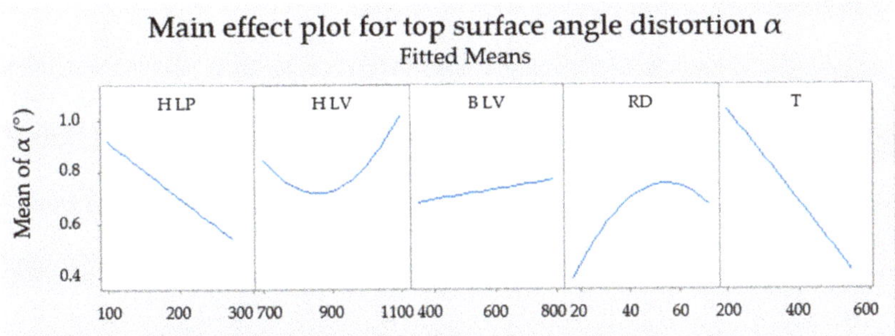

Figure 5. Main effect plot for top surface angle distortion α.

Table 4. ANOVA table for the top surface angle distortion α.

Source	DF	Contribution (%)	Adj SS	Adj MS	F-Value	P-Value
Model	12	91.82	3.44257	0.28688	15.9	0
Linear	5	73.01	2.81581	0.56316	31.21	0
H LP	1	17.22	0.64558	0.64558	35.78	0
H LV	1	3.62	0.13584	0.13584	7.53	0.014
B LV	1	0.59	0.03329	0.03329	1.85	0.192
RD	1	5.26	0.26462	0.26462	14.66	0.001
T	1	46.31	1.73648	1.73648	96.23	0
Square	2	4.73	0.17728	0.08864	4.91	0.021
H LV·H LV	1	1.79	0.17109	0.17109	9.48	0.007
RD·RD	1	2.94	0.11008	0.11008	6.10	0.024
2-Way Interaction	5	14.08	0.52797	0.10559	5.85	0.003
H LP·H LV	1	1.95	0.07309	0.07309	4.05	0.060
H LP·T	1	2.90	0.1089	0.10890	6.04	0.025
H LV·B LV	1	2.68	0.10043	0.10043	5.57	0.031
B LV·RD	1	4.62	0.17310	0.17310	9.59	0.007
B LV·T	1	1.93	0.07244	0.07244	4.01	0.061
Error	17	8.18	0.30677	0.01805	-	-
Lack-of-Fit	14	6.40	0.24014	0.01715	0.77	0.685
Pure Error	3	1.78	0.06662	0.02221	-	-
Total	29	100.00	-	-	-	-

In order to investigate the effect of high energy and high-temperature preheating on the distortion, an additional four bridge samples were made. Those samples were built with increasing H LP according to Table 5.

Table 5. Parameters of samples with increasing laser power and value of α.

Sample Number	H LP (W)	H LV (mm/s)	B LV (mm/s)	RD (s)	T (°C)	H Ed (J·mm^{-3}) [1]	α (°)
31	275	700	350	22	550	65.5	0.363
32	300	700	350	22	550	71.4	0.224
33	325	700	350	22	550	77.4	0.313
34	350	700	350	22	550	83.3	0.098

[1] Calculated as H Ed = H LP·(H LV·Lt·Hs)$^{-1}$, Layer thickness (Lt) = 50 μm, Hatch spacing (Hs) = 120 μm.

3.3. Mathematical Model for Relative Density

Equation (2) represents a mathematical model of significant parameters with an influence on the relative density. Figure 6 shows a Pareto chart of standardized effect for evaluated relative density data. ANOVA results are shown in Figure 7 and Table 6. The correlation of the regression model for relative density value is confirmed by determination coefficients R^2 = 89.38% and adjusted R^2 = 82.29%.

$$\text{Relative density} = 136.67 - 0.1558\ \text{H LP} - 0.0628\ \text{H LV} - 0.00004\ \text{B LV} + 0.00766\ \text{T} + 0.0390\ \text{RD} + 0.000232\ \text{H LP*H LV} \quad (2)$$

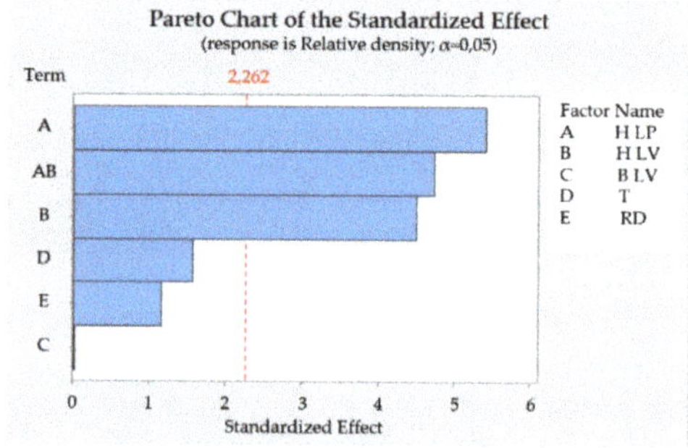

Figure 6. Pareto chart of the standardized effect to relative density.

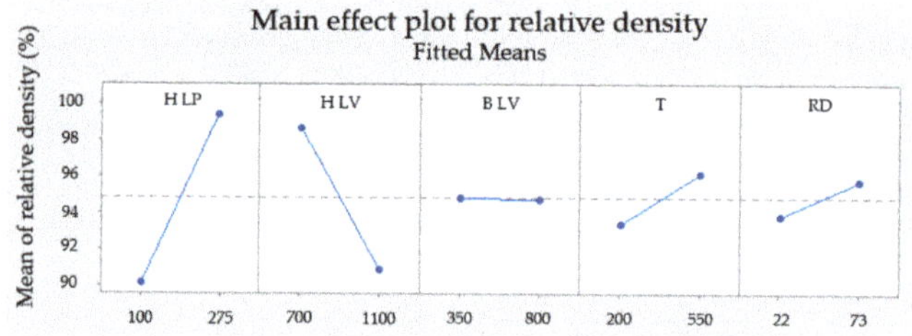

Figure 7. Main effect plot for relative density.

Table 6. ANOVA table for relative density.

Source	DF	Contribution (%)	Adj SS	Adj MS	F-Value	P-Value
Model	6	89.38	893.804	148.967	12.62	0.001
Linear	5	62.96	629.660	125.932	10.67	0.001
H LP	1	34.66	346.611	346.611	29.36	0
H LV	1	23.85	238.471	238.471	20.20	0.002
B LV	1	0	0.0010	0.001	0	0.993
T	1	2.88	28.756	28.756	2.44	0.153
RD	1	1.58	15.821	15.821	1.34	0.277
2-Way Interaction	1	26.41	264.144	264.144	22.38	0.001
H LP·H LV	1	26.41	264.144	264.144	22.38	0.001
Error	9	10.62	106.239	11.804	-	-
Total	15	100.00	-	-	-	-

3.4. Analysis of Used Powder

Figure 8a shows the influence of high-temperature base plate preheating (550 °C) on the powder. The powder significantly changed colour from silver to brown and there is a hint of particle agglomeration. Therefore, the powder used in build job with 550 °C was checked by the SEM microscopy in order to investigate the particle agglomeration and their shape (Figure 8b).

(a) (b)

Figure 8. The powder used in heating unit preheated to the 550 °C (**a**) Build job made with 550 °C; (**b**) SEM microscopy photo of the powder used with 550 °C.

Powder chemistry analysis was done after first build job with preheating to 200 °C and these results were compared with the powder used with 550 °C preheating. Results in Table 7 confirm a rise in oxygen content from 0.12 to 0.33 wt % and in the hydrogen from 0.002 to 0.0168 wt %. Aluminium content also slightly rose from 6.05 to 6.11 wt %, but nitrogen content decreased from 0.017 to 0.0149 wt %. Results are compared with the ASTM B348 Grade 5 titanium requirements and virgin Ti6Al4V powder chemical composition received by the vendor.

Table 7. Chemical composition analysis of the Ti6Al4V powder.

Powder State/Checked Elements	Al (wt %)	O (wt %)	N (wt %)	H (wt %)
ASTM B348 Grade 5	5.50–6.75	Max. 0.20	Max. 0.050	Max. 0.0125
Virgin Ti6Al4V	6.38	0.087	0.0080	0.0020
Ti6Al4V 200 °C	6.05	0.120	0.0170	0.0020
Ti6Al4V 550 °C	6.11	0.330	0.0149	0.0168

4. Discussion

4.1. Top Surface Angle Distortion α

The main contribution of each parameter on the distortion, and therefore the amount of residual stresses, can be derived from the ANOVA (Table 4). Further on the significance of each parameter can be evaluated by a p-value. If the value is lower than 0.05, then the parameter is significant, while the p-value of the lack-of-fit parameter should be high, which shows that the error value is not significant. The p-value of the lack-of-fit parameter 0.685 shows that the regression model for the top surface angle distortion α fits the measured data.

Table 4 and Figure 5 show that the most significant parameters for reduction distortions are T and H LP with 46.31% and 17.22% linear contribution. P-values of those parameters are 0. The RD and H LV are not that significant in comparison with the previous two parameters. Their linear contributions are 5.26% and 3.62%, while p-values are lower than 0.05, therefore parameters are significant. The B LV can be considered as an insignificant parameter with 0.59% linear contribution and the p-value greater than 0.05. Figure 5 indicates that H LP, B LV and T have linear behaviour in contrast to RD and H LV.

From the ANOVA results, it can be concluded that increasing the preheating temperature or laser power causes a reduction in deformation. In contrast, increasing H LV, B LV and RD lead to higher deformations. This can be contributed to the cooling rate, which is lower with slower laser movement, higher preheating and shorter waiting time. Therefore, the thermal gradients, residual stresses and finally distortions are lower [9,11,15,21,22].

Optimal parameters for achieving the lowest distortion can be predicted from the fitted regression model (Figure 9). For reaching the lowest distortion it is predicted to use H LP 275 W, H LV 785 mm/s, B LV 350 mm/s, RD 17 s and T 550 °C. Predicted α is $-0.182°$. In contrast, the lowest distortion predicted with a preheating temperature of 200 °C is 0.139°.

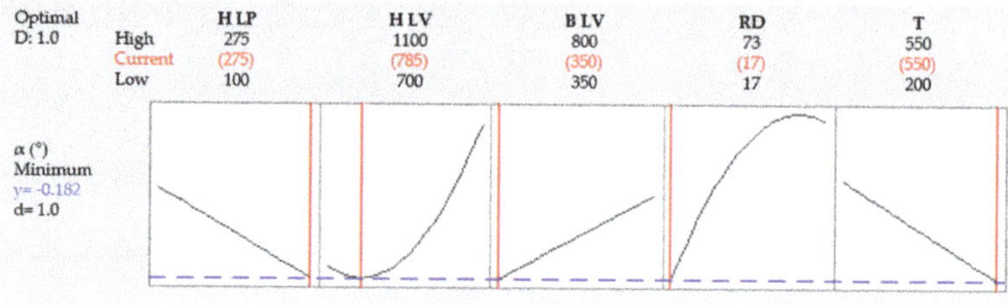

Figure 9. Predicted values for the lowest distortion in the full range of observed parameters.

Figure 10 shows the influence of energy density (ED) on the top surface angle distortion α. Samples used for this comparison were made with the same preheating temperature of 550 °C and 200 °C. The effect of B LV and RD was neglected. The interpolated line for 200 °C samples is constantly dropping with increasing ED and as was measured by Mishurova et al. [7], and this trend constantly continues. In contrast, the interpolated line for 550 °C samples with added high ED samples starts much lower than 200 °C line. The decrease of α value is gradual until 65 J/mm^3 and drops rapidly with higher ED.

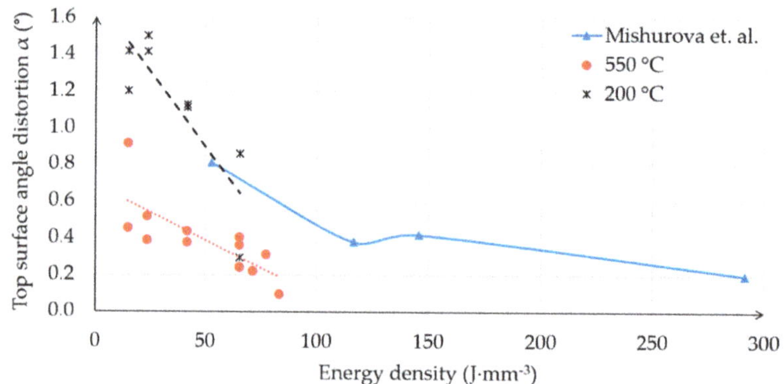

Figure 10. Effect of energy density on the top surface angle distortion α.

4.2. Relative Density

The most significant effect out of involved parameters influencing the relative density are H LP and H LV. The linear contribution is 34.66% for H LP and 23.85% for H LV. They are also significant in their two-way interaction with a contribution of 26.41%. P-values are 0 for H LP and 0.002 for H LV.

Preheating temperature has a linear contribution of 2.88% while its p-value of 0.153 is higher than 0.05. This means that this parameter in the observed range is not that significant for the model due to the high contribution of H LP and H LV. Relative density rose with higher T. Positive influence of preheating was also confirmed on stainless steel M2. It was proved by Kempen et al. [19] that with higher preheating temperature, higher laser velocities can be used while maintaining relative density.

The real delay has a linear contribution of 1.58% and a p-value of 0.277, which means the RD has minimal influence on relative density. Prolonged RD leads to an increase in relative density.

Border laser velocity with a linear contribution of 0 % and p-value 0.993 means that this parameter was not significant for the model, which can be due to the place where porosity was measured.

From the ANOVA (Figure 7) it can be deduced that a sample will have maximum relative density when values of H LP, RD and T are set as the highest and H LV as the lowest. This means the highest energy density is in the hatch.

4.3. Powder Degradation

There is clear evidence that the chemical composition of the powder significantly changed due to oxidation. Titanium alloys suffer high chemical affinity to oxygen leading to form a thin oxide layer even on air room temperature. Exposing titanium to an oxygen-containing atmosphere at elevated temperatures around 550 °C increase diffusion rates through thin oxide layers, and allows penetration of oxygen in the material [23]. After experiments with 550 °C preheating, oxygen content increased against 200 °C preheating from 0.12 to 0.33 wt % which is 0.13% higher than the ASTM B348 requirement for Grade 5 titanium.

Increased oxygen content in the Ti6Al4V causes an increase in yield and ultimate tensile strength, whilst ductility up to 0.19 wt % of the oxygen content remains constant [24]. Ti6Al4V additively manufactured alloy is due to rapid cooling mainly composed from α'martensitic microstructure even with preheating up to 550 °C [15]. Therefore, this is sensitive to oxygen content because of the concentration of oxygen higher than 0.22 wt % leads to the brittleness of the α'martensitic structure. Critical oxygen content for α and β structure is 0.4 wt % [25]. Oxygen concentration above 0.25 wt % leads to change in the typical microstructure, which causes a sharp decrease in ductility of Ti6Al4V [26].

Hydrogen content in the powder used under 550 °C exceeds the approved ASTM B348 limit value of 0.0125 wt % and its content rose from 0.002 to 0.0168 wt %. The diffusion rate of the hydrogen is rapidly increasing at the elevated temperatures [23]. The origin of the hydrogen element is most probably from powder moisture which was kept below 10%. Hydrogen in titanium alloys causes a phenomenon known as hydrogen embrittlement and could lead to part failure [27,28].

It was shown that the critical issues with processing the titanium alloy by SLM at high temperatures are connected with chemical composition changes in the unused powder, although the material was processed under argon atmosphere with oxygen concentration of 0.05% and powder humidity kept below 10%. The measured concentration of oxygen and hydrogen was beyond ASTM B348 requirement for Grade 5 titanium. Therefore, the used powder cannot be used for mechanical stressed parts.

5. Conclusions

Effects of hatch laser power, hatch laser velocity, border laser velocity, preheating temperature and delay time on residual stress and relative density on SLM processed Ti6Al4V samples have been investigated. In addition, the impact of preheating temperature up to 550 °C on Ti6Al4V powder degradation has been discussed. The main findings are the following:

- The preheating temperature has the main effect on the distortion and residual stress out of all observed parameters. With a high preheating temperature of 550 °C, the distortions of the top surface decreased and the relative density increased. The linear contribution effect of preheating was 46.31% on the distortion and 2.88% on the relative density.
- Relative density mainly depends on the hatch laser power and hatch laser velocity.
- Higher energy density decreased the deformations of the BCM samples. The value of the top surface distortion α decreased from 0.363° to 0.098° with increased energy density from 65.5 to 83.3 $J{\cdot}mm^{-3}$.

- Longer delay time negatively influenced distortions, but improved relative density. The linear contribution effect of the delay time was 5.26% on the distortions and 1.58% on the relative density.
- Powder bed preheating to 550 °C led to fast powder degradation. The oxygen and hydrogen content rose beyond the ASTM B348 requirement for Grade 5 titanium after one build job.

Author Contributions: Conceptualization, M.M., D.K.; methodology, M.M., C.H and M.S.; validation, M.M., C.H., B.M., M.S., D.K., R.P., C.S. and D.P.; formal analysis, M.M.; investigation, M.M., B.M. and M.S.; resources, R.P., C.S., D.K. and D.P.; data curation, M.M.; writing—original draft preparation, M.M.; writing—review and editing, M.M., D.K. and M.S.; visualization, M.M.; supervision, M.S., C.H., D.K., R.P. and C.S.; project administration M.M.; funding acquisition, C.S., R.P, D.K. and D.P.

Funding: This research was funded by the ESIF, EU Operational Programme Research, Development and Education within the research project [Architectured materials designed for additive manufacturing] grant number [CZ.02.1.01/0.0/0.0/16_025/0007304] and faculty specific research project FSI-S-17-4144.

Acknowledgments: The authors express their thanks to Philipp Schwemberger from the Institute of Production Engineering, Graz University of Technology, for his help with the samples preparation.

Conflicts of Interest: The authors declare no conflicts of interest. The funders had no role in the design of the study; in the collection, analyses, or interpretation of data; in the writing of the manuscript, or in the decision to publish the results.

References

1. Kruth, J.P.; Froyen, L.; Van Vaerenbergh, J.; Mercelis, P.; Rombouts, M.; Lauwers, B. Selective laser melting of iron-based powder. *J. Mater. Process. Technol.* **2004**, *149*, 616–622. [CrossRef]
2. Schleifenbaum, H.; Meiners, W.; Wissenbach, K.; Hinke, C. Individualized production by means of high power Selective Laser Melting. *CIRP J. Manuf. Sci. Technol.* **2010**, *2*, 161–169. [CrossRef]
3. Gu, D.D.; Meiners, W.; Wissenbach, K.; Poprawe, R. Laser additive manufacturing of metallic components: Materials, processes, and mechanisms. *Laser Addit. Manuf. Mater. Des. Technol. Appl.* **2012**, *6608*, 163–180. [CrossRef]
4. Withers, P.J.; Bhadeshia, H.K.D.H. Residual stress. Part 1—Measurement techniques. *Mater. Sci. Technol.* **2001**, *17*, 355–365. [CrossRef]
5. Kruth, J.P.; Deckers, J.; Yasa, E.; Wauthle, R. Assessing and comparing influencing factors of residual stresses in selective laser melting using a novel analysis method. *Proc. Inst. Mech. Eng. Part B-Journal Eng. Manuf.* **2012**, *226*, 980–991. [CrossRef]
6. Le Roux, S.; Salem, M.; Hor, A. Improvement of the bridge curvature method to assess residual stresses in selective laser melting. *Addit. Manuf.* **2018**, *22*, 320–329. [CrossRef]
7. Mishurova, T.; Cabeza, S.; Artzt, K.; Haubrich, J.; Klaus, M.; Genzel, C.; Requena, G.; Bruno, G. An assessment of subsurface residual stress analysis in SLM Ti-6Al-4V. *Materials* **2017**, *10*. [CrossRef]
8. Sillars, S.A.; Sutcliffe, C.J.; Philo, A.M.; Brown, S.G.R.; Sienz, J.; Lavery, N.P. The three-prong method: A novel assessment of residual stress in laser powder bed fusion. *Virtual Phys. Prototyp.* **2018**, *13*, 20–25. [CrossRef]
9. Ali, H.; Ghadbeigi, H.; Mumtaz, K. Processing Parameter effects on residual stress and mechanical properties of selective laser melted Ti6Al4V. *J. Mater. Eng. Perform.* **2018**, *27*, 4059–4068. [CrossRef]
10. Mugwagwa, L.; Dimitrov, D.; Matope, S.; Yadroitsev, I. Influence of process parameters on residual stress related distortions in selective laser melting. *Procedia Manuf.* **2018**, *21*, 92–99. [CrossRef]
11. Ali, H.; Ghadbeigi, H.; Mumtaz, K. Effect of scanning strategies on residual stress and mechanical properties of Selective Laser Melted Ti6Al4V. *Mater. Sci. Eng. A* **2018**, *712*, 175–187. [CrossRef]
12. Robinson, J.; Ashton, I.; Fox, P.; Jones, E.; Sutcliffe, C. Determination of the effect of scan strategy on residual stress in laser powder bed fusion additive manufacturing. *Addit. Manuf.* **2018**, *23*, 13–24. [CrossRef]
13. Buchbinder, D.; Meiners, W.; Pirch, N.; Wissenbach, K.; Schrage, J. Investigation on reducing distortion by preheating during manufacture of aluminum components using selective laser melting. *J. Laser Appl.* **2014**, *26*. [CrossRef]
14. Vora, P.; Mumtaz, K.; Todd, I.; Hopkinson, N. AlSi12 in-situ alloy formation and residual stress reduction using anchorless selective laser melting. *Addit. Manuf.* **2015**, *7*, 12–19. [CrossRef]

15. Ali, H.; Ma, L.; Ghadbeigi, H.; Mumtaz, K. In-situ residual stress reduction, martensitic decomposition and mechanical properties enhancement through high temperature powder bed pre-heating of Selective laser melted Ti6Al4V. *Mater. Sci. Eng. A* **2017**, *695*, 211–220. [CrossRef]
16. Mertens, R.; Vrancken, B.; Holmstock, N.; Kinds, Y.; Kruth, J.P.; Van Humbeeck, J. Influence of Powder Bed Preheating on Microstructure and Mechanical Properties of H13 tool steel SLM parts. In *Laser Assisted Net Shape Engineering 9 International Conference on Photonic Technologies Proceedings of the Lane, Fürth, Germany, 19–22 September 2016*; Schmidt, M., Vollertsen, F., Arnold, C.B., Eds.; Elsevier: Amsterdam, The Netherlands, 2016; Volume 83, pp. 882–890.
17. Koutny, D.; Palousek, D.; Pantelejev, L.; Hoeller, C.; Pichler, R.; Tesicky, L.; Kaiser, J. Influence of scanning strategies on processing of aluminum alloy EN AW 2618 using selective laser melting. *Materials* **2018**, *11*, 298. [CrossRef]
18. Mertens, R.; Dadbakhsh, S.; Van Humbeeck, J.; Kruth, J.-P. Application of base plate preheating during selective laser melting. *Procedia CIRP* **2018**, *74*, 5–11. [CrossRef]
19. Kempen, K.; Vrancken, B.; Buls, S.; Thijs, L.; Van Humbeeck, J.; Kruth, J.P. Selective Laser melting of crack-free high density M2 High speed steel parts by baseplate preheating. *J. Manuf. Sci. Eng. ASME* **2014**, *136*. [CrossRef]
20. Palousek, D.; Omasta, M.; Koutny, D.; Bednar, J.; Koutecky, T.; Dokoupil, F. Effect of matte coating on 3D optical measurement accuracy. *Opt. Mater. AMST* **2015**, *40*, 1–9. [CrossRef]
21. Ali, H.; Ghadbeigi, H.; Mumtaz, K. Residual stress development in selective laser-melted Ti6Al4V: A parametric thermal modelling approach. *Int. J. Adv. Manuf. Technol.* **2018**, *97*, 2621–2633. [CrossRef]
22. Xu, W.; Lui, E.W.; Pateras, A.; Qian, M.; Brandt, M. In situ tailoring microstructure in additively manufactured Ti-6Al-4V for superior mechanical performance. *Acta Mater.* **2017**, *125*, 390–400. [CrossRef]
23. Lutjering, G.; Williams, J.C. *Titanium*, 2nd ed.; Springer: Berlin, Germany, 2007; ISBN 978-3-540-71397-5.
24. Tang, H.P.; Qian, M.; Liu, N.; Zhang, X.Z.; Yang, G.Y.; Wang, J. Effect of powder reuse times on additive manufacturing of Ti-6Al-4V by Selective electron beam melting. *JOM* **2015**, *67*, 555–563. [CrossRef]
25. Yan, M.; Xu, W.; Dargusch, M.S.; Tang, H.P.; Brandt, M.; Qian, M. Review of effect of oxygen on room temperature ductility of titanium and titanium alloys. *Powder Metall.* **2014**, *57*, 251–257. [CrossRef]
26. Yan, M.; Dargusch, M.S.; Ebel, T.; Qian, M. A transmission electron microscopy and three-dimensional atom probe study of the oxygen-induced fine microstructural features in as-sintered Ti-6Al-4V and their impacts on ductility. *Acta Mater.* **2014**, *68*, 196–206. [CrossRef]
27. Silverstein, R.; Eliezer, D. Hydrogen trapping in 3D-printed (additive manufactured) Ti-6Al-4V. *Mater. Charact.* **2018**, *144*, 297–304. [CrossRef]
28. Tal-Gutelmacher, E.; Eliezer, D. High fugacity hydrogen effects at room temperature in titanium based alloys. *J. Alloys Compd.* **2005**, *404–406*, 613–616. [CrossRef]

 materials

Article

Phase Studies of Additively Manufactured Near Beta Titanium Alloy-Ti55511

Tuerdi Maimaitiyili [1,2,*], Krystian Mosur [3], Tomasz Kurzynowski [3], Nicola Casati [4] and Helena Van Swygenhoven [2,5]

1 Materials and Process Development, Swerim AB, Isafjordsgatan 28A, 16440 Stockholm, Sweden
2 Photons for Engineering and Manufacturing Group, Swiss Light Source, Paul Scherrer Institute-PSI, 5232 Villigen, Switzerland; helena.vanswygenhoven@psi.ch
3 Centre for Advanced Manufacturing Technologies/Fraunhofer Project Center, Wrocław University of Science and Technology, ul. Łukasiewicza 5, 50-371 Wrocław, Poland; krystian.mosur@gmail.com (K.M.); tomasz.kurzynowski@pwr.edu.pl (T.K.)
4 Swiss Light Source, Paul Scherrer Institute-PSI, 5232 Villigen, Switzerland; nicola.casati@psi.ch
5 Neutrons and X-rays for Mechanics of Materials, IMX, Ecole Polytechnique Federale de Lausanne, CH-1012 Lausanne, Switzerland
* Correspondence: Tuerdi.maimaitiyili@swerim.se; Tel.: +46-73-029-3290

Received: 16 March 2020; Accepted: 3 April 2020; Published: 7 April 2020

Abstract: The effect of electron-beam melting (EBM) and selective laser melting (SLM) processes on the chemical composition, phase composition, density, microstructure, and microhardness of as-built Ti55511 blocks were evaluated and compared. The work also aimed to understand how each process setting affects the powder characteristics after processing. Experiments have shown that both methods can process Ti55511 successfully and can build parts with almost full density (>99%) without any internal cracks or delamination. It was observed that the SLM build sample can retain the phase composition of the initial powder, while EBM displayed significant phase changes. After the EBM process, a considerable amount of α Ti-phase and lamella-like microstructures were found in the EBM build sample and corresponding powder left in the build chamber. Both processes showed a similar effect on the variation of powder morphology after the process. Despite the apparent difference in alloying composition, the EBM build Ti55511 sample showed similar microhardness as EBM build Ti-6Al-4V. Measured microhardness of the EBM build sample is approximately 10% higher than the SLM build, and it measured as 348 ± 30.20 HV.

Keywords: titanium alloy; Ti55511; synchrotron; XRD; microscopy; SLM; EBM; EBSD; additive manufacturing; Rietveld analysis

1. Introduction

Additive manufacturing (AM) also known as "3D printing" is an advanced manufacturing technology which allow fabrication of geometrically complex and functional parts directly form the computer-aided design (CAD) model in a short time with limited tooling cost and with almost no material waste [1–6]. Hence, AM is a potential technology in areas where a high degree of customization and on-demand manufacturing is key such as aerospace and medical industry.

Different types of AM techniques have been developed for metallic materials, and they can be categorized into many different subclasses based on energy sources, feed material, and ingredient material feeding approach as described by Liu et al. [7]. Among various AM techniques, the powder bed selective laser melting (SLM) and electron beam melting (EBM) are two common AM techniques, and these systems have been described in detail by Samy et al. [1] and Maimaitiyili et al. [2], respectively. In general, AM processes involve shaping a build plate and selectively melting the raw material (e.g., wire/powder) to form a three-dimensional solid object using a high energy focused laser or electron beam with multi-axis motion. Even though the basic operational principle of the AM method is relatively simple, the actual metal AM process is complex, and the results depend upon different settings of the system such as beam power (current), scanning speed, preheat temperature, etc. These are collectively referred to as processing parameters. In the AM process, processing parameters determine the build environment and cooling conditions and, consequently, affect the phase composition [1,2,8,9], residual stress [2], texture [2,8,10], surface roughness [11], density [7,12,13], and mechanical properties [3,7,12–15] of the as-built part. The lack of understanding of the relationship between process parameters and microstructure hinders the prediction of the properties of the built material and service life.

Titanium-based alloys have been widely used as an engineering material in many industries because of their excellent combination of a high strength/weight ratio and good corrosion resistance [1,2,10,12,14–17]. However, extracting high purity Ti and producing usable Ti-alloy parts are difficult and expensive processes. Therefore, there is a strong interest in using AM, such as SLM and EBM techniques, to process Ti-based materials.

The mechanical properties of Ti-alloys depend on the microstructure, chemical, and phase composition [3,7,12–15]. Based on concentrations of alloying elements, Ti-alloys can be divided into three main classes: α, $\alpha + \beta$, and (meta-stable and stable) β-alloys [17,18]. Depending on alloy composition and heating/cooling rates, Ti-alloys can also contain metastable α' (hcp) and α'' (orthorhombic) phases [1,17–19]. Therefore, it is of fundamental and technological importance to investigate the influence of AM processes on the phase composition and microstructure in Ti-alloys. Many AM studies have been performed on the two-phase ($\alpha + \beta$) Ti-alloy known as Ti-6Al-4V [1,2,5,6,8–11,14,20] ,while other alloy compositions such as Ti–5Al–5Mo–5V–1Cr–1Fe (Ti55511) have been less addressed [13,21].

The Ti55511 is a near β-type alloy with important application in aerospace industries [13,22–24]. Compared to Ti-6Al-4V, the Ti55511 is a superior structural material, as it provides comparable or higher strength with 15%–20% less weight [24]. Most of our knowledge has been derived from conventionally manufactured materials [23,24]. Characterization of phases and microstructure after synthesis using different AM methods has not yet been performed. Here, we report and compare the build quality, microstructure, and phase composition of Ti55511 synthesized by SLM and EBM.

2. Materials and Methods

2.1. Powders

Pre-alloyed Ti–5Al–5Mo–5V–1Cr–1Fe (Ti55511) powder prepared by gas atomization with an average particle size of 43 µm for SLM and 71 µm for EBM was obtained from KAMB Import–Export Warszawa (Nr CAS: 7440-32-6). The chemical compositions (wt%) of the as-received Ti55511 powder was Al 5.17, Mo 4.95, V 4.74, Cr 0.92, Fe 1.01, balanced by Ti.

Laser diffraction particle size analyzer Partica LA-950 V2 system (Horiba, Tokyo, Japan) was used to measure the particle size distribution. For accuracy, each measurement was repeated three times. The results are plotted in Figure 1, and some important parameters are listed in Table 1.

A scanning electron microscope (SEM) equipped with an energy-dispersive X-ray detector system was used to examine the shape, size distribution, surface morphology, and external and internal defects of the powders. In addition, optical microscopy was employed to observe the internal porosity, defects, and cross-section of the powders. Selected SEM images of powders before and after processing are shown in Figure 2. Figure 3 shows the internal defect of the powders. The size distribution determined by SEM agrees well with data presented in Figure 1 and Table 1.

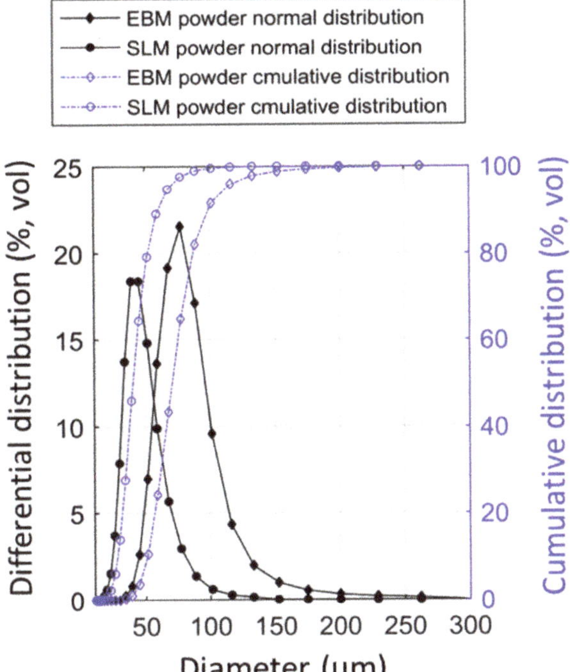

Figure 1. Particle size distribution of Ti55511 powders.

Table 1. Powder size distribution before and after processing measured by the laser diffraction method.

	Ti55511					
	SLM			EBM		
	Before	**After**	**% Change**	**Before**	**After**	**% Change**
D_{10} (μm)	27.96	30.42	8.80	50.88	59.74	17.41
D_{50} (μm)	40.42	44.82	10.89	70.41	85.32	21.18
D_{90} (μm)	60.32	68.58	13.68	99.23	127	28.02

"D_{10}", "D_{50}", and "D_{90}" mean the particle sizes at 10 vol%, 50 vol%, and 90 vol%, respectively.

2.2. Sample Build Processes

The Ti55511 blocks with dimension of 2.5 × 2.5 × 5 cm were fabricated using both SLM and EBM methods in 90° (long side of the sample was along with the build direction), 0° (long side of the sample was in the build plane), and 45° (long side of the sample was 45° to the build plane) orientations. All microstructure related results presented were obtained from 90° samples.

The SLM samples were made with ReaLizer 250 II SLM machine (ReaLizer GmbH, Borchen, Germany) equipped with a 400 W fiber laser. The laser melting process was carried under a protective argon atmosphere with O_2 content less than 0.1 vol.%. The laser power (P) was 200 W, the scan speed (v) 330 mm/s, the hatch spacing (h) 0.21 mm, and layer thickness (d) 50 μm. To reduce thermal residual stress and elemental segregation, the samples were built on a Ti-6Al-4V substrate pre-heated to 250 °C prior to the building process.

The EBM specimens were built on an Arcam A2 machine (Arcam AB, Mölndal, Sweden) with a layer thickness of 50 μm. The processing parameters, such as spot size and scan velocity, were defined by the Arcam A2 process control algorithm. The scanning speed was 4530 mm/s, current 15mA, focus offset 3mA, and preheat temperature 650 °C. To investigate only the process effect, no post-treatment was applied to the specimens.

The scan strategy used in SLM was the "island" scan strategy in which each layer was divided into 3 mm × 3 mm square islands. Scan tracks in each island were exposed at the same orientation with respect to the neighboring island and rotated 90° among alternating layers.

In the EBM, a bidirectional scan strategy was used in which all tracks are made with alternating direction, i.e., left-to-right, then right-to-left.

The scanning strategies used are those that resulted in this material lowest residual stress and porosity which determined after cube print tests.

2.3. Characterization Methods

All samples used for microscopic studies were prepared by using standard metallographic preparation routines. Examination of microstructure was performed using a Visible Light Microscope (VLM, Leica DMRX + SpeedXT Core5, Wetzlar, Germany) and ZEISS NVision40 scanning electron microscope equipped with an energy-dispersive X-ray spectrometer (EDS) analysis system from Oxford Instruments (Oberkochen, Germany). To obtain phase and texture related information from the as-built material, electron backscatter diffraction (EBSD) investigations were performed using a field emission gun scanning electron microscope (FEG SEM) ZEISS ULTRA 55 equipped with an EDAX Hikari Camera (Oberkochen, Germany) operated at 20 kV in a high current mode with 120 μm aperture.

To identify the phase composition of powders, synchrotron X-ray powder diffraction were carried out at the Material Science (MS) beamline X04SA-MS4 of the Swiss Light Source (Paul Scherrer Institute, Villigen, Switzerland) using the MYTHEN II detector. All measurements were made at room temperature with 25.1 keV (λ = 0.4940 Å) X-ray beam and 60s exposure.

Diffraction data of as built material were acquired using a D500 X-ray diffractometer (XRD) from Bruker–Siemens (Karlsruhe, Germany) with Cu Kα radiation (λ = 0.15406 nm) operating at 40 mA and 40 kV. The step size and the acquisition time were 0.01° and 1 s respectively. All measurements were conducted at room temperature at the center of each test blocks cut from each sample faces (xy-, xz- and yz-planes) from the top- and bottom-half of the sample. The quantitative phase analysis was performed with a Topas-Academic software package.

Porosity was characterized using the Archimedes technique and microscopy. Cuboid samples were sectioned at different depths, ground and polished, and inspected in SEM. On average, 120 images were captured from each sectioned part and stitched with the functions in IMAGIC IMS V17Q4. These color images were then converted into 8 bit black-and-white images using ImageJ. To understand the shape of the pores, a circularity of the pores was also calculated with this program.

3. Results and Discussions

3.1. Powder Characterization

Figure 1 and Table 1 present the results of the powder size distribution (PSD) of different powder samples obtained using a Partica LA-950 V2 laser particle size analyzer (Kyoto, Japan). Figures 2 and 3 show the morphology and external/internal defect of the powders, respectively. Both SLM and EBM powder particles predominantly in spherical shape with limited quantity of non-spherical particles and spherical imperfections. Both SLM and EBM powders have a nearly normal size distribution (Figure 1). The SLM powder had a size distribution between 28 (D10) and 60 μm (D90) with mean volume diameter around 40 μm. The EBM powder had a size distribution between 51 (D10) and 99 μm (D90) with mean volume diameter around 70 μm.

Figure 2. SEM images showing the morphologies of ingredient powder before (**a**,**c**) and after (**b**,**d**) the EBM (**a**,**b**) and SLM (**c**,**d**) processes.

It has been reported that a spherical powder with narrow PSD has a positive effect in both mechanical properties and finishing surface for the sample [25]. Because of charging problems associated with electron beam, commonly, a larger powder is used in the EBM [2]. According to the literature [2,4,5], measured PSD of the EBM and SLM was in the suggested PSD range for respective methods. Hence, both types of powders are ideal for processing with corresponding methods and all observation reported here can be directly related to the alloy and the manufacturing methods in use.

After processing, the powder remains predominantly spherical as shown in Figure 2b,d. Table 1 shows, however, that both the SLM and EBM process caused changes in powder size distribution. The averaged sizes tended to increase which can be due to the powder agglomeration or powder recoating. Occasionally, broken powder particles were observed as shown in Figure 2d. It is, however, clear that both process routes have limited effect on the powder quality.

Figure 2 shows cross-sections of embedded powder after etching before and after usage. The initial microstructure (before usage) of the SLM and EBM powders was very similar. The powder used in the SLM process was not different, but the powder used in EBM showed a lamellae microstructure with a mixture of very fine and coarser α and β phases. Similar lamella microstructure is reported by Li et al. [26] for thermally treated conventional Ti55511 which first solution treated at 920 °C for 120 min and annealed at 700 °C for 60 min. Therefore, it is believed that the high build plate temperature used in the EBM process together with heat dissipated from the melt zone during electron beam scanning is the main cause of such significant phase transformation observed in the remaining powders left in the build chamber.

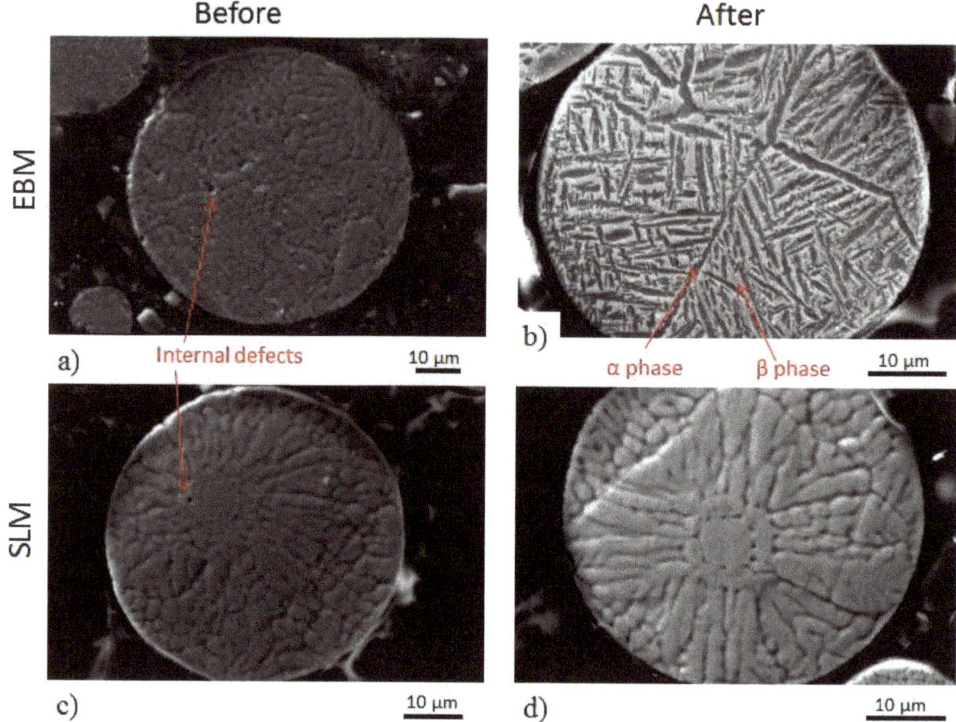

Figure 3. Internal microstructure of EBM powder (**a**) before and (**b**) after processing. Microstructure of SLM powder (**c**) before and (**d**) after processing.

The chemical composition of powder samples before and after the process was assessed by EDS analysis. As in Figure 3, the elemental map of powders before SLM and EBM process and as well as after SLM process do not show any distinct regions, and they all seemed homogeneous and featureless. However, there are two distinct regions in the elemental map of powder after the EBM process: Mo dense and Mo depleted region. Such a difference indicates a difference in phases. In Ti-alloy, Mo and V work as a β stabilizers, and Al is an α stabilizer [17]. Therefore, during phase formation, these elements will preferentially partition to the respective phases.

The diffraction patterns of powders before and after the process shown in Figure 4 confirm a microstructure only composed of the β Ti-phase. Remaining powder in the SLM build chamber after the process consisted of comparable phase composition as ingredient powder, however, the remaining powder after the EBM process showed the clear presence of α Ti-phase in addition to β Ti-phase.

It is well known that titanium and its alloys have a strong affinity for oxygen, and they can react to form detrimental oxides at elevated temperatures which potentially degrade the quality of the build parts [7]. However, from Figure 4, one can see that all powders are free from oxides. In addition, all powders did not show any observable color changes after the process. Therefore, it is safe to say that both EBM and SLM process are equally effective in preventing oxygen contamination, and powder degradation related to oxygen from both methods are minimal.

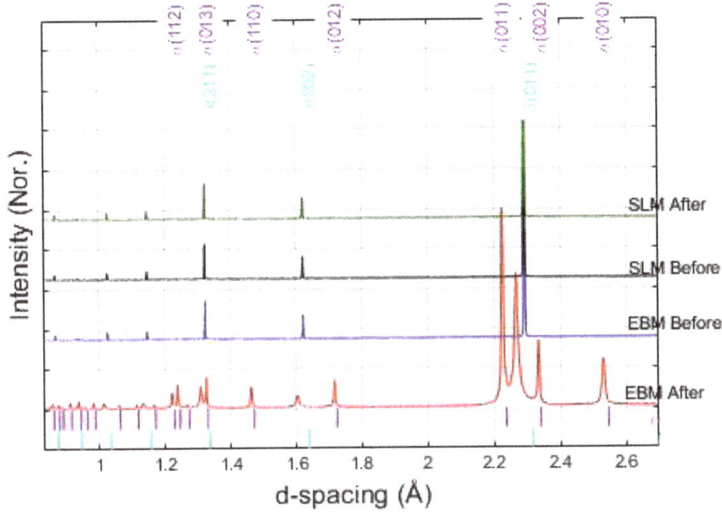

Figure 4. Diffraction pattern of powders before and after the process. These color-coded tick marks under the diffractogram correspond to the expected peak positions of α- and β-Ti phases reported in the Inorganic Crystal Structure Database (ICSD) [27] (ICSD reference number of α-phase is 191187 and β-phase is 653278).

It is commonly reported that the powder morphology [25], oxygen content [7], and PSD [25] have significant impact on the final build material quality, and for that reason these parameters are often used/discussed in the literature for evaluating the impact of an additive manufacturing process to the powder degradation behavior. However, it is not clear whether the phase composition of the ingredient powder has any influence on the porosity of the build parts. As different phases have a different crystal structure and each phase mixtures can have specific microstructures, it can be expected that the thermal/chemical properties of the powders with phase transformation can be different from the standard/initial powders. Therefore, it might be also important to include powder phase composition in the discussion of powder degradation evaluations together with other parameters.

3.2. Surface Roughness

All builds from both methods were successful, with no warping, distortion, lifting from the base and no macro/micro-scale cracking. The physical appearance of vertically built samples in the 90° orientation from SLM and EBM are shown in Figures 5a and 5b, respectively. Clear band-like patterns were observed in the SLM-built specimen (Figure 5a). A limited number of rough spikes sticking to the sides of the EBM-built specimen, as shown with arrows in Figure 5b, were observed. It is believed that these spikes might be caused by fallen agglomerated powders/melts which spatter out during scanning processes. As the distribution of these spikes is random, the occurrence is limited and can be removed relatively easily, they are excluded from the surface roughness evaluations. The roughness was measured with a Veeco Dektak 8 (NY, USA) profilometer and values of 12.27 µm and 38.05 µm were obtained for SLM and EBM, respectively. The surface roughness of the additively manufactured parts was mainly influenced by the powder particle size and build layer thickness [28,29]. In general, the smaller the particle size, the thinner the layer thickness and so the higher the surface quality. The size of the powder particles used in EBM was twice the size of SLM, so a higher surface roughness was expected for EBM.

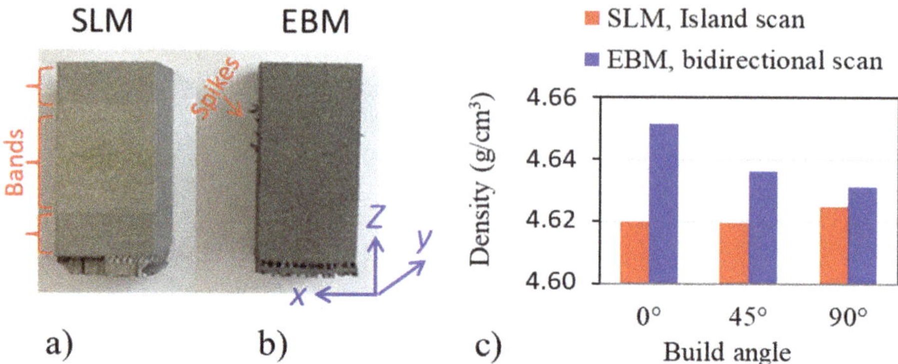

Figure 5. Photography of (**a**) SLM and (**b**) EBM samples build on 90° orientation. (**c**) Density comparison of SLM (red) and EBM (blue) samples built on three different orientations obtained from the Archimedes principle.

3.3. Porosity

Figure 5c shows the Archimedes density measurement results of samples built in three different orientations. The standard deviations of density measurement were less than 0.005 g/cm^3 for all measurements. As can be observed, EBM in general gave higher density than SLM irrespective of sample build orientation, but the difference between these two was less than one percent (0.39%). With respect to reported densities in literature, both methods can produce an almost fully dense structure (99.38 and 99.77% for SLM and EBM, respectively). Results of the porosity after image analysis from the xy- and xz-planes agreed well with the results presented in Figure 5c. The SLM sample had more pores than the EBM samples. The size of the pored ranged between 5 and 300 µm in SLM and 5 and 70 µm in EBM. The distribution of the porosity in the xy planes can be observed in Figure 6. The figures confirm the difference in porosity but also reveal differences in their distribution.

Generally, two types of pores exist in powder-based AM: spherical, gas-induced pores, and irregular-shaped, process-induced pores [7,30]. The first can occur due to the presence of entrapped gas in the powder particles during atomization, the latter mainly associated with non-optimal process parameters [30]. According to Figure 6 and results from the image analysis, pores in the EBM sample were mostly gas pore type, while in SLM both types were observed.

An interesting point to notice on the micrographs presented in Figure 6 were the pore patterns. In the EBM build sample, pores seemed to form randomly at both build direction and build planes, while in SLM, there was a clear tendency to form preferentially in both directions. In SLM most pores tended to form linearly in the build direction. In the build plane, pores occurred mostly at or near to the corners of a scanning "island" (Figure 6b,c). The reason for having such a pore formation pattern in SLM may be related to the scanning strategies in use. When the microstructure was observed after etching in the xy-plane of SLM build sample (Figure 6c), one can identify the used scanning strategy. The approximate size and locations of one "island" is shown in Figure 6c by a red square. As seen in Figure 6c, pores were indeed most prevalent at the "island" intersection or corner regions. One of the possible reasons for this is when the laser reaches the "island" border and starts to melt the next scan vector, a process associated pores, such as keyhole and lack of fusion might be formed. This problem can be mitigated or eliminated by (1) increasing the overlap between "islands", (2) introducing a shift and tilt between layers, (3) adding a contour scan with lower energy density in each "island" just before or after its completion, (4) changing the scanning speed when the scan vector approaches the "island" border.

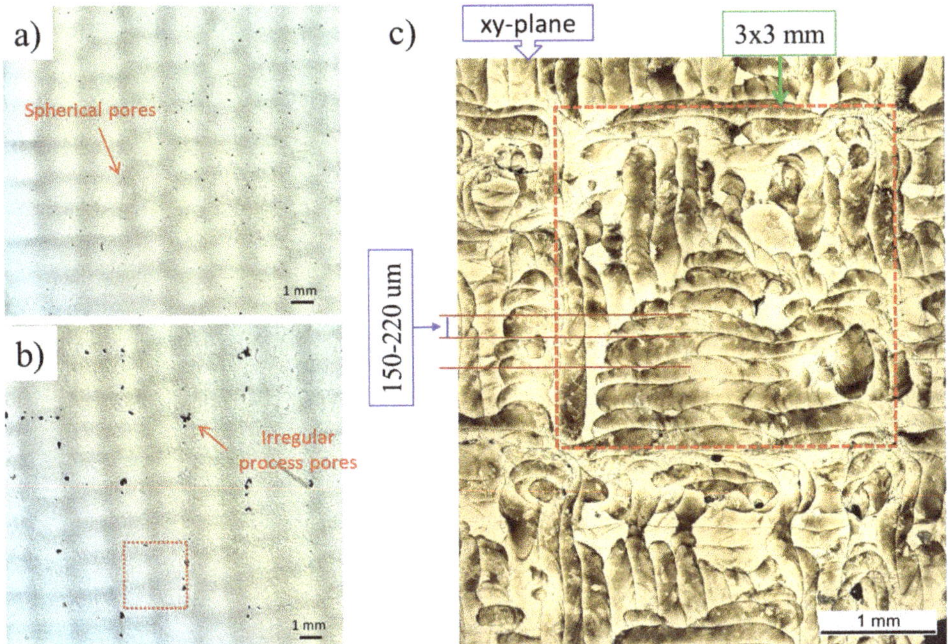

Figure 6. VLM image of (**a**) EBM and (**b**) SLM samples before etching. (**c**) VLM image of SLM sample after etching. All images are taken from the xy-plane (or build plane) of samples.

3.4. Microstructures of the Build Material

Figure 7 illustrates typical microstructural features from three perpendicular planes of SLM (first row) and EBM (second row) processed Ti55511 alloy. The orientation of the planes is indicated, and the axes are given in Figure 5b. In the xz surface plane of the SLM sample (Figure 7a), individual scan tracks and molten pool boundaries with the typical arc-shaped configuration are observed. The tracks are about 60–110 μm in thickness and 100–200 μm in width and are produced by the Gaussian-like energy distribution of the laser.

Parallel to the building direction (Z) columnar grains are visible in both SLM and EBM samples. These columns are much larger than the layer thickness. The width of the columnar grains is smaller in the SLM built sample than in the EBM (Figure 7c,f). In the xy-plane the cross-section of the columns was equiaxed in both samples (Figure 7b,e). Despite significant alloy composition differences, in the EBM sample fine grains with lamella and Widmanstätten-like structure (Figure 7d–f) similar to EBM, processed Ti-6Al-4V can be observed within the large columns. A similar microstructure is observed in the used powder (Figure 3b). Figure 8 shows an EBSD phase map and an inverse pole figure taken from the xz-plane of the EBM sample revealing the presence of about 67 wt% α phase and 33 wt% β phase. Areas of similar orientation that were likely to have originated from the same parent β grain can be recognized.

This microstructure and phase differences between EBM and SLM build samples can be ascribed to the difference in build plate temperature and cooling rates applied. Because of faster scanning speed and a high chamber temperature of the EBM, the cooling rate will be slower than the SLM. Therefore, the average temperature of the melt pool and heat affected zone in EBM will be relatively higher than in SLM which will consequently lead to grain growth and $\beta \rightarrow \alpha$ transformation.

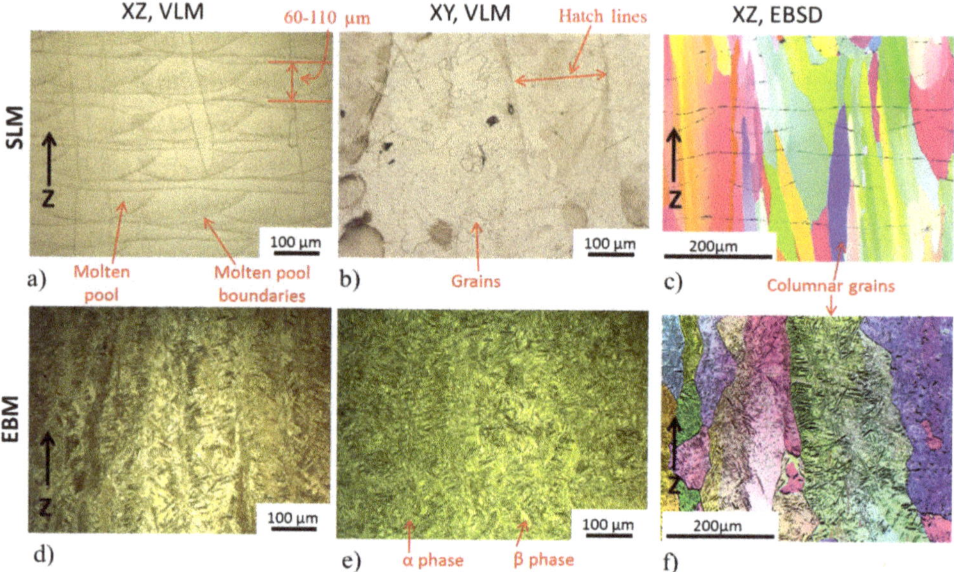

Figure 7. Microstructure of SLM (three images in first/top row) and EBM (all three images in second/bottom row) build samples. (**a**,**d**) are VLM image from the xz-plane; (**b**,**e**) are from the xy-plane; (**c**,**f**) are EBSD pattern from the xz-plane.

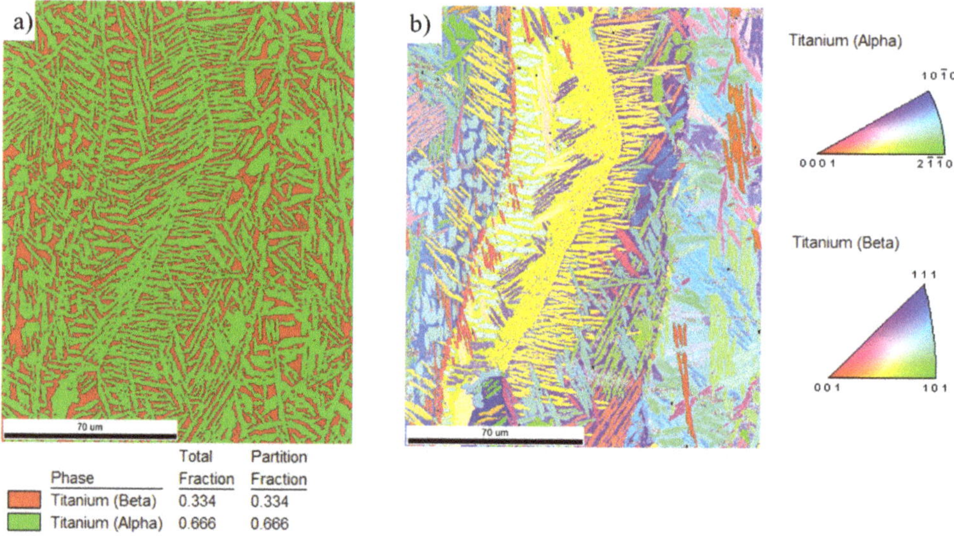

Figure 8. High-resolution EBSD (**a**) phase map and (**b**) IPF of EBM build sample.

Similar to the powder analysis, the chemical composition of the as-built samples was assessed by EDS analysis. The elemental map of the as-built SLM sample does not show any distinct regions and they all seem homogeneous and featureless. In the as-built EBM sample, however, there are two distinct regions in the elemental map like what was observed in the EBM powders after processing.

The XRD analysis, performed on the xy-plane of the as-built samples, confirmed that the SLM consists predominantly of beta phase but reveals also the presence of a limited amount of α or α' phase (<3 wt%) as shown in Figure 9a. Because of limited quantity and the almost identical unit cell parameters of α and α' phase, it is not possible to distinguish between them. According to the literature, the α' phase commonly associated with the extreme temperature change is usually observed in the SLM build Ti-6Al-4V [1,8,17]. The α phase, on the other hand, is associated with an isothermal cooling condition [2,10,17,31] and is commonly observed in the EBM build Ti-6Al-4V. Therefore, it is believed that that the minority phase present in the SLM build sample might be α' phase.

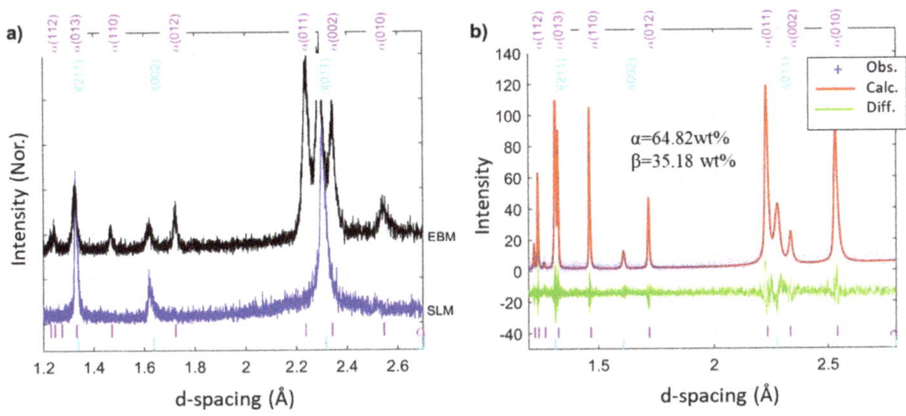

Figure 9. (**a**) Diffraction pattern of as-built samples. These color-coded tick marks under the diffractogram correspond to the expected peak positions of reported α- and β-Ti phases. (**b**) Rietveld refinement result of XRD data from EBM build sample.

The XRD pattern of the EBM sample confirms the presence of α and β phases. Rietveld analysis (Figure 9b) was performed and the lattice constants of hexagonal close packed α phase determined to be a = 2.931(7) Å and c = 4.658(8) Å, respectively, with c/a ratio of 1.5891. For the body-centered cubic β phase, a = 3.230(9) Å. As shown in the figure, the phase fraction of the α phase is 64.82 wt% and the β phase is 35.18%. This result supports the observation from the SEM.

The mechanical properties of a Ti-alloys are strongly determined by their microstructure and phase composition. Measured microhardness of various samples seemed to match with literature. The mean microhardness profiles of the SLM and EBM build samples were 315 ± 3.41 HV and 348 ± 30.20 HV, respectively. Increased microhardness of the EBM build sample can be attributed to the amount of α phase present. From the values of standard deviations of microhardness, it is evident that the SLM build sample has relatively consistent microhardness while the EBM samples show a larger deviation. This variation indicates that the SLM sample has a uniform microstructure throughout while the EBM sample has an inhomogeneous microstructure. This agrees well with the microstructure presented in Figures 7 and 8. Here it is also important to point out that despite apparent differences in alloy chemical composition, the microhardness of the EBM build sample is very similar to α phase dominated EBM build Ti-6Al-4V reported by Neikter et al. [32]. Such similarity can be explained by the similar $\alpha + \beta$ microstructure observed in both types of alloys after the EBM process. The microhardness of the EBM build sample seemed to slightly lower (but same in standard deviation) than what Kurzynowski et al. [13] reported for EBM build Ti55511, previously.

As hardness of the material is the resistance of the material to plastic deformation, one can expect that the EBM build sample may show higher strength than the SLM build samples. However, this needs further investigation.

4. Conclusions

The Ti55511 parts were successfully fabricated using SLM and EBM techniques. The results indicate that both methods can process Ti55511 and achieve almost full density with limited porosity. For the process parameters used in this study, the SLM process gives a slightly lower density and better surface quality. The shape of the pores in the EBM production appears to be mostly spherical while more random shapes are observed in SLM. The dominant β phase in the original powder becomes a minority phase after EBM processing, while there is almost no phase transformation in the SLM. Because of high build temperature and relatively slow cooling rate, the EBM build samples show a lamella and Widmanstätten-like structure similar to the microstructure observed in EBM processed Ti-6Al-4V despite it is alloy composition. Because of the lamella microstructure, the EBM build sample showed about 10% higher microhardness than the SLM build samples and it measured as 348 ± 30.20 HV. To achieve near-β phase composition in Ti55511 after EBM processing, the current processing route needs to be optimized or the build part needs additional post-heat treatments.

Author Contributions: Conceptualization, T.M.; Data curation, T.M.; Formal analysis, T.M. and H.V.S.; Funding acquisition, H.V.S.; Investigation, T.M. and K.M.; Methodology, T.M., K.M. and N.C.; Resources, K.M., T.K. and H.V.S.; Supervision, T.K. and H.V.S.; Visualization, T.M.; Writing—original draft, T.M. and H.V.S. All authors have read and agreed to the published version of the manuscript.

Funding: This research was funded by the project PREMPA, a Strategic Focus Area project of the ETH board in Switzerland.

Acknowledgments: Tuerdi Maimaitiyili and Van Swygenhoven Helena thank the financial support of the project PREMPA, a Strategic Focus Area project of the ETH board. Tuerdi Maimaitiyili thanks Haydous Fatima, Martin Elsener, and Miroslav Smid at the Paul Scherrer Institut (PSI) for their help with surface roughness, PSD, and EBSD measurements, respectively. Tuerdi Maimaitiyili also thanks Christof W. Schneider at PSI for the XRD measurements of as-built specimens.

Conflicts of Interest: The authors declare no conflict of interest.

References

1. Hocine, S.; Swygenhoven, H.V.; Petegem, S.V.; Chang, C.S.T.; Maimaitiyili, T.; Tinti, G.; Sanchez, D.F.; Grolimund, D.; Casati, N. Operando X-ray diffraction during laser 3D printing. *Mater. Today* **2019**. [CrossRef]
2. Maimaitiyili, T.; Woracek, R.; Neikter, M.; Boin, M.; Wimpory, R.C.; Pederson, R.; Strobl, M.; Drakopoulos, M.; Schafer, N.; Bjerken, C. Residual Lattice Strain and Phase Distribution in Ti-6Al-4V Produced by Electron Beam Melting. *Materials* **2019**, *12*, 667. [CrossRef] [PubMed]
3. Yap, C.Y.; Chua, C.K.; Dong, Z.L.; Liu, Z.H.; Zhang, D.Q.; Loh, L.E.; Sing, S.L. Review of selective laser melting: Materials and applications. *Appl. Phys. Rev.* **2015**, *2*, 041101. [CrossRef]
4. Herzog, D.; Seyda, V.; Wycisk, E.; Emmelmann, C. Additive manufacturing of metals. *Acta Mater.* **2016**, *117*, 371–392. [CrossRef]
5. Murr, L.E.; Esquivel, E.V.; Quinones, S.A.; Gaytan, S.M.; Lopez, M.I.; Martinez, E.Y.; Medina, F.; Hernandez, D.H.; Martinez, E.; Martinez, J.L.; et al. Microstructures and mechanical properties of electron beam-Rapid manufactured Ti–6Al–4V biomedical prototypes compared to wrought Ti–6Al–4V. *Mater. Charact.* **2009**, *60*, 96–105. [CrossRef]
6. Harun, W.S.W.; Manam, N.S.; Kamariah, M.S.I.N.; Sharif, S.; Zulkifly, A.H.; Ahmad, I.; Miura, H. A review of powdered additive manufacturing techniques for Ti-6al-4v biomedical applications. *Powder Technol.* **2018**, *331*, 74–97. [CrossRef]
7. Liu, S.; Shin, Y.C. Additive manufacturing of Ti6Al4V alloy: A review. *Mater. Des.* **2019**, *164*, 107552. [CrossRef]

8. Al-Bermani, S.S.; Blackmore, M.L.; Zhang, W.; Todd, I. The Origin of Microstructural Diversity, Texture, and Mechanical Properties in Electron Beam Melted Ti-6Al-4V. *Metall. Mater. Trans. A* **2010**, *41*, 3422–3434. [CrossRef]
9. Kenel, C.; Grolimund, D.; Li, X.; Panepucci, E.; Samson, V.A.; Sanchez, D.F.; Marone, F.; Leinenbach, C. In situ investigation of phase transformations in Ti-6Al-4V under additive manufacturing conditions combining laser melting and high-Speed micro-X-Ray diffraction. *Sci. Rep.* **2017**, *7*, 16358. [CrossRef]
10. Neikter, M.; Woracek, R.; Maimaitiyili, T.; Scheffzük, C.; Strobl, M.; Antti, M.-L.; Åkerfeldt, P.; Pederson, R.; Bjerkén, C. Alpha texture variations in additive manufactured Ti-6Al-4V investigated with neutron diffraction. *Addit. Manuf.* **2018**, *23*, 225–234. [CrossRef]
11. Safdar, A.; He, H.Z.; Wei, L.; Snis, A.; Chavez, D.P. Effect of process parameters settings and thickness on surface roughness of EBM produced Ti-6Al-4V. *Rapid Prototyp. J.* **2012**, *18*, 401–408. [CrossRef]
12. Phutela, C.; Aboulkhair, T.N.; Tuck, J.C.; Ashcroft, I. The Effects of Feature Sizes in Selectively Laser Melted Ti-6Al-4V Parts on the Validity of Optimised Process Parameters. *Materials* **2019**, *13*, 117. [CrossRef] [PubMed]
13. Kurzynowski, T.; Madeja, M.; Dziedzic, R.; Kobiela, K. The Effect of EBM Process Parameters on Porosity and Microstructure of Ti-5Al-5Mo-5V-1Cr-1Fe Alloy. *Scanning* **2019**, *2019*, 2903920. [CrossRef] [PubMed]
14. Neikter, M.; Colliander, M.; Schwerz, C.d.A.; Hansson, T.; Åkerfeldt, P.; Pederson, R.; Antti, M.-L. Fatigue Crack Growth of Electron Beam Melted Ti-6Al-4V in High-Pressure Hydrogen. *Materials* **2020**, *13*, 1287. [CrossRef]
15. Xie, Z.; Dai, Y.; Ou, X.; Ni, S.; Song, M. Effects of selective laser melting build orientations on the microstructure and tensile performance of Ti-6Al-4V alloy. *Mater. Sci. Eng. A* **2020**, *776*, 139001. [CrossRef]
16. Shipley, H.; McDonnell, D.; Culleton, M.; Coull, R.; Lupoi, R.; O'Donnell, G.; Trimble, D. Optimisation of process parameters to address fundamental challenges during selective laser melting of Ti-6Al-4V: A review. *Int. J. Mach. Tools Manuf.* **2018**, *128*, 1–20. [CrossRef]
17. Gerd, L.; James, C.W. *Titanium*, 2nd ed.; Springer: Heidelberg, Germany, 2007.
18. Polmear, I.; StJohn, D.; Nie, J.; Qian, M. Light Alloys: Metallurgy of the Light Metals, Chapter 7-Titanium Alloys. In *Light Alloys: Metallurgy of the Light Metals*, 5th ed.; Butterworth-Heinemann: Boston, MA, USA, 2017; pp. 369–460. [CrossRef]
19. Tan, X.; Kok, Y.; Toh, W.Q.; Tan, Y.J.; Descoins, M.; Mangelinck, D.; Tor, S.B.; Leong, K.F.; Chua, C.K. Revealing martensitic transformation and α/β interface evolution in electron beam melting three-Dimensional-Printed Ti-6Al-4V. *Sci. Rep.* **2016**, *6*, 26039. [CrossRef]
20. Zhao, C.; Fezzaa, K.; Cunningham, R.W.; Wen, H.; De Carlo, F.; Chen, L.; Rollett, A.D.; Sun, T. Real-Time monitoring of laser powder bed fusion process using high-Speed X-Ray imaging and diffraction. *Sci. Rep.* **2017**, *7*, 3602. [CrossRef]
21. Wang, K.; Bao, R.; Zhang, T.; Liu, B.; Yang, Z.; Jiang, B. Fatigue crack branching in laser melting deposited Ti55511 alloy. *Int. J. Fatigue* **2019**, *124*, 217. [CrossRef]
22. Boyer, R.R.; Briggs, R.D. The use of beta titanium alloys in the aerospace industry. *J. Mater. Eng. Perform.* **2005**, *14*, 681–685. [CrossRef]
23. Ahmed, M.; Savvakin, D.G.; Ivasishin, O.M.; Pereloma, E.V. The effect of cooling rates on the microstructure and mechanical properties of thermo-Mechanically processed Ti–Al–Mo–V–Cr–Fe alloys. *Mater. Sci. Eng. A* **2013**, *576*, 167–177. [CrossRef]
24. Ran, C.; Chen, P.; Li, L.; Zhang, W. Dynamic shear deformation and failure of Ti-5Al-5Mo-5V-1Cr-1Fe titanium alloy. *Mater. Sci. Eng. A* **2017**, *694*, 41–47. [CrossRef]
25. Levy, G.; Herres, N.; Spierings, A.B. Influence of the particle size distribution on surface quality and mechanical properties in AM steel parts. *Rapid Prototyp. J.* **2011**, *17*, 195–202. [CrossRef]
26. Li, C.; Zhang, X.; Zhou, K.; Peng, C. Relationship between lamellar α evolution and flow behavior during isothermal deformation of Ti–5Al–5Mo–5V–1Cr–1Fe near β titanium alloy. *Mater. Sci. Eng. A* **2012**, *558*, 668–674. [CrossRef]
27. Belsky, A.; Hellenbrandt, M.; Karen, V.L.; Luksch, P. New developments in the Inorganic Crystal Structure Database (ICSD): Accessibility in support of materials research and design. *Acta Crystallogr. Sect. B* **2002**, *58*, 364–369. [CrossRef] [PubMed]
28. Santos, E.C.; Shiomi, M.; Osakada, K.; Laoui, T. Rapid manufacturing of metal components by laser forming. *Int. J. Mach. Tools Manuf.* **2006**, *46*, 1459–1468. [CrossRef]

29. Algardh, J.K.; Horn, T.; West, H.; Aman, R.; Snis, A.; Engqvist, H.; Lausmaa, J.; Harrysson, O. Thickness dependency of mechanical properties for thin-Walled titanium parts manufactured by Electron Beam Melting (EBM)®. *Addit. Manuf.* **2016**, *12 Pt A*, 45–50. [CrossRef]
30. Sames, W.J.; List, F.A.; Pannala, S.; Dehoff, R.R.; Babu, S.S. The metallurgy and processing science of metal additive manufacturing. *Int. Mater. Rev.* **2016**, *61*, 315–360. [CrossRef]
31. Nan, Y.; Ning, Y.; Liang, H.; Guo, H.; Yao, Z.; Fu, M.W. Work-Hardening effect and strain-Rate sensitivity behavior during hot deformation of Ti–5Al–5Mo–5V–1Cr–1Fe alloy. *Mater. Des.* **2015**, *82*, 84–90. [CrossRef]
32. Neikter, M.; Åkerfeldt, P.; Pederson, R.; Antti, M.-L.; Sandell, V. Microstructural characterization and comparison of Ti-6Al-4V manufactured with different additive manufacturing processes. *Mater. Charact.* **2018**, *143*, 68–75. [CrossRef]

Article

Can Appropriate Thermal Post-Treatment Make Defect Content in as-Built Electron Beam Additively Manufactured Alloy 718 Irrelevant?

Sneha Goel [1,*], Kévin Bourreau [2], Jonas Olsson [1], Uta Klement [3] and Shrikant Joshi [1]

[1] Department of Engineering Science, University West, Trollhättan 46186, Sweden; jonas.olsson@hv.se (J.O.); shrikant.joshi@hv.se (S.J.)

[2] Specialty Materials, University of Limoges, Limoges 87000, France; kevinbourreau@live.fr

[3] Department of Industrial and Materials Science, Chalmers University of Technology, Gothenburg 41296, Sweden; uta.klement@chalmers.se

* Correspondence: sneha.goel@hv.se

Received: 5 December 2019; Accepted: 21 January 2020; Published: 23 January 2020

Abstract: Electron beam melting (EBM) is gaining rapid popularity for production of complex customized parts. For strategic applications involving materials like superalloys (e.g., Alloy 718), post-treatments including hot isostatic pressing (HIPing) to eliminate defects, and solutionizing and aging to achieve the desired phase constitution are often practiced. The present study specifically explores the ability of the combination of the above post-treatments to render the as-built defect content in EBM Alloy 718 irrelevant. Results show that HIPing can reduce defect content from as high as 17% in as-built samples (intentionally generated employing increased processing speeds in this illustrative proof-of-concept study) to <0.3%, with the small amount of remnant defects being mainly associated with oxide inclusions. The subsequent solution and aging treatments are also found to yield virtually identical phase distribution and hardness values in samples with vastly varying as-built defect contents. This can have considerable implications in contributing to minimizing elaborate process optimization efforts as well as slightly enhancing production speeds to promote industrialization of EBM for applications that demand the above post-treatments.

Keywords: additive manufacturing; electron beam melting; defects; microstructure; hardness; alloy 718; hot isostatic pressing; post-treatment

---

1. Introduction

Electron beam melting (EBM) is a powder bed fusion-based metal additive manufacturing (AM) technique bearing capability to produce components with high design flexibility. The material systems which can significantly benefit from EBM technology include high-performance materials such as Ni-based superalloys, e.g., Alloy 718, and Ti-alloys [1,2]. In the beginning, a vast majority of the research activity relating to EBM was focused on Ti and its alloys, particularly Ti-6Al-4V [3,4]. Since incorporation of defects (typically gas and shrinkage porosity, and lack of fusion) is a concern during EBM production, there have been several reported efforts with emphasis on process understanding and optimization to curtail defect formation in EBM manufactured Ti-6Al-4V [5–8]. Moreover, use of unoptimized process parameters has been shown to result in extensive defect formation [8].

Such defect formation in EBM Alloy 718 can also have particularly adverse consequences, as defects can degrade mechanical behavior of the EBM-built material. For instance, anisotropy in tensile behavior of heat-treated EBM Alloy 718 has been attributed to remnant defects [9,10]. The formation of the above defects is not only attributable to the use of unoptimized process parameters but can also occur due to stochastic instabilities in the process. These can include a temporary decrease in electron beam power, non-uniform spread of powder, poor sintering of the powder prior to melting, spattering of powder from the build layer due to excessive electrostatic forces in an event called "smoking", etc. [11–13]. All of these have been reported to cause formation of lack of fusion defects [13]. Consequently, a set of post-treatments are typically considered to enhance the properties of the as-built material [14–16] and are deemed particularly necessary in the case of the demanding applications that Alloy 718 is routinely employed for.

In this context, even with use of optimized parameters, defects, to some degree, are inherent to EBM processed Alloy 718 [17]. Consequently, the as-built material is subjected to hot isostatic pressing (HIPing) to close the defects present in the material [18]. In addition, the HIP'ed material is typically subjected to solution treatment and aging to achieve the required phase composition. In this context, it is pertinent to mention that the typical recommended parameters for HIPing are 1120–1185 °C, 100 MPa, 4 h, to be followed by a 1 h solutionizing treatment and a two-step aging involving a 8 h treatment for each step, which is perhaps "borrowed" from what has been the practice with wrought Alloy 718 [19]. Therefore, a holistic approach of optimizing the process, together with applied post-treatments, is relevant from a technology industrialization standpoint. In this context, prior work in the authors' group has shown that the aging treatment can potentially be shortened considerably to (4 + 1) h instead of (8 + 8) h [20]. Reduction in processing time and costs are also pivotal to industrial exploitation of the EBM technology. Sames et al. [18] have reported that a major limitation is the processing speed. However, faster production speeds are known to lead to poor surface finish and/or increased defect content [18,21]. Specifically, in cases where such post-treatment involving HIPing is deemed inevitable, there could be an opportunity to increase the EBM processing speed without being constrained by the need to eliminate or minimize defects.

The present study explores the potential of the combined post-treatments involving HIPing, solutionizing and aging treatment to test the hypothesis of whether they can render the as-built defect content in EBM Alloy 718 irrelevant. The defect content in the samples was intentionally tailored by systematically increasing the hatch line spacing (LS), which refers to the spacing between raster scanning lines during EBM processing. The extent of defects thus incorporated far exceeded those typically encountered in EBM builds for the sake of this illustrative proof-of-concept study.

2. Experimental Procedure

Samples were EBM built with different defect contents by varying LS to assess their response to identical post-treatments, particularly HIPing. Plasma atomized Alloy 718 powder (AP&C, Québec, Canada) [5] recycled ~20 times was used. Six different samples (15 × 15 × 15 mm^3) (Figure 1), henceforth referred to as AB#1 to AB#6, were built using EBM (Model A2X, Arcam AB, Gothenburg, Sweden) with varying LS corresponding to different processing speeds (expressed as relative melting time and relative melted area per unit time by normalizing with the "standard" values), as summarized in Table 1. All the samples were melted by a snake hatch method during which the electron beam was moved in a back-and-forth raster scan pattern [18]. All the other process parameters were kept constant as per the EBM process theme version 4.2.205, and some of the key process parameters associated with this theme are listed in Table 2. It is worth mentioning that the parameter speed function dynamically adjusts the electron beam scan speed depending on the scan length [22].

Selected samples were further subjected to two different HIPing treatments (HIP1, HIP2). The choice of parameters for HIP1 (1120 °C/100 MPa/4 h/rapid cooling) and HIP2 (1185 °C/100 MPa/4 h/rapid cooling) was based on recommendations of the ASTM (F3055) standard for post-treatment of powder bed fusion-produced Alloy 718 [19]. For one of the HIPing conditions selected based on ensuing results, the samples were also subjected to solution treatment (954 °C/1 h/water cooling), and double aging (740 °C/4 h, cool at 55 °C/h to 635 °C, held at 635 °C/1 h/air cooling) called HIP1 + STA. Details of the above mentioned post-treatments are summarized in Table 3. It is relevant to mention here that the aging protocol reflected above corresponds to a shortened heat treatment compared to the schedule recommended in the above ASTM standard and is based on prior work carried out in this group [20].

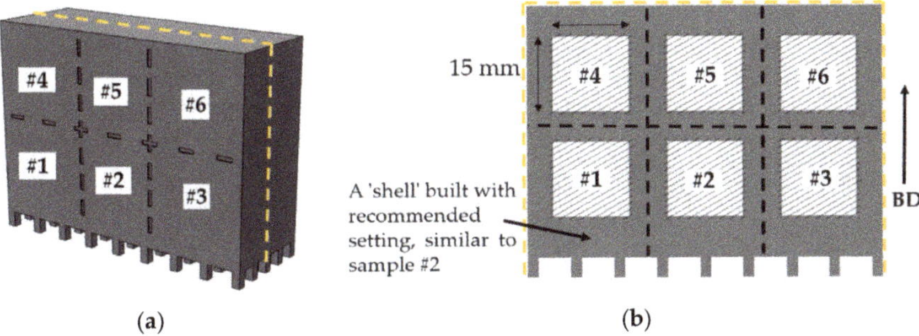

Figure 1. Schematic of the CAD model of one of multiple identical blocks comprising the electron beam melting (EBM) build (**a**), and cross-section view of the block revealing the cubes (marked by hatched lines) processed with different line spacing (**b**). The arrow on the right shows the build direction, BD.

Table 1. EBM process parameters.

Sample Nomenclature	Hatch Line Spacing (LS, μm)	Relative Melting Time Per Layer [2]	Relative Melted Area Per Unit Time [2]
#1	75	167%	0.6
#2 [1]	125 *	100%	1 [1]
#3	175	71%	1.4
#4	225	56%	1.8
#5	275	45%	2.2
#6	325	38%	2.6

[1] Arcam recommended setting. [2] Values normalized with respect to the Arcam recommended setting.

Table 2. Process parameters in the Inco 4.2.205 theme for EBM Alloy 718.

Parameter	Value
Layer thickness (μm)	75
Acceleration voltage (kV)	60
Hatch scan rotation (°)	72
Pre-heat temperature (°C)	1025
Speed function	63
Focus offset (mA)	15
Max. beam current (mA)	18

Table 3. Details of post-treatments.

Designation	Cycle
HIP1	1120 °C/100 MPa/4 h/RC
HIP2	1185 °C/100 MPa/4 h/RC
HIP1 + STA	HIP1: 1120°C/100 MPa/4 h/RC +STA: 954 °C/1 h/WC to RT, 740° C/4 h/cool at 55 °C/h to 635 °C, hold at 635 °C/1 h/AC to RT

Note: The abbreviations RC, WC, AC, and RT denote rapid cooling, water cooling, air cooling, and room temperature, respectively.

All the investigated samples were sectioned along the build direction using an alumina cutting blade. The sample cross-sections, each 15 × 15 mm^2, were hot mounted, ground, and polished. Microstructural investigation of the prepared samples was carried out using an Olympus BX60 M optical microscope (OM), and HITACHI TM3000 (equipped with energy-dispersive X-ray spectroscopy (EDS)) and LEO 1550 Gemini scanning electron microscopes (SEMs). Quantification of defects was done using OM micrographs of unetched polished samples processed using ImageJ software. In each case, more than 10 images were analyzed to get a representative value of the defect content (magnification: 50×, analyzed area: ~21 mm^2). For characterization of secondary phases using SEM imaging, the selected samples were electrolytically etched using 50% diluted (in ethanol) Kalling's 2 reagent (2–3 V applied for 3–5 s). Vickers micro-hardness testing (HMV-2, Shimadzu Corp., Japan) on selected samples was performed using a 500 g load which was applied for 15 s in ambient conditions. Up to 25 random indents were made on each of the samples. Since some of the defect sizes in the intentionally prepared high-defect-content builds were very large (as shown in Figure 2) and exceeded the indent size, few extreme outliers exhibiting "zero" hardness were noted whenever the indent happened to coincide with or be in the immediate vicinity of a very large defect. Average hardness values (excluding only such extreme outliers) are reported herein.

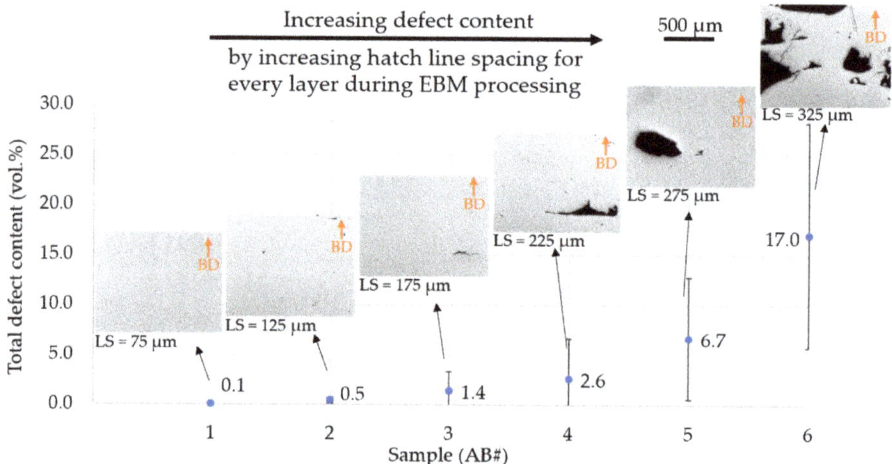

Figure 2. Defect content in as-built samples intentionally built with varied hatch line spacing (LS). The size and amount of defects are visualized in the insets. Arrows in the inset indicate build direction.

3. Results and Discussion

Since the main theme of the paper was to assess whether defects in the as-built condition matter if the subsequent post-treatment steps involve HIPing in any case, results pertaining to as-built conditions are presented first, followed by their response to post-treatments.

The as-built microstructures of samples processed with varying LS corresponding to different melting speeds (Table 1) are depicted in Figure 2. Increase in LS above the "recommended" value resulted in a rise in the defect content, as shown in Figure 2. The defect contents stated represent the sum of all types of defects observed in the as-built samples, i.e., gas and shrinkage porosity, and lack of fusion. Incidentally, these defects were also present in samples produced at the recommended LS = 125 μm (AB#2) as well as at lower LS = 75 μm (AB#1). Samples intentionally produced with higher LS exhibited defect contents up to 17%, as shown in Figure 2 and also visualized in accompanying micrographs. In a word, a systematic increase in line offset (processing speed) from 75 to 325 μm resulted in an increased defect content ranging from 0.1% to 17% in the investigated samples. It is worth mentioning that large defects such as those observed in samples AB#5 (LS = 275 μm) and AB#6 (LS = 325 μm) are not typical of an EBM built even with not fully optimized processing conditions [23]. However, these samples were produced for the present 'proof of concept' investigation regarding the extent of defect closure enabled through HIPing.

Samples with distinct defect contents, i.e., AB#2, AB#4, and AB#6 were subjected to two HIPing treatments to investigate the efficacy of defect closure. The defect contents in as-built samples AB#2, AB#4, and AB#6 (corresponding to LS = 125, 225, and 325 μm, respectively) were 0.5%, 2.6%, and 17%, respectively. After both the HIPing treatments, the defect content in all samples was significantly reduced, regardless of the original extent, as summarized in Table 4. Moreover, the difference in the extent of defect closure accomplished at 1120 °C (HIP1) and 1185 °C (HIP2) was not significant. The considerable densification of samples achieved by HIPing is also visualized from the microstructures in Figure 3. Furthermore, the small amount of remnant defects in all HIP'ed samples discussed were found to be typically associated with the presence of Al-rich oxide, which can inhibit complete defect closure. This is clearly evident from the representative EDS results given in Figure 4. It is worth mentioning that Al-rich oxides are thermodynamically very stable and not removed by HIPing, as previously observed in HIP'ed EBM Alloy 718 [24].

Table 4. Defect content (vol.%) in various samples in as-built and HIP'ed conditions.

Sample	Condition		
	As-built	HIP1	HIP2
#2 (LS = 125 μm)	0.5 ± 0.3	0.1	0.1
#4 (LS = 225 μm)	2.6 ± 4.1	0.2 ± 0.1	0.3 ± 0.2
#6 (LS = 325 μm)	16.9 ± 11.3	0.2 ± 0.1	0.2 ± 0.1

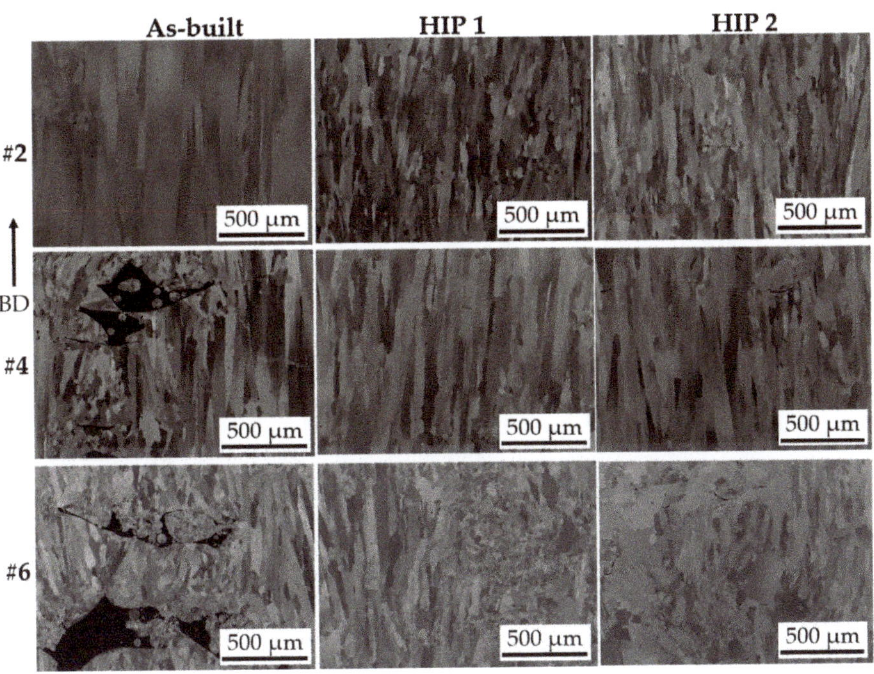

Figure 3. SEM micrographs showing defects in samples #2 (LS = 125 μm), #4 (LS = 225 μm), and #6 (LS = 325 μm) in as-built and HIP'ed conditions. The arrow indicates the build direction.

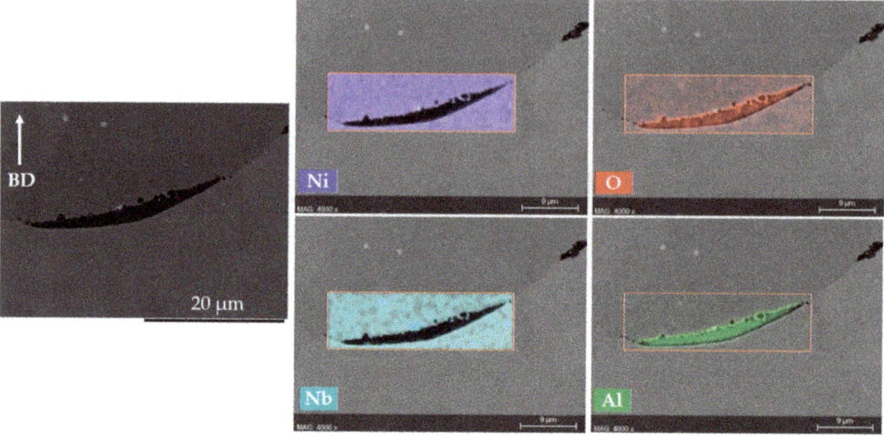

Figure 4. Representative EDS analysis showing inclusions typically associated with many remnant defects in sample #4 (LS = 225 μm) after HIP1. The arrow indicates the build direction.

It is pertinent to note that concerns regarding thermally induced porosity (TIP), i.e., regrowth of apparently "closed" defects during subsequent heat treatment, have been reported [25,26]. Therefore, the samples HIP1#2, #4, and #6 were also subjected to heat treatment at 1180° C/4 h under vacuum, as previously studied by Benn et al. [26]. The results showed no significant difference in defect content before and after the heat treatment. Therefore, the TIP phenomenon was not discernible in the present HIP'ed EBM built Alloy 718 samples, at least to a statistically discernible limit, thereby indicating the efficacy of defect closure achieved by HIPing. It is worth mentioning that TIP plausibly did not happen in the present case because of insignificant gas entrapment in the defects owing to the vacuum conditions employed during EBM processing [27]. This is in contrast to LPBF processing, during which there have been concerns related to process gas infiltration because the builds are carried out in an inert atmosphere, typically using argon [28].

Such extensive and efficient reduction in the defect content by orders of magnitude (as evident from Table 4 and Figure 3) *prima facie* provides an opportunity to also potentially increase productivity—at least to some reasonable extent—during EBM processing. It is worth mentioning that the melting step typically comprises ~35% (the exact value depends on the build design [22]) of the layer heating time (preheat + melt + post-heat), which corresponds to 15% of the total layer processing time (also including lowering of build platform + powder raking) spent during the EBM process. Therefore, decreasing the melting time (by increasing the LS, as done in this study) would also have a corresponding effect on the overall EBM processing time.

Consistent with the idea of taking a holistic approach to optimize the processing chain for EBM-built Alloy 718 components, the present study also sought to evaluate the role of shortened solutionizing and aging treatment (mentioned earlier) following HIPing. For this purpose, the samples HIP1#2, HIP1#4, and HIP1#6 were first subjected to a standard solution treatment followed by a shorter aging treatment (Table 3) in comparison to ASTM F3055 [19] specification, as previously suggested in a recent study carried out in the authors' group [20]. The shortened aging treatment refers to a (740 °C/4 h cool at 55 °C/h to 635 °C, hold at 635 °C/1 h/AC to RT) cycle compared to the conventional (740 °C/8 h cool at 55 °C/h to 635 °C, hold at 635 °C/8 h/AC to RT) cycle suggested in the ASTM F3055 [19] specification.

Since the solution treatment was carried out at 954 °C, all the HIP1 + STA samples #2, #4, and #6 (corresponding to LS = 125, 225, and 325 μm, respectively) exhibited a virtually similar distribution of grain boundary δ phase, as shown in Figure 5a–c. This is in accordance with the reported time–temperature-transformation diagram of Alloy 718 [29], and has also been previously observed in solution-treated EBM Alloy 718 that was not subjected to prior HIPing [10]. The effect of aging treatment can be readily seen from the similar distribution of γ'' phase as shown in Figure 6a–c. These microstructural similarities in the three samples (HIP1 + STA #2, #4, #6) were also reflected in the microhardness results, as described below.

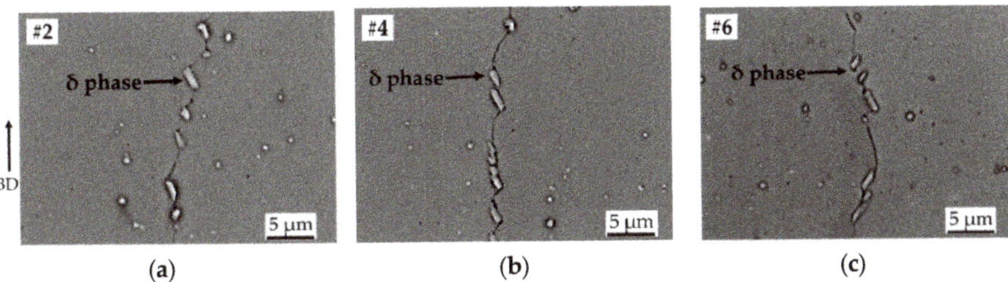

Figure 5. SEM micrographs of HIP1 + STA samples (**a**) LS = 125 μm, (**b**) LS = 225 μm, and (**c**) LS = 325 μm showing similar distribution of the δ phase. The arrow indicates the build direction.

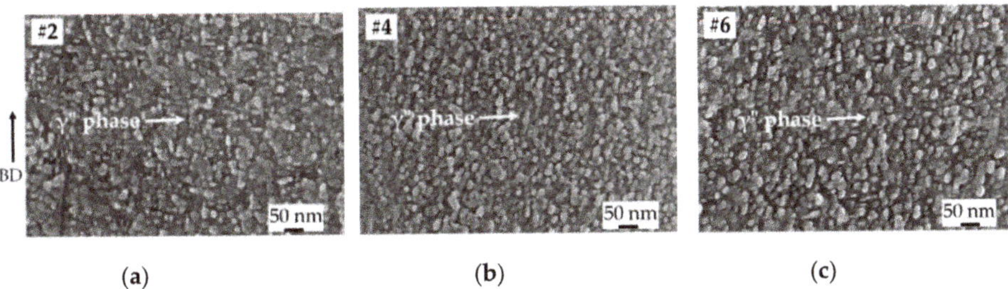

Figure 6. SEM micrographs of HIP1 + STA samples (**a**) LS = 125 μm, (**b**) LS = 225 μm, and (**c**) LS = 325 μm showing similar distribution of the γ'' phase. The arrow indicates the build direction.

The average microhardness values of the as-built and post-treated samples, determined as previously described in the experimental procedure, are shown in Figure 7. It can be seen that, although the defect contents in the as-built samples were vastly different, they exhibit similar hardness values of ~430 kgf/mm^2. This can be explained by the similarity in local melting conditions for the three as-built specimens, as the electron beam parameters were unchanged, and only the line spacing was varied. This contrasts a previous reported study by Lee et al. [23], where hardness was observed to vary (405–450 kgf/mm^2) with defect content (4.5%–0.5%) of the EBM Alloy 718 samples. In the study by Lee et al. the local melting conditions were varied between the different specimens by changing the focus offset, thereby influencing energy density. The HIP1 treatment caused a reduction in the hardness of all the samples to nearly identical values of ~200 kgf/mm^2. This is consistent with the previous observations on EBM and LPBF Alloy 718 [30,31]. After HIP1 + STA, all the samples exhibited similar hardness values (~475 kgf/mm^2), which were higher than in the as-built condition. This value is close to the reported peak hardness (~490 kgf/mm^2) of Alloy 718 [10]. Previous studies on EBM Alloy 718 have also reported an increase in hardness in the as-built condition from 410 kgf/mm^2 to 470 kgf/mm^2 after aging [32–34]. Thus, the foregoing results amply reveal that, regardless of the vastly different defect content in the as-built condition, the final post-treated condition exhibits a very similar density (Figure 3), phase constitution in terms of δ phase and γ'' phase distribution (Figures 5 and 6), as well as microhardness values (Figure 7).

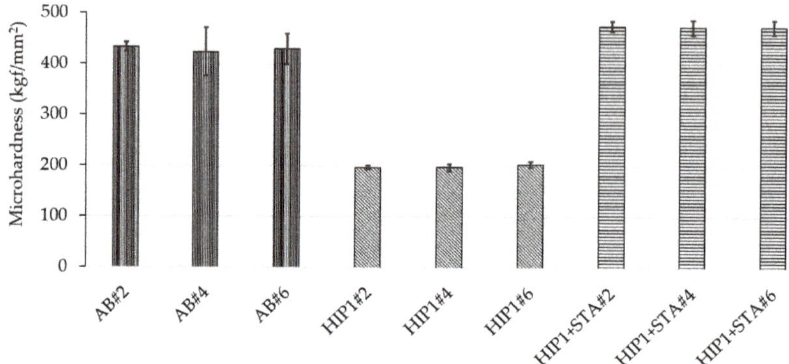

Figure 7. Microhardness values of the as-built and post-treated samples #2 (LS = 125 μm), #4 (LS = 225 μm), #6 (LS = 325 μm).

4. Conclusions

The present study on EBM Alloy 718 intentionally built with varied defect contents (by increasing hatch line spacing) clearly shows that regardless of the quantum of defects in the as-built condition, after HIPing, the material exhibits a nearly completely densified structure. The results are suggestive of the possibility of appropriate thermal post-treatments nullifying the influence of defects incorporated during EBM processing, thereby reducing elaborate process optimization efforts. A spin-off prospect of these findings can also be an opportunity to EBM print parts somewhat faster compromising defect content, if the production protocol is to involve an ensuing HIPing step in particular. In addition, subsequent solution treatment and 'shortened' aging compared to the conventional long aging yields a phase constitution and hardness that are virtually identical despite a vast difference in as-built defect contents. The resultant hardness was close to the peak hardness of Alloy 718. The above demonstrates the ability of thermal post-treatments involving HIPing, solutionizing and short aging treatments to render the defect content in as-built EBM Alloy 718 irrelevant. Such optimization of the build process and post-treatment together can have significant implications on industrialization of EBM technique.

Author Contributions: Conceptualization, S.G. and S.J.; Methodology, S.G., J.O.; Formal Analysis, S.G., K.B., S.J.; Investigation, S.G., K.B.; Resources, J.O., S.J.; Data Curation, S.G., K.B.; Writing—Original Draft Preparation, S.G.; Writing—Review & Editing, S.G., K.B., U.K., S.J.; Visualization, S.G., K.B., S.J.; Supervision, S.G., U.K., S.J.; Project Administration, S.G., S.J.; Funding Acquisition, S.J. All authors have read and agreed to the published version of the manuscript.

Funding: This research was supported by the STIFTELSEN FÖR KUNSKAPS-OCH KOMPETENSUTVECKLING, Sweden, grant number 20160281 for the SUMAN-Next Project.

Acknowledgments: Thanks to Johannes Gårdstam (Quintus Technologies AB, Sweden) for the HIPing treatments.

Conflicts of Interest: The authors declare no conflict of interest.

References

1. Kirka, M.M.; Greeley, D.A.; Hawkins, C.; Dehoff, R.R. Effect of Anisotropy and Texture on the Low Cycle Fatigue Behavior of Inconel 718 Processed via Electron Beam Melting. *Int. J. Fatigue* **2017**, *105*, 235–243. [CrossRef]
2. Smith, C.J.; Derguti, F.; Nava, E.H.; Thomas, M.; Tammas-Williams, S.; Gulizia, S.; Fraser, D.; Todd, I. Dimensional Accuracy of Electron Beam Melting (EBM) Additive Manufacture with Regard to Weight Optimized Truss Structures. *J. Mater. Process. Technol.* **2016**, *229*, 128–138. [CrossRef]
3. Körner, C. Additive Manufacturing of Metallic Components by Selective Electron Beam Melting—A Review. *Int. Mater. Rev.* **2016**, *61*, 361–377. [CrossRef]

4. Tammas-Williams, S.; Withers, P.J.; Todd, I.; Prangnell, P.B. The Effectiveness of Hot Isostatic Pressing for Closing Porosity in Titanium Parts Manufactured by Selective Electron Beam Melting. *Metall. Mater. Trans. A* **2016**, *47*, 1939–1946. [CrossRef]
5. Gong, H.; Rafi, K.; Starr, T.; Stucker, B. The Effects of Processing Parameters on Defect Regularity in Ti-6Al-4V Parts Fabricated by Selective Laser Melting and Electron Beam Melting. In Proceedings of the 24th International SFF Symposium—An Additive Manufacturing Conference, SFF., Austin, TX, USA, 12–14 August 2013; pp. 424–439.
6. Gong, H.; Rafi, K.; Gu, H.; Starr, T.; Stucker, B. Analysis of Defect Generation in Ti–6Al–4V Parts Made Using Powder Bed Fusion Additive Manufacturing Processes. *Addit. Manuf.* **2014**, *1–4*, 87–98. [CrossRef]
7. Gong, H.; Rafi, K.; Gu, H.; Ram, G.D.J.; Starr, T.; Stucker, B. Influence of Defects on Mechanical Properties of Ti-6Al-4V Components Produced by Selective Laser Melting and Electron Beam Melting. *Mater. Des.* **2015**, *86*, 545–554. [CrossRef]
8. Murr, L.E.; Gaytan, S.M.; Medina, F.; Martinez, E.; Hernandez, D.H.; Martinez, L.; Lopez, M.I.; Wicker, R.B.; Collins, S. Effect of Build Parameters and Build Geometries on Residual Microstructures and Mechanical Properties of Ti-6Al-4V Components Built by Electron Beam Melting (EBM). In Proceedings of the 20th Annual International Solid Freeform Fabrication Symposium, SFF 2009, Austin, TX, USA, 3–5 August 2009; pp. 374–397.
9. Strondl, A.; Palm, M.; Gnauk, J.; Frommeyer, G. Microstructure and Mechanical Properties of Nickel Based Superalloy IN718 Produced by Rapid Prototyping with Electron Beam Melting (EBM). *Mater. Sci. Technol.* **2011**, *27*, 876–883. [CrossRef]
10. Deng, D.; Moverare, J.; Peng, R.L.; Söderberg, H. Microstructure and Anisotropic Mechanical Properties of EBM Manufactured Inconel 718 and Effects of Post Heat Treatments. *Mater. Sci. Eng. A* **2017**, *693*, 151–163. [CrossRef]
11. Cordero, Z.C.; Meyer, H.M.; Nandwana, P.; Dehoff, R.R. Powder Bed Charging during Electron-Beam Additive Manufacturing. *Acta Mater.* **2017**, *124*, 437–445. [CrossRef]
12. Sigl, M.; Lutzmann, S.; Zaeh, M.F. Transient Physical Effects in Electron Beam Sintering. In Proceedings of the Solid Freeform Fabrication Symposium, Austin, TX, USA; 2006; pp. 464–477.
13. Polonsky, A.T.; Echlin, M.P.; Lenthe, W.C.; Dehoff, R.R.; Kirka, M.M.; Pollock, T.M. Defects and 3D Structural Inhomogeneity in Electron Beam Additively Manufactured Inconel 718. *Mater. Charact.* **2018**, *143*, 171–181. [CrossRef]
14. Raghavan, S.; Zhang, B.; Wang, P.; Sun, C.-N.; Nai, M.L.S.; Li, T.; Wei, J. Effect of Different Heat Treatments on the Microstructure and Mechanical Properties in Selective Laser Melted INCONEL 718 Alloy. *Mater. Manuf. Process.* **2017**, *32*, 1588–1595. [CrossRef]
15. Sun, S.-H.; Koizumi, Y.; Saito, T.; Yamanaka, K.; Li, Y.-P.; Cui, Y.; Chiba, A. Electron Beam Additive Manufacturing of Inconel 718 Alloy Rods: Impact of Build Direction on Microstructure and High-Temperature Tensile Properties. *Addit. Manuf.* **2018**, *23*, 457–470. [CrossRef]
16. Kirka, M.M.; Medina, F.; Dehoff, R.; Okello, A. Mechanical Behavior of Post-Processed Inconel 718 Manufactured Through the Electron Beam Melting Process. *Mater. Sci. Eng. A* **2017**, *680*, 338–346. [CrossRef]
17. Gong, X.; Anderson, T.; Chou, K. Review on Powder-Based Electron Beam Additive Manufacturing Technology. *Manuf. Rev.* **2014**, *1*, 1–12. [CrossRef]
18. Sames, W.J.; List, F.A.; Pannala, S.; Dehoff, R.R.; Babu, S.S. The Metallurgy and Processing Science of Metal Additive Manufacturing. *Int. Mater. Rev.* **2016**, *61*, 315–360. [CrossRef]
19. American Society for Testing and Materials. *ASTM F3055−14a. Standard Specification for Additive Manufacturing Nickel Alloy (UNS N07718) with Powder Bed Fusion*; ASTM: West Conshohocken, PA, USA, 2014.
20. Zaninelli, E. Effects of Post-Processing on EBM Fabricated Inconel 718. Master's Thesis, University of Modena and Reggio Emilia, Modena, Italy, 2018.
21. Bauereiß, A.; Scharowsky, T.; Körner, C. Defect Generation and Propagation Mechanism During Additive Manufacturing by Selective Beam Melting. *J. Mater. Process. Technol.* **2014**, *214*, 2522–2528. [CrossRef]
22. Sames, W.J. Additive Manufacturing of Inconel 718 Using Electron Beam Melting: Processing, Post-Processing, & Mechanical Properties. Ph.D. Thesis, Texas A&M University, College Station, TX, USA, 2015.

23. Lee, H.J.; Kim, H.K.; Hong, H.U.; Lee, B.S. Influence of the Focus Offset on the Defects, Microstructure, and Mechanical Properties of an Inconel 718 Superalloy Fabricated by Electron Beam Additive Manufacturing. *J. Alloys Compd.* **2019**, *781*, 842–856. [CrossRef]
24. Balachandramurthi, A.R.; Moverare, J.; Dixit, N.; Pederson, R. Influence of Defects and As-Built Surface Roughness on Fatigue Properties of Additively Manufactured Alloy 718. *Mater. Sci. Eng. A* **2018**, *735*, 463–474. [CrossRef]
25. Tammas-Williams, S.; Withers, P.J.; Todd, I.; Prangnell, P.B. Porosity Regrowth during Heat Treatment of Hot Isostatically Pressed Additively Manufactured Titanium Components. *Scr. Mater.* **2016**, *122*, 72–76. [CrossRef]
26. Benn, R.C.; Salva, R.P. Additively Manufactured Inconel® Alloy 718. In Proceedings of the 7th International Symposium on Superalloy 718 and Derivatives, Pittsburgh, PA, USA, 10–13 October 2010; pp. 455–469.
27. Ding, X.; Koizumi, Y.; Aoyagi, K.; Kii, T.; Sasaki, N.; Hayasaka, Y.; Yamanaka, K.; Chiba, A. Microstructural Control of Alloy 718 Fabricated by Electron Beam Melting with Expanded Processing Window by Adaptive Offset Method. *Mater. Sci. Eng. A* **2019**, *764*, 138058. [CrossRef]
28. Schaak, C.; Tillmann, W.; Schaper, M.; Aydinöz, M.E. Process Gas Infiltration in Inconel 718 Samples During SLM Processing. *RTeJournal—Fachforum für Rapid Technol.* **2016**.
29. Nandwana, P.; Kirka, M.; Okello, A.; Dehoff, R. Electron Beam Melting of Inconel 718: Effects of Processing and Post-Processing. *Mater. Sci. Technol.* **2018**, *34*, 612–619. [CrossRef]
30. Goel, S.; Gundgire, T.; Varghese, J.; Rajulapati, K.V.; Klement, U.; Joshi, S. Role of HIPing and Heat Treatment on Properties of Alloy 718 Fabricated by Electron Beam Melting. In Proceedings of the Euro PM2019, Maastricht, The Netherlands, 13–16 October 2019.
31. Aydinöz, M.E.; Brenne, F.; Schaper, M.; Schaak, C.; Tillmann, W.; Nellesen, J.; Niendorf, T. On the Microstructural and Mechanical Properties of Post-Treated Additively Manufactured Inconel 718 Superalloy Under Quasi-Static and Cyclic Loading. *Mater. Sci. Eng. A* **2016**, *669*, 246–258. [CrossRef]
32. Al-Juboori, L.A.; Niendorf, T.; Brenne, F. On the Tensile Properties of Inconel 718 Fabricated by EBM for As-Built and Heat-Treated Components. *Metall. Mater. Trans. B* **2018**, *49B*, 2969–2974. [CrossRef]
33. Balachandramurthi, A.R.; Moverare, J.; Dixit, N.; Deng, D.; Pederson, R. Microstructural influence on fatigue crack propagation during high cycle fatigue testing of additively manufactured Alloy 718. *Mater. Charact.* **2019**, *149*, 82–94. [CrossRef]
34. Goel, S.; Ahlfors, M.; Bahbou, F.; Joshi, S. Effect of different post-treatments on the microstructure of EBM-built Alloy 718. *J. Mater. Eng. Perform.* **2018**, *28*, 673–680. [CrossRef]

Article

Encapsulation of Electron Beam Melting Produced Alloy 718 to Reduce Surface Connected Defects by Hot Isostatic Pressing

Yunus Emre Zafer [1,*], Sneha Goel [1], Ashish Ganvir [2], Anton Jansson [3] and Shrikant Joshi [1]

1 Department of Engineering Science, University West, 461 86 Trollhättan, Sweden; sneha.goel@hv.se (S.G.); shrikant.joshi@hv.se (S.J.)
2 Research & Technology, Department of Process Engineering, GKN Aerospace Engine Systems AB, 461 81 Trollhättan, Sweden; ashish.ganvir@gknaerospace.com
3 School of Science and Engineering, Örebro University, 701 82 Örebro, Sweden; anton.jansson.mail1@gmail.com
* Correspondence: yunusemrezafer@gmail.com

Received: 12 February 2020; Accepted: 4 March 2020; Published: 9 March 2020

Abstract: Defects in electron beam melting (EBM) manufactured Alloy 718 are inevitable to some extent, and are of concern as they can degrade mechanical properties of the material. Therefore, EBM-manufactured Alloy 718 is typically subjected to post-treatment to improve the properties of the as-built material. Although hot isostatic pressing (HIPing) is usually employed to close the defects, it is widely known that HIPing cannot close open-to-surface defects. Therefore, in this work, a hypothesis is formulated that if the surface of the EBM-manufactured specimen is suitably coated to encapsulate the EBM-manufactured specimen, then HIPing can be effective in healing such surface-connected defects. The EBM-manufactured Alloy 718 specimens were coated by high-velocity air fuel (HVAF) spraying using Alloy 718 powder prior to HIPing to evaluate the above approach. X-ray computed tomography (XCT) analysis of the defects in the same coated sample before and after HIPing showed that some of the defects connected to the EBM specimen surface were effectively encapsulated by the coating, as they were closed after HIPing. However, some of these surface-connected defects were retained. The reason for such remnant defects is attributed to the presence of interconnected pathways between the ambient and the original as-built surface of the EBM specimen, as the specimens were not coated on all sides. These pathways were also exaggerated by the high surface roughness of the EBM material and could have provided an additional path for argon infiltration, apart from the uncoated sides, thereby hindering complete densification of the specimen during HIPing.

Keywords: electron beam melting; additive manufacturing; Alloy 718; surface defects; encapsulation; coating; hot isostatic pressing

---

1. Introduction

Additive manufacturing (AM), also commonly known as three-dimensional (3D) printing, is a rapidly growing technology for generating complex geometrical products layer by layer from a 3D computer-aided design (CAD) model data [1,2]. This technology offers significant design freedom compared to conventional manufacturing methods such as forging, casting, etc. [3]. Electron beam melting (EBM) is one of the AM techniques; it utilizes an electron beam as the heat source [4]. Manufacturing of difficult-to-machine nickel-based superalloys, such as Alloy 718, by using EBM is being explored and has attracted significant interest in the aerospace industry due to reduced raw material wastage, which leads to lower costs and buy-to-fly-ratio, and less contamination due to vacuum conditions during the process [5].

However, EBM-manufactured Alloy 718 is typically characterized by the presence of inevitable defects such as lack of fusion, gas porosity, and shrinkage porosity, which can be detrimental to the mechanical properties of the material [6]. Therefore, thermal post-treatment involving hot isostatic pressing (HIPing) is typically employed to reduce such defects in the AM manufactured material [3,7–9]. In this context, it is pertinent to state that micro-computed tomography is used to obtain detailed information about the location and size of defects present in the material in 3D [10–12]. It has been reported that HIPing can heal most defects in AM manufactured specimens, except the surface-connected defects [11,13]. The presence of these defects becomes vital, as they are not affected by HIPing and are found to be potential crack initiation sites leading to fracture of the material under fatigue loading [6]. Therefore, an idea of encapsulating the surface-connected "open" defects through deposition of a thin film/coating on the as-built specimen surface was explored in the literature for laser-based AM techniques [13]. However, this approach has not yet been widely investigated, and in particular no such efforts have yet been reported in case of EBM-manufactured material, where retained surface-connected defects after HIPing can be of concern. Therefore, the challenges presented in the above led to a hypothesis that is tested in the present work. The hypothesis is that encapsulation of these surface defects by applying a coating on the EBM-built Alloy 718 could allow HIPing to effectively heal all defects present, including those that are surface-connected.

In the present study, the coating technique utilized to explore the encapsulation hypothesis is high-velocity air fuel (HVAF), which is one of the thermal spray coating techniques [14,15]. Some of the EBM-manufactured Alloy 718 specimens were coated. Afterward, the coated and uncoated specimens were subjected to HIPing to enable a comparison of extent of defect closure in the two conditions. A detailed investigation of the defects in the specimens is carried out by using light optical microscopy (LOM), scanning electron microscopy (SEM), and X-ray computed tomography (XCT). In addition, surface roughness of the as-built specimen is measured using white light interferometry.

2. Materials and Methods

2.1. EBM Processing

The feedstock material used for EBM production was plasma-atomized Alloy 718 powder supplied by Advanced Powders and Coating (Québec, Canada). The nominal powder particle size range was 45–105 μm, and its chemical composition, as provided by the supplier, is given in Table 1. The powder was recycled several times before usage.

Table 1. Nominal chemical composition of Alloy 718 powder used during electron beam melting (EBM).

Element	Ni	C	Cr	Mo	Ti	Al	Fe	Nb	Ta
Wt %	54.10	0.03	19.00	2.90	1.00	0.50	Bal.	4.90	<0.01

An Arcam A2X EBM machine was used to produce several rectangular specimens of dimensions 53 × 45 × 5 mm, as shown in Figure 1a. Typical process parameter settings recommended by Arcam AB (Mölndal, Sweden) were used (EBM control software version 4.2.76). The build process started after preheating the stainless-steel build plate to ~1025 °C. A layer thickness of 75 μm and an acceleration voltage of 60 kV were used. The periphery of the specimens, also termed as contour, was melted using multispot melting. The interior of the specimens, known as hatch, was processed using back and forth raster scanning. During hatch melting, the scan direction was rotated by 72° for every added layer.

2.2. Encapsulation Concept

The feedstock material used for coating was Alloy 718 powder manufactured by Praxair Surface Technologies (Indianapolis, USA). The powder particle size range was 15–45 μm, and its chemical composition, as provided by the supplier, is listed in Table 2. Half of the as-built EBM Alloy 718 specimens were coated on the back and front sides (53 × 45 mm faces), leaving the remaining sides uncoated, as shown in Figure 1b. The coating was deposited using an HVAF M3 system (Uniquecoat, Richmond, USA) with a target coating thickness of about 500 μm. Although substrates to be HVAF sprayed are typically grit blasted to a roughness of about 5–10 μm arithmetic mean roughness (Ra), it is important to point out that there was no surface preparation done prior to coating deposition on the EBM specimens, which already had an average surface roughness of about 80 μm arithmetic mean height (Sa) in as-built condition.

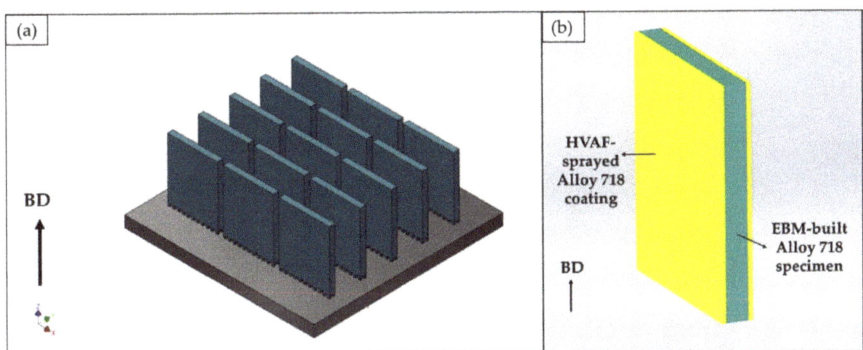

Figure 1. (**a**) Image of the computer-aided design (CAD) model of the entire electron beam melting (EBM) build, and (**b**) illustration of the coated specimen. The arrow indicates the build direction (BD).

Table 2. Nominal chemical composition of Alloy 718 powder used for coating.

Element	Ni	C	Cr	Mo	Ti	Al	Fe	Nb + Ta
Wt %	50.00–55.00	0.02–0.08	17.00–21.00	2.80–3.30	0.70–1.10	0.03–0.70	15.00–21.00	4.70–5.50

2.3. Hot Isostatic Pressing

Some of the specimens in as-built and coated conditions were HIPed in a QIH21 model molybdenum HIP furnace at Quintus Technologies (Västerås, Sweden). HIPing was carried out at a temperature of 1,120 °C, and pressure of 100 MPa was applied for 4 h. Argon was used as an inert process gas. After the dwell time, the specimens were rapidly cooled.

2.4. Metallographic Preparation and Characterization

For microstructural investigation, samples were sectioned along and perpendicular to the build direction with an alumina abrasive cutting disc mounted on a Struers Secotom 10 machine. The sectioned samples were hot mounted using a Buehler SimpliMet 3000 press. The mounted samples were grinded using silicon carbide paper ranging from P360 to P1200 grits, followed by polishing with 9 μm and 3 μm diamond suspensions, and thereafter with 0.05 μm colloidal silica suspension. Grinding and polishing were done using a semi-automatic Buehler Ecomet 300 Pro machine.

The defects, particularly their type, amount, and distribution, were analyzed at suitable magnifications using Zeiss Axio light optical microscope. The defect content in the hatch and contour regions were distinctly evaluated using the LOM micrographs, which were processed by using image analysis software ImageJ. In each case, no less than 15 LOM images captured along the cross-section (see Figure 2) were analyzed to obtain representative defect content. Hitachi TM3000 scanning electron microscope was employed to perform a high magnification analysis of the coating. The porosity of the coating, before and after HIPing, was evaluated by image analysis software ImageJ, using no less than 15 SEM micrographs captured along the cross-section to get a representative value of the porosity content.

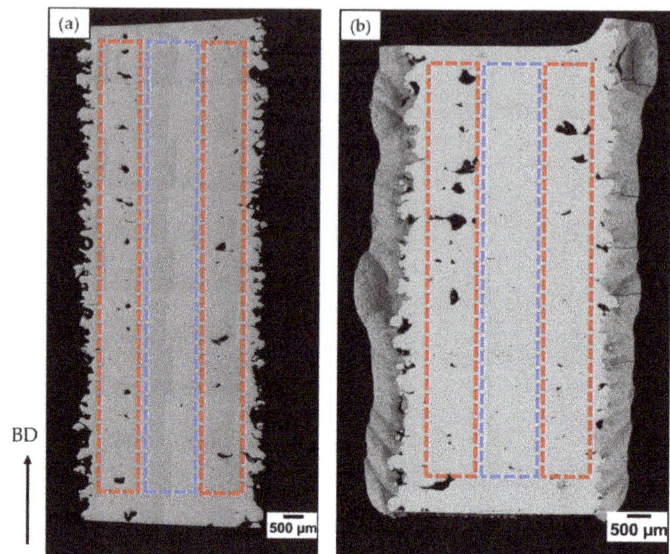

Figure 2. Light optical microscopy (LOM) micrographs of the specimen cross-sections in the (**a**) uncoated and (**b**) coated conditions. The dotted red and blue lines indicate the regions of contour and hatch, respectively, distinctly analyzed for defect quantification. The arrow indicates the build direction.

For a further precise analysis of the defects, XCT was employed. For this, a sample of dimensions 10 × 10 × 6 mm was sectioned out. The defects present in the specimen, including their size, shape, and location, were characterized using a Zeiss Xradia Versa 520 XCT system. XCT scans were performed over the entire volume of the same coated specimen (10 × 10 × 6 mm) before and after HIPing.

The surface roughness of the as-built specimen was characterized through white light interferometry using a profilm3D (Filmetrics, San Diego, CA, USA). The surface roughness parameter, Sa, was evaluated from the 3D topography maps of the specimen. For each measurement point, an area of 1 mm by 1 mm was analyzed. In total, six such measurements were performed over the as-built specimen.

3. Results and Discussion

3.1. Uncoated Condition

The as-built EBM Alloy 718 specimens were characterized by the presence of defects. Three types of defects were observed, i.e., lack of fusion, shrinkage porosity, and gas porosity, and they are visualized in Figure 3. Such defects have also been previously observed by Goel et al. [3], and the influence of the defects on mechanical properties has been elaborated elsewhere [16,17]. Round-shaped gas porosities (refer Figure 3a) were randomly distributed, and they are attributable to entrapped gas inside the virgin powder, which could have found its way into the EBM build [1]. The shrinkage porosities, shown in Figure 3b, were typically aligned along the build direction, and are reportedly known to form as a result of interdendritic shrinkage after solidification [18]. The lack of fusion defects were primarily concentrated in the contour regions and often contained partially molten powder particles. The main reason for the formation of lack of fusion defects is expected to be inappropriate energy input. Low energy input can result in incomplete bonding between the layers [19], and, as a result, partially molten particles can be observed in these defects (see Figure 3c). On the other hand, high energy input can cause the melt to spatter away [19].

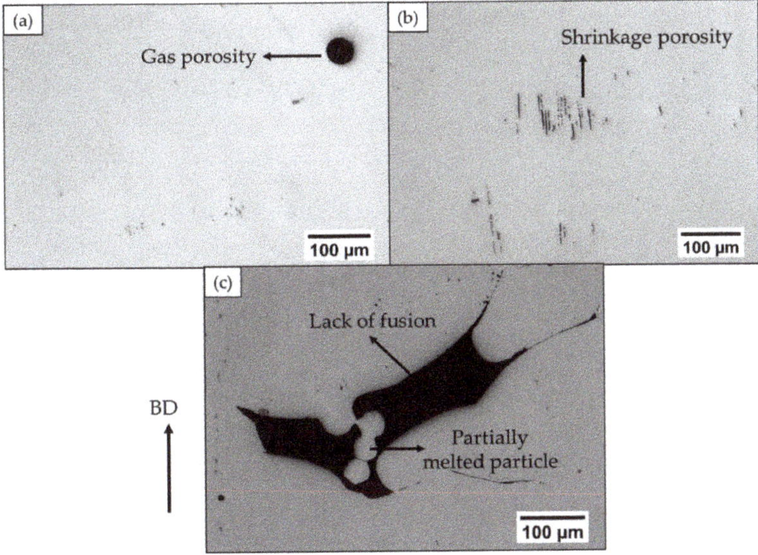

Figure 3. LOM micrographs showing the different kinds of defects present in the as-built EBM Alloy 718: (**a**) gas porosity, (**b**) shrinkage porosity, and (**c**) lack of fusion. The arrow on the left indicates the build direction.

The as-built specimen was also subjected to HIPing, which caused a significant reduction in defect content. A closer investigation of the entire HIPed specimen revealed that nearly all the shrinkage porosities were healed. However, some of the gas porosities and lack of fusion defects were present after HIPing, as shown in Figure 4. A possible explanation for remnant gas porosities after HIPing could be entrapped argon gas, which exerts opposite pressure and can hinder complete closure [6]. However, the defects of typically more serious concern are lack of fusions [6]. A vast majority of remnant lack of fusion defects were found to be surface-connected (open), as exemplified in Figure 4b; therefore, these were not healed after HIPing. This is consistent with several other studies by Tammas-Williams et al. [11] and Tillmann et al. [13], in which XCT was utilized to investigate the effects of HIPing on the defects in EBM-built materials, and it was observed that the remnant lack of fusion defects open to the surface were not closed after HIPing. The reason for this is pressure equalization inside and outside the defect, which can prevent it from closing [13]. In the case of EBM-built Alloy 718, Kotzem et al. [20] have also observed the presence of surface-connected defects. Therefore, to close such defects in EBM Alloy 718, the specimens were coated and HIPed.

3.2. Coated Condition

The as-built EBM Alloy 718 specimens (~53 mm × 45 mm × 5 mm) were coated in an effort to enclose the surface-connected defects, as shown in Figure 5a. Some of the coated specimens were then subjected to HIPing to heal the defects present in EBM Alloy 718. The volume fraction of defects in hatch and contour regions in the coated condition before and after HIPing, as evaluated by image analysis, is given in Figure 6. It can be seen that the defect content was significantly reduced in both the hatch and contour regions after HIPing. However, full densification after HIPing could not be achieved as hypothesized at the beginning of this study. Some of the lack of fusion defects, specifically those in the contour and close to the surface, and gas porosities were found to persist even after HIPing the coated specimen. A post mortem study revealed that the coating seemed unable to completely seal the EBM specimen surface, as several defects at the EBM material-coating interface were observed (see Figure 5b).

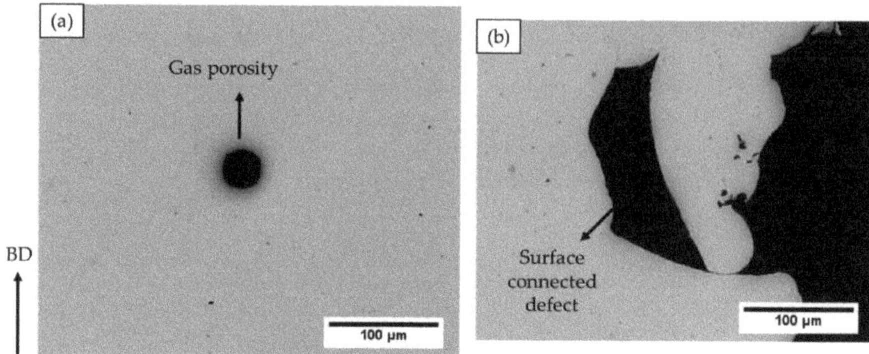

Figure 4. LOM micrographs revealing the remnant defects in the hot isostatic pressed (HIPed) EBM Alloy 718: (**a**) gas porosity and (**b**) lack of fusion. The arrow on the left indicates the build direction.

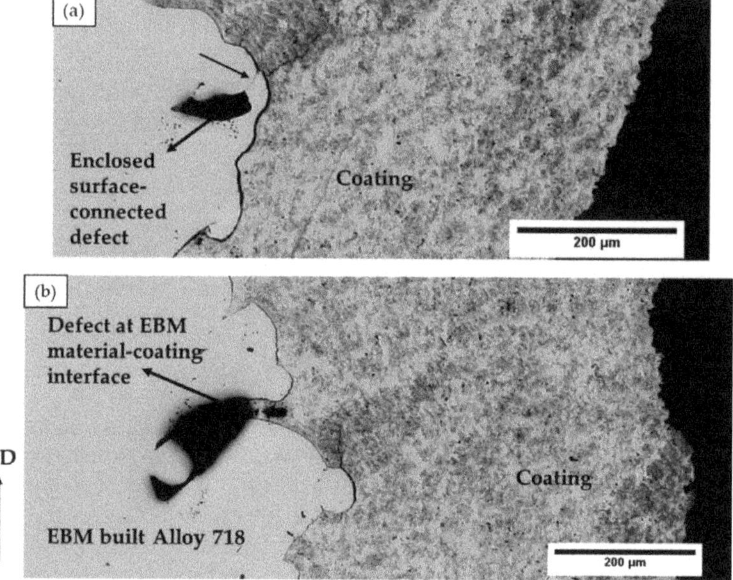

Figure 5. LOM micrographs showing (**a**) an enclosed defect in the EBM material and (**b**) a defect at the EBM material-coating interface in the as-built condition. The arrow on the left indicates the build direction.

In order to get a more detailed understanding of the above observation, especially to identify surface-connected defects, further XCT analysis was carried out. A coated sample was analyzed before and after HIPing to precisely track the defects in the two conditions through XCT. The larger lack of fusion defects, which had a connection to the EBM specimen-coating interface, were separately analyzed, as shown in Figure 7. The defect close to the EBM specimen surface marked with the blue dotted circle in Figure 7a appeared to be fully closed after HIPing, or at least reduced to a size below the resolution limit of XCT. This is attributable to the successfully complete encapsulation of this otherwise surface-connected defect, which could be healed during HIPing. However, some other defects appeared to be not fully closed. This could be possibly attributed to incomplete encapsulation of the defect, as discussed later. The defects were further individually analyzed using 2D slices from the XCT data.

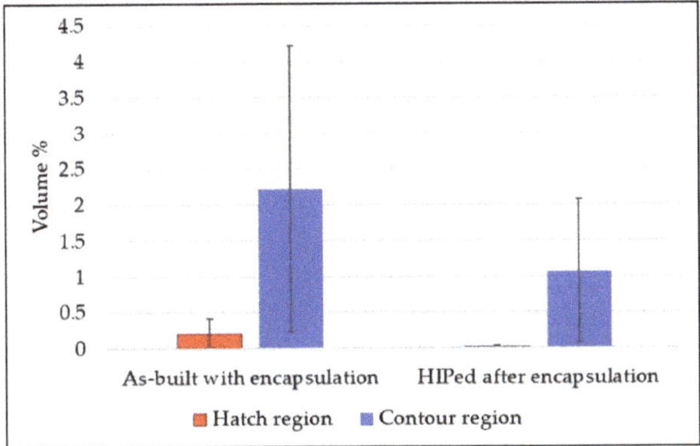

Figure 6. Defect content in the coated condition measured in the hatch and contour regions along the build direction.

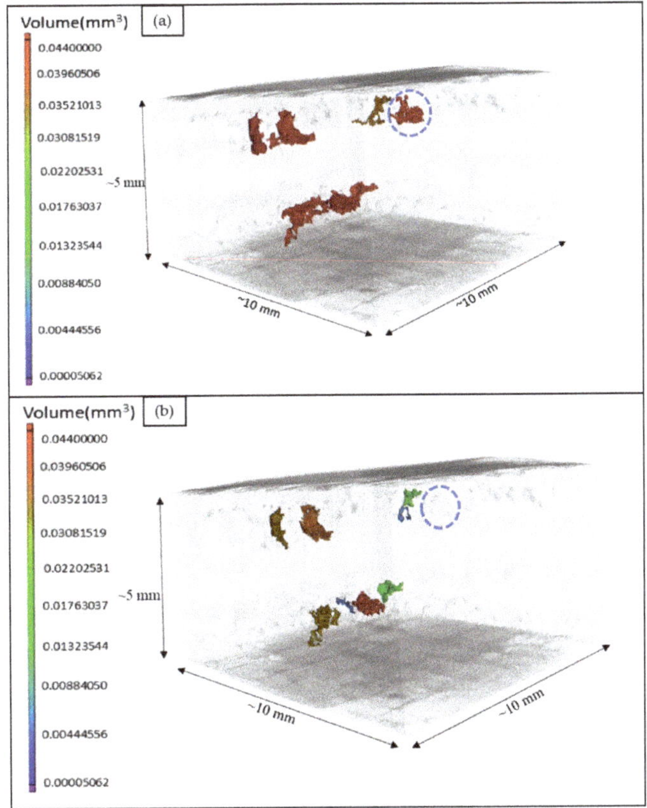

Figure 7. Three-dimensional (3D) views of large defects in the same coated sample (**a**) before and (**b**) after hot isostatic pressing (HIPing), obtained from x-ray computed tomography (XCT) analysis.

Figure 8 shows a comparison of defects present at identical cross-sections of the same coated EBM specimen before and after HIPing as obtained from the XCT analysis. It is important to mention that specimens were only coated on the front and back sides as shown in Figure 1b. Some defects, for instance, the one marked with red dotted lines in Figure 8, can be clearly seen to be open to the uncoated side. It was observed that most of the lack of fusion defects that appeared to have a connection to the EBM specimen-coating interface did not close after HIPing (highlighted with yellow dotted lines in the figure). Nevertheless, some of the lack of fusion defects that were close to the interface seemed to be healed (marked with green dotted lines). Moreover, most of the gaps at the interface of EBM-built material and coating remained unhealed (marked with blue dotted lines in Figure 8). This can be explained.

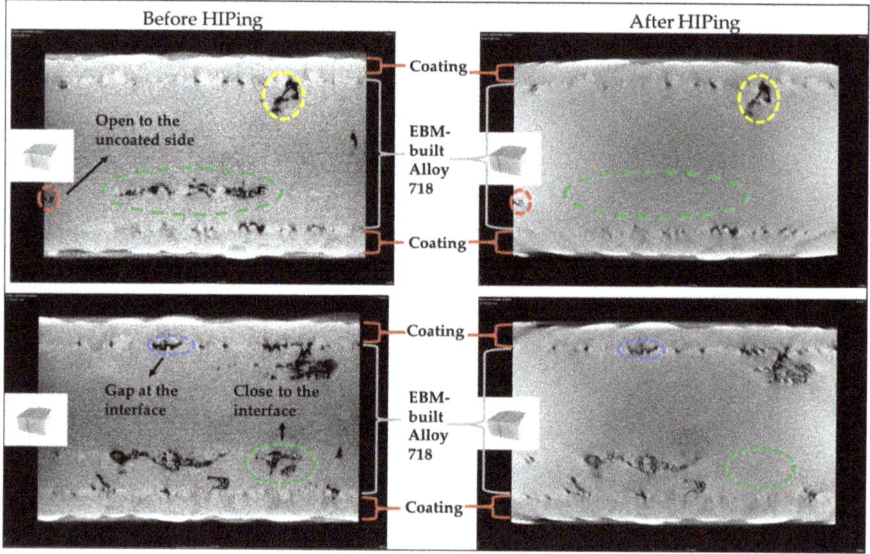

Figure 8. Two-dimensional (2D) sliced images from XCT analysis showing the "tracked" defects at the same cross-sections in the entire coated sample before and after HIPing.

The interface between EBM specimen and coating exhibited gaps, as shown in Figure 5b, which could be attributed to the high surface roughness of the as-built EBM Alloy 718. It is pertinent to note that the coating was applied on the as-built surface without any prior surface preparation. The surface roughness value of the as-built EBM specimen was nearly 80 µm Sa, as measured by white light interferometry, and such high roughness could have hindered complete sealing of the specimen surface. In this context, it is worth mentioning that, although it is necessary to have an appropriately rough substrate to be used for any thermal spray coating process, to achieve better coating-substrate adhesion, an excessively high surface roughness can introduce defects at the substrate-coating interface by not permitting the molten splats to completely fill the "valleys" in the existing surface asperities (see Figure 5b). In the present study, the coating was deposited on a rather rough surface (Sa ~ 80 µm) compared to what is typically reported in the literature, i.e., about 10 µm [21–24]. In addition to these interface defects, several through-thickness vertical cracks were observed in both the coated samples, i.e., HIPed and not HIPed condition, as shown in Figure 9, which could also be possibly attributed to the high surface roughness. The total porosity content of the coatings before and after HIPing, as evaluated by image analysis, is shown in Figure 10. It can be seen from the figure that HIPing resulted in porosity reduction by an order of magnitude. However, some amount of porosity was present after HIPing. The vertical cracks, gaps at the EBM specimen-coating interface, and the porosity in the coating could have provided the path for argon gas infiltration inside the EBM specimen during HIPing. This could possibly explain why some of the defects, mainly lack of fusions connected to the EBM specimen-coating interface, and the gaps at the interface were still present in the coated and HIPed EBM specimen.

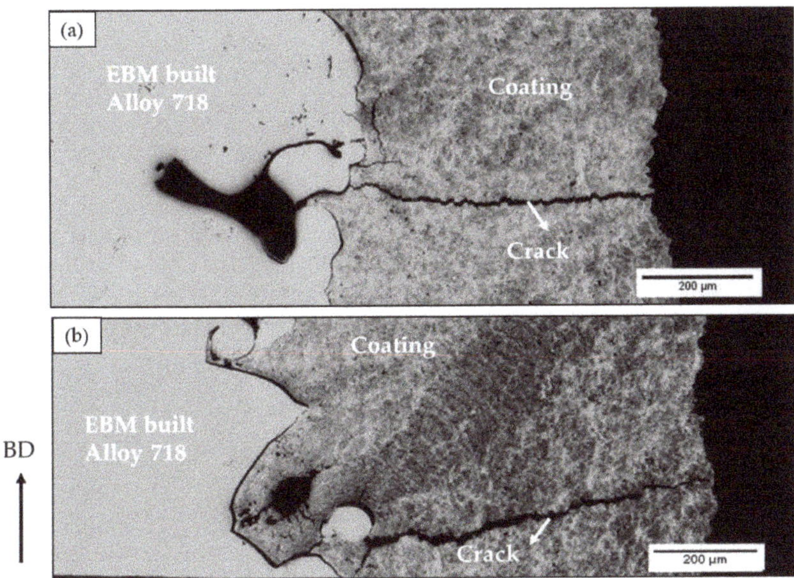

Figure 9. LOM micrographs showing (**a**) a crack in the as-built sample and (**b**) a remnant crack present in the HIPed sample.

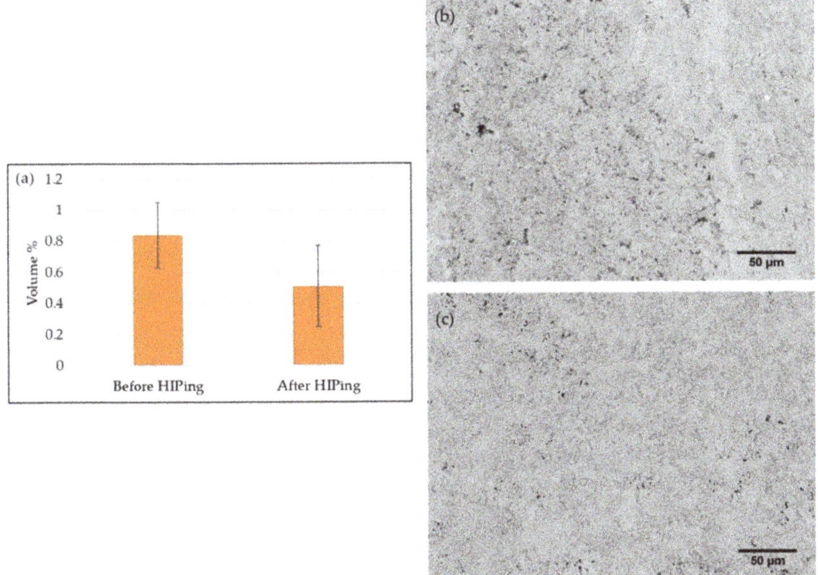

Figure 10. (**a**) Porosity content in the coatings before and after HIPing, as determined by image analysis, SEM micrographs of the coating (**b**) before HIPing, and (**c**) after HIPing.

4. Summary and Conclusions

In this study, the efficacy of encapsulating the as-built EBM Alloy 718 specimens was investigated to eventually close surface-connected defects after HIPing. The major findings of the study based on the obtained results are as follows:

- The hypothesis that encapsulation of EBM specimens through coatings can eliminate surface-connected defects during subsequent HIPing presents a novel idea. However, it could only be partly tested in this paper due to (a) very large surface roughness of the as-built EBM 718 specimen used for this study and (b) only two sides of the as-built specimen being coated.
- Few of the surface-connected defects were closed after subjecting the coated EBM-built specimen to HIPing. However, some of the lack of fusion defects and gaps at the EBM specimen-coating interface remained after HIPing.
- The presence of defects in the coated and HIPed specimen was rationalized as follows: the high surface roughness of the as-built specimens caused only partial "sealing" of defects, as gaps were observed between the EBM specimen and the coating. In addition, the through-thickness cracks resulting during coating on very rough substrate surfaces could have also connected the defects to the surface, despite the application of coating.
- The specimens were coated on only the two larger faces, leaving the remaining sides uncoated. This could have provided an additional path for HIP process gas infiltration from the uncoated sides.

It is inferred that the surface roughness of the EBM specimen, prior to coating deposition, should be reduced to enable complete sealing of the surface-connected defects. Thus, before coating, prior surface preparation by mechanical post-treatment techniques, such as shot peening, machining, and grit blasting, can be used in cases where the as-built surface roughness is very large. Encapsulation of EBM-built materials should be done on all the sides of the specimens, as it can enable more effective defect closure during HIPing.

Author Contributions: Y.E.Z. performed all the experimental investigations, analyzed all the results, and wrote the paper; A.J. performed the XCT data processing; S.J., S.G., and A.G. have contributed in defining the problem, planning the experimental approach, and reviewing analysis of the results. All authors discussed the results and finalized the paper. All authors have read and agreed to the published version of the manuscript.

Funding: This research was funded by KK foundation, grant number 20160281.

Acknowledgments: The support from GKN Aerospace Engine Systems AB is highly acknowledged. The assistance of research engineers Jonas Ölsson and Stefan Björklund from University West for building the EBM specimens and spraying the coating, respectively, is gratefully acknowledged.

Conflicts of Interest: Authors declare no conflict of interest.

Data Availability: The raw data required to reproduce these findings cannot be shared at this time, since the data also form part of an ongoing study in the author's research group.

References

1. Sames, W. Additive Manufacturing of Inconel 718 using Electron Beam Melting: Processing, Post-Processing, & Mechanical Properties. Ph.D. Thesis, Texas A&M University, College Station, TX, USA, 2015.
2. Gibson, I.; Rosen, D.; Stucker, B. *Additive Manufacturing Technologies: 3D Printing, Rapid Prototyping, and Direct Digital Manufacturing*, 2nd ed.; Springer: New York, NY, USA, 2015; ISBN 978-1-4939-2112-6. [CrossRef]
3. Goel, S.; Olsson, J.; Ahlfors, M.; Klement, U.; Joshi, S. The Effect of Location and Post-treatment on the Microstructure of EBM-Built Alloy 718. In *Proceedings of the 9th International Symposium on Superalloy 718 & Derivatives: Energy, Aerospace, and Industrial Applications*; Ott, E., Liu, X., Andersson, J., Bi, Z., Bockenstedt, K., Dempster, I., Groh, J., Heck, K., Jablonski, P., Kaplan, M., et al., Eds.; Springer International Publishing: Cham, Switzerland, 2018; pp. 115–129. [CrossRef]

4. Karimi, P.; Sadeghi, E.; Deng, D.; Gruber, H.; Andersson, J.; Nylén, P. Influence of build layout and orientation on microstructural characteristics of electron beam melted Alloy 718. *Int. J. Adv. Manuf. Technol.* **2018**, *99*, 2903–2913. [CrossRef]
5. English, C.L.; Tewari, S.K.; Abbott, D.H. An overview of Ni base additive fabrication technologies for aerospace applications. In Proceedings of the 7th International Symposium on Superalloy, Pittsburgh, PA, USA, 10–13 October 2010; Volume 718, pp. 399–412.
6. Balachandramurthi, A.R.; Moverare, J.; Dixit, N.; Pederson, R. Influence of defects and as-built surface roughness on fatigue properties of additively manufactured Alloy 718. *Mater. Sci. Eng. A* **2018**, *735*, 463–474. [CrossRef]
7. Goel, S.; Sittiho, A.; Charit, I.; Klement, U.; Joshi, S. Effect of post-treatments under hot isostatic pressure on microstructural characteristics of EBM-built Alloy 718. *Addit. Manuf.* **2019**, *28*, 727–737. [CrossRef]
8. Goel, S.; Ahlfors, M.; Bahbou, F.; Joshi, S. Effect of Different Post-treatments on the Microstructure of EBM-Built Alloy 718. *J. Mater. Eng. Perform.* **2019**, *28*, 673–680. [CrossRef]
9. Goel, S.; Bourreau, K.; Olsson, J.; Klement, U.; Joshi, S. Can Appropriate Thermal Post-Treatment Make Defect Content in as-Built Electron Beam Additively Manufactured Alloy 718 Irrelevant? *Materials* **2020**, *13*, 536. [CrossRef] [PubMed]
10. Tammas-Williams, S.; Withers, P.J.; Todd, I.; Prangnell, P.B. Porosity regrowth during heat treatment of hot isostatically pressed additively manufactured titanium components. *Scr. Mater.* **2016**, *122*, 72–76. [CrossRef]
11. Tammas-Williams, S.; Withers, P.J.; Todd, I.; Prangnell, P.B. The Effectiveness of Hot Isostatic Pressing for Closing Porosity in Titanium Parts Manufactured by Selective Electron Beam Melting. *Metall. Mater. Trans. A* **2016**, *47*, 1939–1946. [CrossRef]
12. du Plessis, A.; Rossouw, P. Investigation of Porosity Changes in Cast Ti6Al4V Rods After Hot Isostatic Pressing. *J. Mater. Eng. Perform.* **2015**, *24*, 3137–3141. [CrossRef]
13. Tillmann, W.; Schaak, C.; Nellesen, J.; Schaper, M.; Aydinöz, M.E.; Niendorf, T. Functional encapsulation of laser melted Inconel 718 by Arc-PVD and HVOF for post compacting by hot isostatic pressing. *Powder Metall.* **2015**, *58*, 259–264. [CrossRef]
14. Zeng, Z.; Sakoda, N.; Tajiri, T.; Kuroda, S. Structure and corrosion behavior of 316L stainless steel coatings formed by HVAF spraying with and without sealing. *Surf. Coat. Technol.* **2008**, *203*, 284–290. [CrossRef]
15. Guo, R.Q.; Zhang, C.; Chen, Q.; Yang, Y.; Li, N.; Liu, L. Study of structure and corrosion resistance of Fe-based amorphous coatings prepared by HVAF and HVOF. *Corros. Sci.* **2011**, *53*, 2351–2356. [CrossRef]
16. Fousová, M.; Vojtěch, D.; Doubrava, K.; Daniel, M.; Lin, C.-F. Influence of Inherent Surface and Internal Defects on Mechanical Properties of Additively Manufactured Ti6Al4V Alloy: Comparison between Selective Laser Melting and Electron Beam Melting. *Materials* **2018**, *11*, 537. [CrossRef] [PubMed]
17. Gong, H.; Rafi, K.; Gu, H.; Janaki Ram, G.D.; Starr, T.; Stucker, B. Influence of defects on mechanical properties of Ti–6Al–4V components produced by selective laser melting and electron beam melting. *Mater. Des.* **2015**, *86*, 545–554. [CrossRef]
18. Deng, D.; Moverare, J.; Peng, R.L.; Söderberg, H. Microstructure and anisotropic mechanical properties of EBM-manufactured Inconel 718 and effects of post heat treatments. *Mater. Sci. Eng. A* **2017**, *693*, 151–163. [CrossRef]
19. Karimi Neghlani, P. Electron Beam Melting of Alloy 718: Influence of Process Parameters on the Microstructure. Ph.D. Thesis, University West, Trollhättan, Sweden, 2018.
20. Kotzem, D.; Arold, T.; Niendorf, T.; Walther, F. Influence of specimen position on the build platform on the mechanical properties of as-built direct aged electron beam melted Inconel 718 alloy. *Mater. Sci. Eng. A* **2020**, *772*, 138785. [CrossRef]
21. Singh, R.; Rauwald, K.-H.; Wessel, E.; Mauer, G.; Schruefer, S.; Barth, A.; Wilson, S.; Vassen, R. Effects of substrate roughness and spray-angle on deposition behavior of cold-sprayed Inconel 718. *Surf. Coat. Technol.* **2017**, *319*, 249–259. [CrossRef]
22. Pershin, V.; Lufitha, M.; Chandra, S.; Mostaghimi, J. Effect of substrate temperature on adhesion strength of plasma-sprayed nickel coatings. *J. Therm. Spray Technol.* **2003**, *12*, 370–376. [CrossRef]

23. Staia, M.H.; Ramos, E.; Carrasquero, A.; Roman, A.; Lesage, J.; Chicot, D.; Mesmacque, G. Effect of substrate roughness induced by grit blasting upon adhesion of WC-17% Co thermal sprayed coatings. *Thin Solid Films* **2000**, *377*, 657–664. [CrossRef]
24. Wang, Y.-Y.; Li, C.-J.; Ohmori, A. Influence of substrate roughness on the bonding mechanisms of high velocity oxy-fuel sprayed coatings. *Thin Solid Films* **2005**, *485*, 141–147. [CrossRef]

Article

Influence of Scanning Speed on Microstructure and Properties of Laser Cladded Fe-Based Amorphous Coatings

Xiangchun Hou [1], Dong Du [1], Baohua Chang [1,*] and Ninshu Ma [2]

[1] State Key Laboratory of Tribology, Department of Mechanical Engineering, Tsinghua University, Beijing 100084, China; houxc16@mails.tsinghua.edu.cn (X.H.); dudong@tsinghua.edu.cn (D.D.)

[2] Joining and Welding Research Institute, Osaka University, 11-1 Mihogaoka, Ibaraki, Osaka 567-0047, Japan; ma.ninshu@jwri.osaka-u.ac.jp

* Correspondence: bhchang@tsinghua.edu.cn

Received: 3 March 2019; Accepted: 17 April 2019; Published: 18 April 2019

Abstract: Fe-based amorphous alloys with excellent mechanical properties are suitable for preparing wear resistant coatings by laser cladding. In this study, a novel Fe-based amorphous coating was prepared by laser cladding on 3Cr13 stainless steel substrates. The influence of scanning speeds on the microstructures and properties of the coatings was investigated. The microstructure compositions and phases were analyzed by scanning electron microscope, electron probe microanalyzer, and x-ray diffraction respectively. Results showed that the microstructures of the coatings changed significantly with the increase of scanning speeds. For a scanning speed of 6 mm/s, the cladding layer was a mixture of amorphous and crystalline regions. For a scanning speed of 8 mm/s, the cladding layer was mainly composed of block grain structures. For a scanning speed of 10 mm/s, the cladding layer was composed entirely of dendrites. Different dilution rates at the bonding zones were the main reasons for the microstructure change for different claddings. For all three scanning speeds, the coatings had higher hardness and wear resistance when compared with the substrate; as the scanning speed increased, the hardness and wear resistance of the coatings gradually decreased due to the change in microstructure.

Keywords: Fe-based amorphous coating; laser cladding; microstructure; property

---

1. Introduction

Amorphous alloys have unique properties such as high hardness, high strength, high wear resistance, and high corrosion resistance because of their short-range order and long-range disordered structure [1–3]. Their unique properties make amorphous alloys suitable for surface coatings. Different processes have been used to prepare surface coatings such as laser cladding, plasma spray, spark cladding, and high velocity oxygen fuel (HVOF) [4–8]. Among them, laser cladding can meet the requirements of preparing amorphous alloys thanks to its rapid cooling characteristics. In addition, the laser cladded coatings may have good metallurgical bonding with substrates [9,10]. Therefore, laser cladding has attracted more and more attention in the fabrication of amorphous surface coatings. Since Yoshioka et al. [11] successfully deposited a Ni-Cr-P-B amorphous coating by laser cladding on low carbon steel in 1975, different types of amorphous alloy coatings have been prepared by laser cladding such as Zr-based, Ni-based, Cu-based, Al-based, Co-based, etc. [12–16]. Compared with the above amorphous alloy systems, Fe-based amorphous coatings have comparable mechanical properties but a much lower cost [17], and therefore have great potential in engineering applications. So far, efforts have been made to prepare different kinds of Fe-based amorphous coatings using laser cladding. Sahasrabudhe et al. [10] deposited Fe-Cr-Mo-W-C-Mn amorphous coatings on a Zr substrate by laser cladding where the coating had an amorphous and crystalline composite structure and the crystal phase was distributed in the amorphous region. The coatings had much higher hardness and wear resistance than the substrate. Zhu et al. [18] laser cladded Fe-Co-B-Si-Nb coatings on low carbon steel at different scanning speeds of 6, 17, and 50 m/s and found that the main factors that affected the microstructure and the fraction of the amorphous phase were dilution ratio and scanning speed. Wu et al. [19] prepared Fe-Co-Ni-Zr-Si-B amorphous coatings on AISI 1045 steel by laser cladding where the thickness of the coating was 1.2 mm and had a high hardness of 1270 HV as well as good corrosion resistance.

In the present paper, an Fe-based amorphous alloy with the chemical composition of $Fe_{46.8}Mo_{22.7}Cr_{13.6}Co_{7.6}C_{4.8}B_{2.3}Y_{1.2}Si_{1.0}$ (wt.%), which had high glass forming ability, was used to prepare coatings by laser cladding. To our knowledge, the powders have not previously been used in laser cladding. The main objective of this study was to investigate the effects of scanning speeds on microstructure and properties of the coatings.

2. Materials and Methods

2.1. Materials

3Cr13 stainless steel with dimensions of 80 mm × 50 mm × 6 mm was used as the substrate material. The chemical composition of 3Cr13 stainless steel is shown in Table 1. The chemical composition of the Fe-based amorphous alloy powders was $Fe_{46.8}Mo_{22.7}Cr_{13.6}Co_{7.6}C_{4.8}B_{2.3}Y_{1.2}Si_{1.0}$ (wt.%). Pure elemental powders of Fe, Mo, Cr, Co, C, Y, Si, and B with 99.8 to 99.9 wt.% in purity were used to prepare powders by vacuum gas atomization. The particle size of the powders was 40–75 μm.

Table 1. Chemical composition of the 3Cr13 stainless steel (wt.%).

Element	C	Mn	Ni	Si	P	S	Cr	Fe
wt.%	0.26–0.35	≤1.00	≤0.60	≤1.00	≤0.035	≤0.03	12.00–14.00	Balance

2.2. Laser Cladding

A fiber laser (YLS-2000 IPG) equipped with a lateral powder feeder system was used in laser cladding. The laser beam with a Gaussian energy distribution was 3 mm in diameter. Ar gas was employed to transport powders and protect the bath from oxidation. The gas flow rate was 10 L/min. The laser cladding was carried out at the scanning speeds of 6, 8 and 10 mm/s. The detailed laser parameters used in this study are listed in Table 2.

Table 2. Laser processing parameters.

Laser Power (W)	Scanning Speed (mm/s)	Powder Feeding Rate (g/min)	Heat Input (J/mm)	Overlapping Rate (%)
1000	6	20	166.7	30
1000	8	20	125.0	30
1000	10	20	100.0	30

2.3. Microanalysis

The laser cladding samples were cut along cross sections with a wire electric discharge machining (EDM) machine, then were polished and etched by aqua regia for microscopy observations and hardness tests. A scanning electron microscope (SEM, QUANTA 200 FEI, Hillsboro, OR, USA) with energy dispersive spectroscopy (EDS) and an electron probe microanalyzer (EPMA SHIMADZU, Kyoto, Japan) was used to analyze the microstructure and element distributions. The phase composition of the laser-processed samples were characterized by an x-ray diffractometer (D8 Bruker, Billerica, MA, USA) with Cu–Ka (λ = 0.154060 nm) radiation. The X-ray diffraction (XRD) system was operated at

40 kV and 200 mA in a 2θ range of 20–80°. The step size was 0.02° and the scanning speed was 4° per minute. Before the XRD test, the laser cladding samples with dimensions of 8 mm × 8 mm × 6 mm were ground to a position of 0.3 mm in a thickness direction from the top surface.

2.4. Microhardness and Tribological Tests

The microhardness of the samples was measured using a Vickers hardness tester (HM-800 Mitutoyo, Takatsu-ku, Kawasaki, Japan) under a load of 200 g and a dwell time of 15 s. The samples with dimensions of 15 mm × 15 mm × 6 mm were cut from the laser cladding samples, and then were ground to keep the surface roughness less than Ra3.2. A ball-on-disc tribometer (UMT-3 Bruker, Billerica, MA, USA) was used to measure the tribological properties of the substrate and the coatings. A silicon nitride ball with a diameter of 4 mm and hardness of 1550 $HV_{0.1}$ was used as the upper specimen. The wear experiment was performed at a load of 3 N and a sliding speed of 100 r/min, in a circle with the diameter of 10 mm for 30 min. A precision analytical balance with an accuracy of 0.001 mg was used to measure the wear mass losses and then calculate the volume losses. The worn morphologies and compositions were measured by scanning electron microscopy (SEM, QUANTA 200 FEI) with energy dispersive spectroscopy (EDS FEI).

3. Results and Discussion

3.1. Phases

The x-ray diffraction patterns of the coatings produced at different laser scanning speeds are shown in Figure 1. The diffraction patterns varied significantly with the laser scanning speed. When the scanning speed was 6 mm/s, a broad halo peak characterizing an amorphous phase could be observed at the diffraction angle around 44°, on which sharp diffraction peaks corresponding to the carbides $M_{23}(B,C)_6$ and M_7C_3 superposed. Here, M is Fe and Cr. When the scanning speed was increased to 8 mm/s and 10 mm/s, the broad halo peak was no longer apparent and the diffraction patterns were composed of sharp peaks of crystalline phases. As the scanning speed increased, the intensity of the crystallization peak increased significantly, indicating an increase in crystallinity and a suppression of amorphous formation. There was a definite change in the phase constituents when the scanning speed was increased from 6 mm/s to 8 mm/s. Besides the carbides $M_{23}(B,C)_6$, M_7C_3, and Mo_2C, diffraction peaks from the solid solution γ-(Fe,Cr) and intermetallic compound Co_7Mo_6 were also identified. The phases at the scanning speed of 10 mm/s were mainly carbides $M_{23}(B,C)_6$, M_7C_3, solid solution γ-(Fe,Cr), and intermetallic compound Co_7Mo_6, which were the same as those for 8 mm/s.

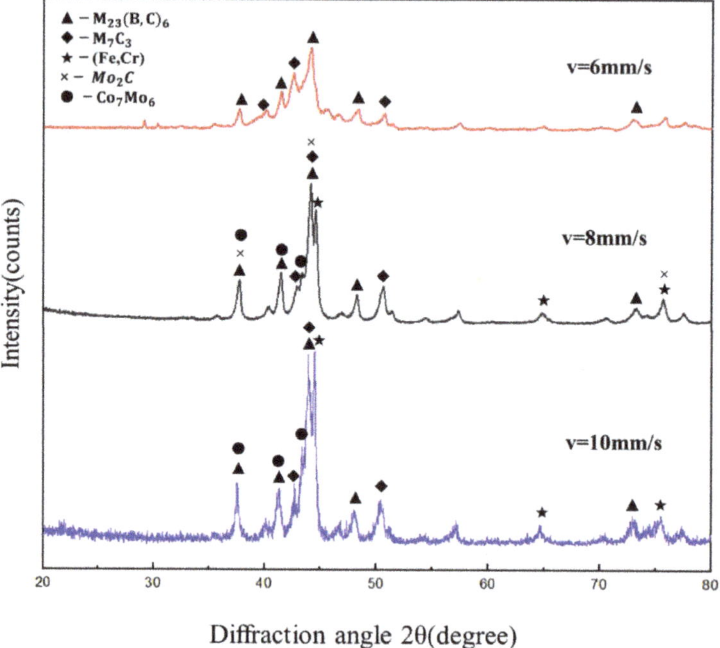

Figure 1. XRD patterns of the cladding layers at different scan speeds.

3.2. Geometrical Morphologies and Microstructures

The cross sections of coatings obtained at different laser scanning speeds are shown in Figure 2. The cladding layers were dense and clear interface lines could be seen between the cladding layers and the substrates for all three different scanning speeds, which indicated good metallurgical bonding was achieved in laser cladding. Cracks may have existed in some cladding layers (Figure 1). The curvature of the interface decreased as the heat input increased, indicating the change in dilution rates. The geometric parameters on cross-sections of coatings at different scanning speeds are shown in Table 3. The dilution rates were calculated using Equation (1) [20].

$$\text{Dilution rate } \mu = \frac{S_2}{S_1 + S_2} \times 100\%, \quad (1)$$

where S_1 is the area of cladding layer and S_2 represents the area of molten substrate on the cross-section of a specimen. The average geometric parameters and the dilution rates are presented in Table 3.

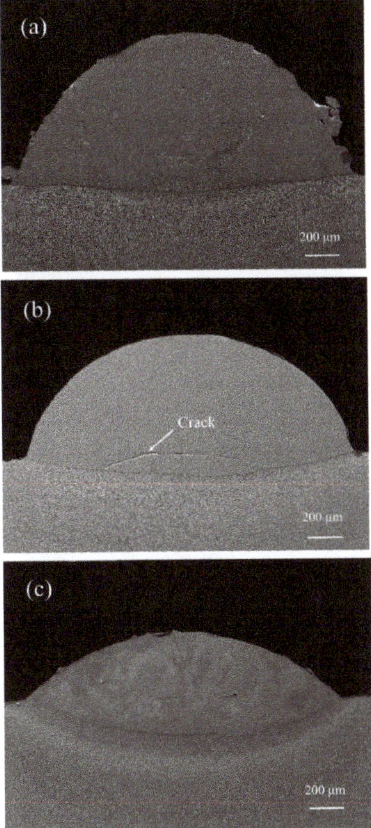

Figure 2. Cross sections of the laser cladding coatings for different scanning speeds: (**a**) 6 mm/s; (**b**) 8 mm/s; (**c**) 10 mm/s.

Table 3. Average geometric parameters and dilution rates for the different scanning speeds.

Scanning Speed (mm/s)	Width of the Cladding Layer (μm)	Height of the Cladding Layer (μm)	Depth of the Molten Substrate (μm)	Dilution Rate
6	2139.2	906.9	68.3	7.0%
8	2070.7	749.6	120.6	13.9%
10	1965.8	409.2	230.6	36.0%

The width and height of the cladding layers decreases with an increase in scanning speeds under comparable laser cladding conditions because the amount of powder fed into the molten pool per unit length is reduced as the scanning speed increases. On the other hand, increasing scanning speed leads to an increase in the depth of the melted substrate, and in turn, the dilution rate. This can be attributed to the weaker "heat shielding effect" [21] of the powders and more heat absorbed by the substrates at higher scanning speeds.

The typical microstructures of coatings made at different scanning speeds are presented in Figures 3–5. Figure 3 shows the SEM photographs of the laser cladding for a scanning speed of 6 mm/s. Figure 3a is an overall view of the cross section, where the cladding layer shows a layered structure. The substrate under the interface exhibited a martensite structure, and a columnar crystal zone with a thickness of about 40–60 μm formed above the interface as shown in Figure 3b. Strip-shaped precipitates could be observed between columnar grains. Figure 3c is a magnification of area C, and it can be seen that above the columnar crystal region was a fine equiaxed crystal region surrounded by a featureless phase structure. In the upper part of the cladding layer, all of the microstructures were the featureless structure shown in Figure 3d.

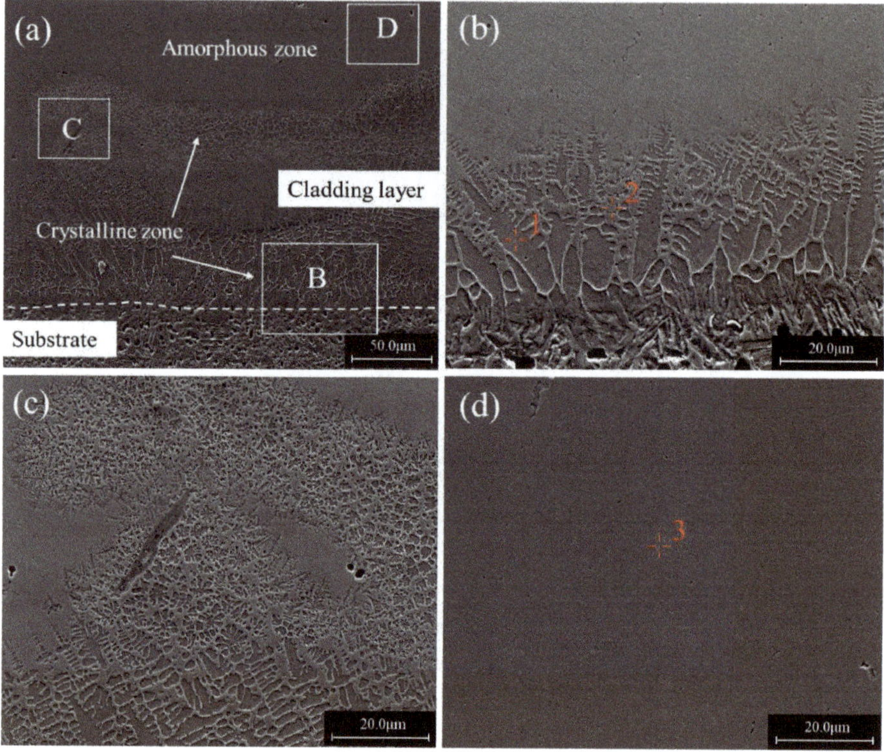

Figure 3. Cross sectional morphologies of a single-bead cladded specimen produced at the scanning speed of 6 mm/s: (**a**) Overall view; (**b**) an enlarged view of region B in (**a**); (**c**) an enlarged view of region C in (**a**); (**d**) an enlarged view of region D in (**a**).

EPMA analysis revealed that the featureless phase structure (Point 3 in Figure 3d) had an average composition of $Fe_{47.3}Mo_{22.5}Cr_{13.9}Co_{7.2}C_{5.1}B_{2.2}Y_{1.0}Si_{0.8}$ (wt.%), which was very close to the nominal composition of the powder. Combined with the XRD results, it can be inferred that this featureless structure zone was mainly composed of the amorphous phase. The EPMA at point 1 showed that the primary arm of the columnar dendrites at the bottom of the cladding layer had a composition of $Fe_{81.5}Mo_{1.2}Cr_{11.6}Co_{1.3}C_{2.6}B_{0.3}Y_{1.3}Si_{0.5}$, where a large amount of Fe element was detected. The composition of the strip-shaped precipitates between columnar grains at Point 2 was $Fe_{63.9}Mo_{11.9}Cr_{18.5}Co_{2.4}C_{5.7}B_{2.0}Y_{0.2}Si_{0.3}$ (wt.%), which had a relatively high carbon content and could mainly be Fe, Cr carbides. There was a higher Fe content and lower Mo, Cr contents in the columnar crystal region near the substrate than those in the upper amorphous region, which demonstrates the elements' migration caused by the dilution effect.

The microstructures of the specimen with a scanning speed of 8 mm/s is shown in Figure 4. Figure 4a is an overall view of the cross section. Figure 4b is an enlarged backscattered photograph around the bonding region of the cladding layer and the substrate. There was a martensite structure under the interface between the cladding layer and the substrate, above which was a layer of planar crystal. Above the plane crystal was a layer of dendrites with a thickness of about 20–30 μm. Above the dendritic structure, the coating was composed of a block grain structure as shown in Figure 4c. Figure 4d is a further enlargement of the block grain structure in Figure 4c.

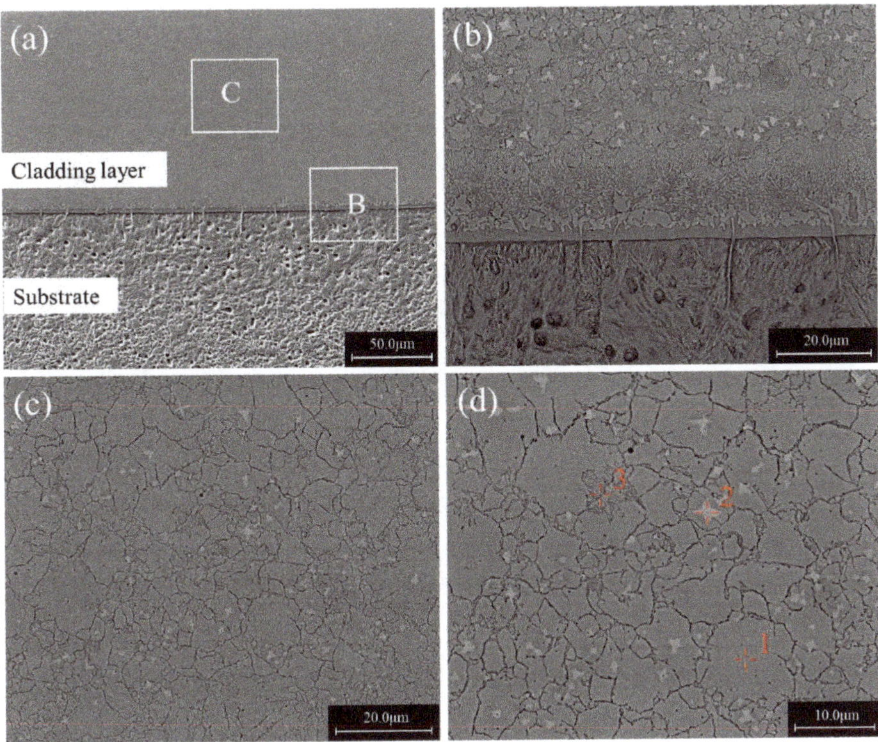

Figure 4. Cross sectional morphologies of a single-bead cladded specimen produced at the scanning speed of 8 mm/s: (**a**) Overall view; (**b**) an enlarged view of region B in (**a**); (**c**) an enlarged view of region C in (**a**); (**d**) an enlarged view of (**c**).

EPMA results showed that the chemical compositions of point 1 inside the block grain in Figure 4d was $Fe_{47.4}Mo_{22.6}Cr_{15.5}Co_{7.6}C_{5.6}B_{2.4}Y_{0.1}Si_{0.6}$ (wt.%), which was close to the nominal composition of Fe-based amorphous alloy powder. The composition of the block grain boundary at point 3 was $Fe_{44.9}Mo_{22.9}Cr_{14.6}Co_{8.4}C_{7.5}B_{0.5}Y_{0.5}Si_{0.7}$ (wt.%), which had a higher C content than the inside. In Figure 4d, a star-like white phase could be observed with a composition of $Fe_{33.0}Mo_{38.5}Cr_{11.3}Co_{6.8}C_{8.2}B_{1.8}Y_{0.1}Si_{0.3}$ (wt.%) (Point 2), having high Mo and C content. Combined with the XRD phase analysis results, it can be inferred that it could be mainly composed of Mo_2C and $M_{23}(B,C)_6$.

Figure 5 shows the SEM photographs of the specimen obtained at the scanning speed of 10 mm/s. It can be noted that dendritic structures existed throughout the cladding layer. In the vicinity of the interface between cladding and substrate, the columnar crystals grew upwards, and were perpendicular to the interface because the temperature gradient is large near the bonding zone, and columnar crystals grow along the temperature gradient direction [22]. It can also be seen that a large number of network-like precipitates existed between the dendrites. Figure 6 shows the elemental distributions obtained by EDS map scanning. It can be observed that the precipitates between the dendrites were rich in Mo, Cr, and C. In contrast, the dendrites were rich in Fe and C, with a trace of Cr, and a lack of Mo. There was no significant difference in the content of Co between the dendrites and the inner precipitates.

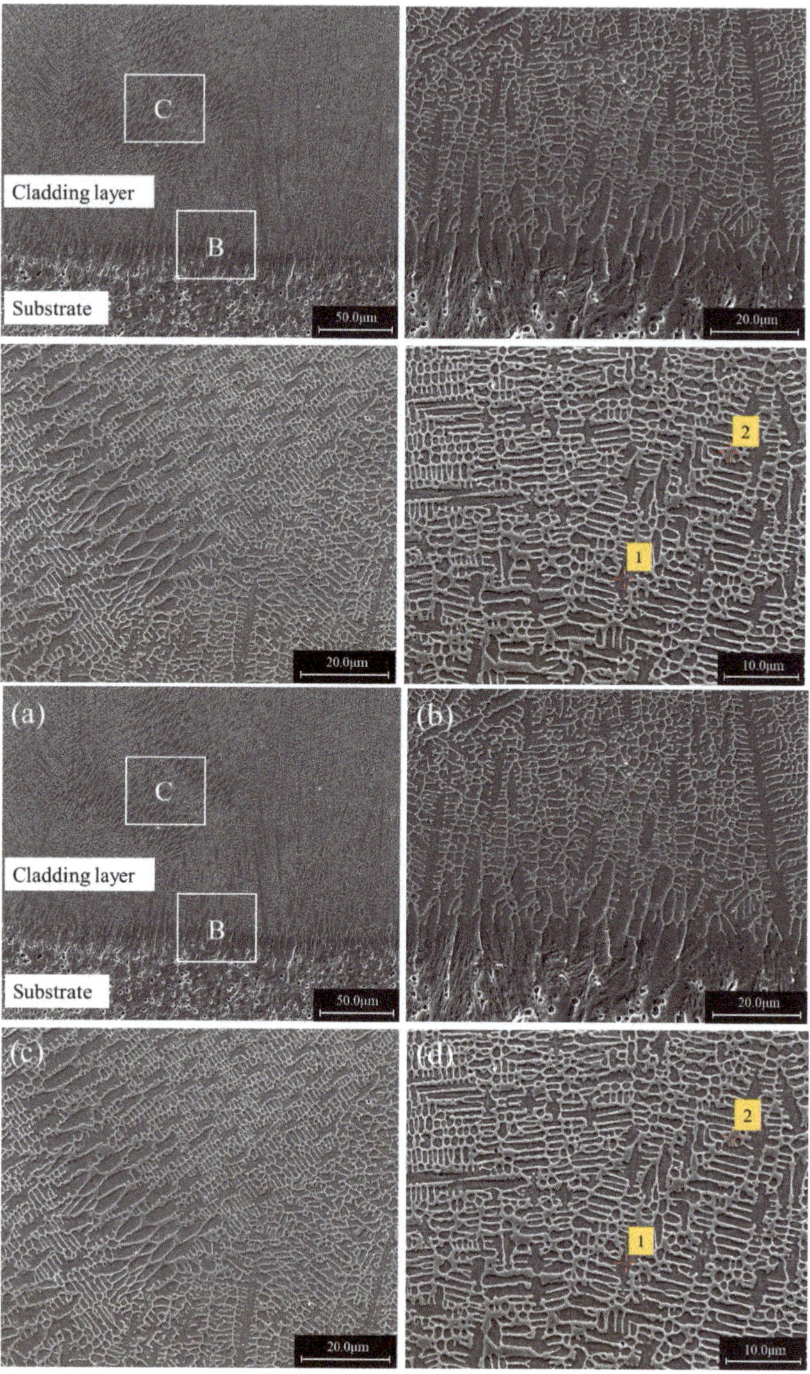

Figure 5. Cross sectional morphologies of a single-bead cladded specimen produced at the scanning speed of 10 mm/s: (**a**) Overall view; (**b**) an enlarged view of region B in (**a**); (**c**) an enlarged view of region C in (**a**); (**d**) an enlarged view of (**c**).

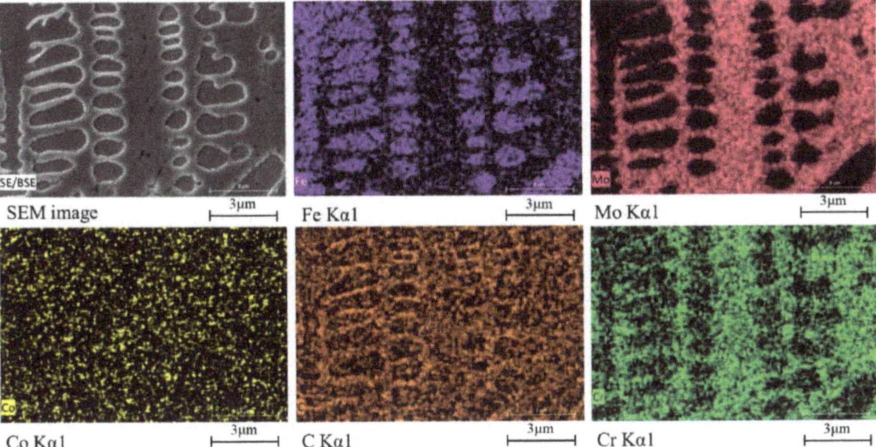

Figure 6. Element distributions of the cladding layer obtained at the scanning speed of 10 mm/s.

EPMA point scanning results showed that the chemical compositions of the dendrites at point 1 in Figure 6 was $Fe_{76.7}Mo_{6.9}Cr_{14.7}Co_{2.8}C_{3.5}B_{0.5}Y_{0.2}Si_{0.8}$ (wt.%), which was dominated by Fe and Cr, with traces of Mo, C, and a little B. The chemical composition of the precipitates at point 2 in Figure 6 was $Fe_{52.4}Mo_{19.3}Cr_{15.0}Co_{3.5}C_{6.5}B_{1.9}Y_{0.8}Si_{0.6}$ (wt.%). Combined with the XRD phase analysis results, it can be inferred that the dendrites were mainly γ-(Fe, Cr) solid solution and the network-like precipitates were composed of carbides $M_{23}(B,C)_6$ and M_7C_3. Zeisig et al. obtained a similar carbide network precipitation when laser cladding the Fe-Cr-Mo-V-C alloy [23].

In summary, as the scanning speeds increased, the microstructures of the laser cladding layers exhibited three distinct microstructure characteristics. For the scanning speed of 6 mm/s, the cladding layer was a mixture of amorphous and crystalline regions. When the scanning speed was increased to 8 mm/s, the cladding layer was mainly composed of a block grain structure. As the scanning speed was further increased to 10 mm/s, the cladding layer was composed entirely of dendrites. Combined with the XRD test results, it can be found that the glass forming ability decreases as the scanning speed increases. A large number of studies have shown that there are two main factors affecting the ability to form amorphous phases, i.e., the composition and the cooling rate [24–26]. The absence of amorphous structures can be attributed either to a cooling rate experienced by the coated layer, which is lower than the critical value required to form a glassy microstructure, or to a substantial solute redistribution or compositional changes within the molten pool or across the coating-substrate interface by dissolution or inter-diffusion.

The cooling rate $\partial T/\partial t$ during the laser cladding process can be approximately estimated using the Rosenthal solution proposed by Steen and Mazumder [27]:

$$\frac{\partial T}{\partial t} = 2\pi k\left[\frac{v}{P}\right](T - T_0)^2, \tag{2}$$

where k is thermal conductivity; v is scanning speed; P is power density; and T_0 is the initial temperature of the substrate. From the formula, it can be inferred that the cooling rate increases as the scanning speed increases. For a more quantitative study, the cooling rates were numerically computed in the present study using the finite element software JWRIAN developed by JWRI [28,29]. Figure 7 presents the thermal histories of points at the deposit/substrate interface and at the top surface of deposits for three scanning speeds. At the deposit/substrate interface, the cooling rates from 1000 °C to 500 °C for the three scanning speeds (6 m/s, 8 m/s, and 10 m/s) were 819.7, 1087.0, and 1388.9 °C/s, respectively. At the top surface of the deposits, the cooling rates from 1000 °C to 500 °C for the three scanning speeds were 847.5, 1111.1, and 1388.9 °C/s, respectively. Obviously, at both locations, the predicted cooling rates increased for the increased scanning speeds, which was the same as the analytical solution. Hence, increasing scanning speeds is theoretically favorable to the formation of an amorphous structure. Jin et al. [30] experimentally demonstrated that a higher scanning speed favored the formation of an amorphous phase in their work. However, the results in this study showed the opposite to the above deduction, where an amorphous microstructure tended to form at relatively lower scanning speed (6 mm/s) while dendrites formed at a higher speed (10 mm/s). Therefore, the main factor affecting the microstructures and amorphous forming ability of the deposits is more likely to be the change in the composition, but not scanning speed. As the laser scanning speed increased from 6 mm/s to 10 mm/s, the dilution ratios of the cladding layers increased significantly (by more than four times) as presented in Table 2. Consequently, more elements migrated from the substrates to the cladding layers, causing the compositions of the cladding layers to deviate from the nominal compositions of the Fe-based amorphous alloy powder, resulting in a decrease in the ability to form an amorphous phase. Li et al. [20] also showed that a coating with a lower dilution rate exhibited a larger volume fraction of the amorphous phase.

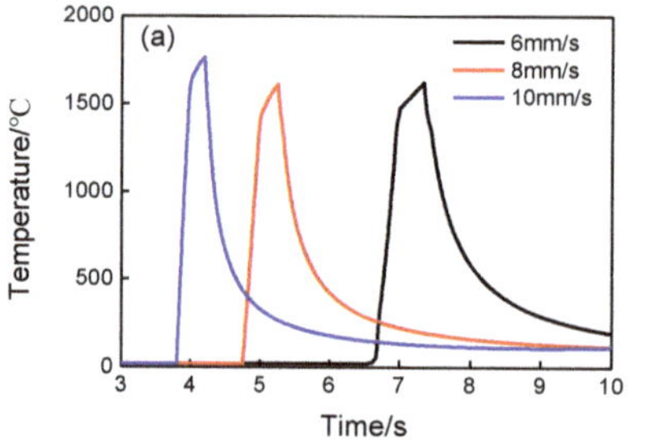

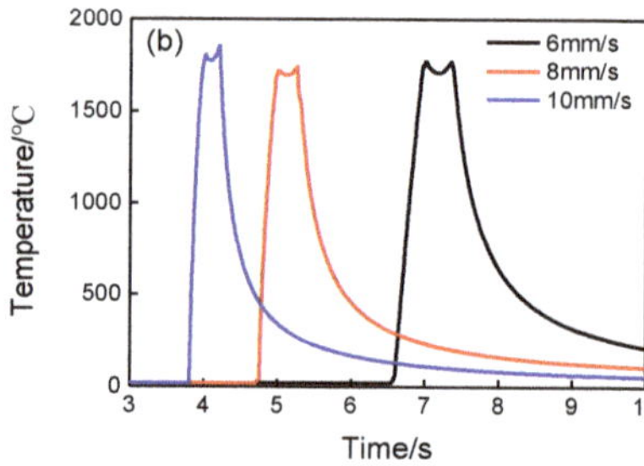

Figure 7. Thermal histories of points at: (**a**) deposit/substrate interface; (**b**) top surface of the deposits for the three scanning speeds obtained by finite element analysis.

3.3. Microhardness

The hardness distributions in the thickness direction for the cladding layers with different scanning speeds are shown in Figure 8. It can be found that the hardness decreased from the cladding layer to the heat affected zone and then to the substrate for all three scanning speeds. The cladding layer had an extremely high hardness of about 1300 $HV_{0.2}$ for the scanning speeds 6 mm/s and 8 mm/s, and a lower high hardness of 700 $HV_{0.2}$ for 10 mm/s. When the scanning speed was 6 mm/s, there was a sudden decrease in hardness to about 700 $HV_{0.2}$ at 200 μm from the substrate, which was because the test point was located just in the crystallization zone in the cladding layer. Overall, the hardness of the cladding layers was significantly improved when compared to the substrate, and the hardness distribution in the cladding layer was quite stable. Both the amorphous phase and the block grain structure had extremely high hardness. The high hardness of the block grain structure formed at 8 mm/s may mainly be attributed to the formation of Mo_2C and other carbides [31].

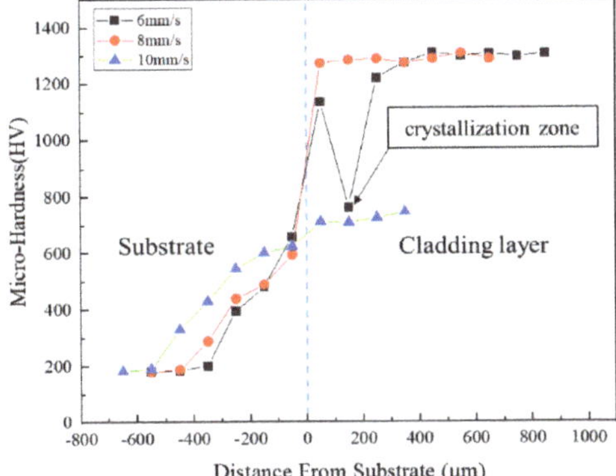

Figure 8. Variation of microhardness in the thickness direction of laser claddings produced with different scanning speeds.

3.4. Wear Resistance

Figure 9 shows the wear volume losses of the substrate and the coatings obtained at different scanning speeds. It is clear that the wear volume losses of the cladding layers increased gradually with an increase of the scanning speeds. The wear volume losses of coatings at the three scanning speeds were much smaller than that of the substrate, indicating that the cladding layers significantly improved the wear resistance of the substrate. With the scanning speed of 6 mm/s, the cladding layer had the smallest wear volume loss of 0.71 mm^3, while that for the substrate was 5.60 mm^3. The wear resistance of the substrate was increased by 7.9 times.

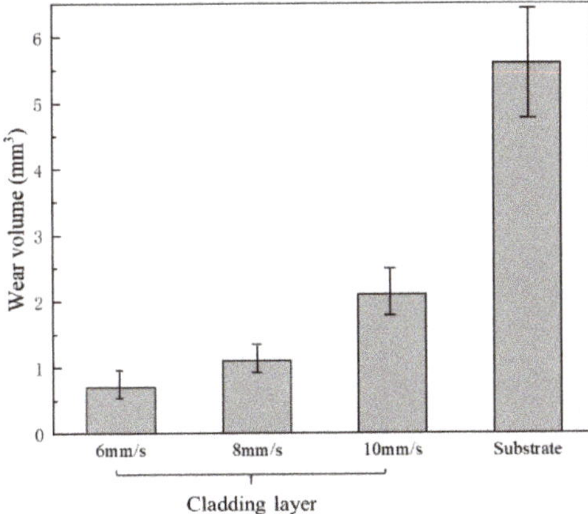

Figure 9. Volume losses in the wear tests for the substrate and the cladding layers produced with different scanning speeds.

Figure 10 shows the morphologies of the worn surfaces of the substrate and the coatings obtained at different scanning speeds. Deep furrows with a peeling of large oxide patches could be observed in the worn surface of the substrate. The worn substrate surface (given in Figure 11a) had a composition of mainly Fe, O, Cr, Si elements using the EDS spectrum results in Figure 11. Significantly, a higher silicon content indicates that silicon was transferred from the upper sample to the substrate, indicating that adhesive wear had occurred. It can be inferred that the main wear mechanism of the substrate was adhesive wear and oxidative wear. At the scanning speed of 6 mm/s, there was only a small number of light grooves and a small amount of oxide particles on the surface of the cladding layer (Figure 10b). Its main wear mechanism was abrasive wear. The formation of the high hardness amorphous phase gave it the highest wear resistance. At the scanning speed of 8 mm/s, the surface of the cladding layer was uniform, and deeper grooves than those at 6 mm/s could be observed (Figure 10c). The wear resistant was high because of the presence of the high hardness Mo_2C, M_7C_3 carbides in the cladding layer. This is mainly due to abrasive wear, and the carbide particles falling off from the cladding layer may be the source of the abrasive. When the scanning speed increased to 10 mm/s, the wear resistance of the cladding layer was notably reduced due to its lower hardness when compared to those obtained at scanning speeds of 6 mm/s and 8 mm/s (Figure 10d). Large oxides (of which the EDS results are given in Figure 11b) and furrows appeared on the worn surface, which indicated that adhesion and oxidative wear occurred on the cladding layer surface.

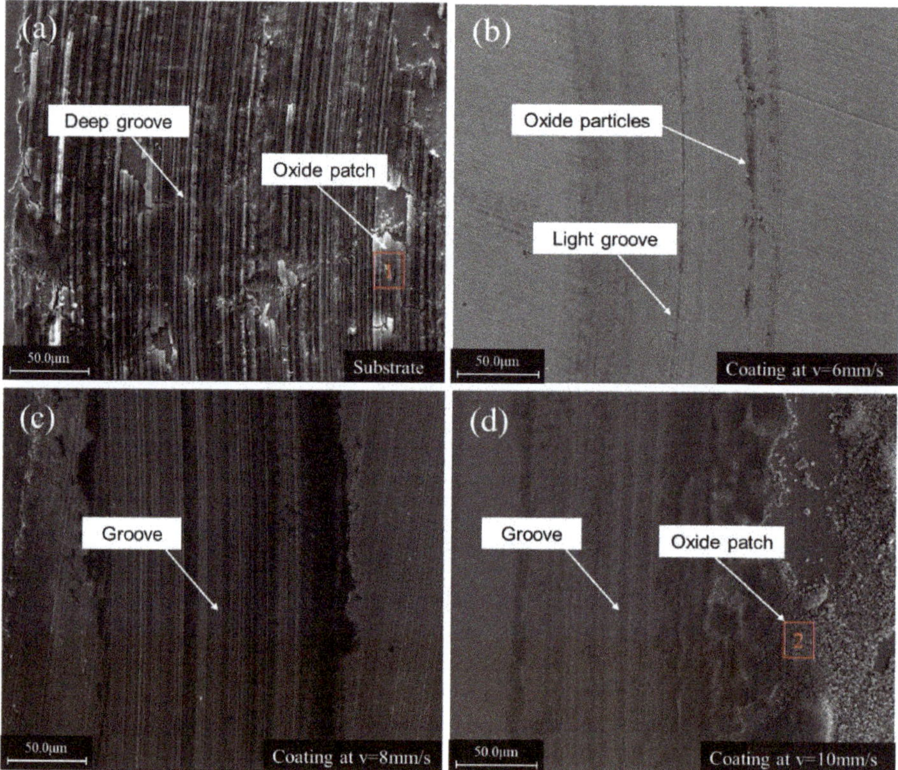

Figure 10. SEM micrographs of the worn surfaces of the substrate and the claddings obtained with different scanning speeds: (**a**) Substrate; (**b**) cladding at 6 mm/s; (**c**) cladding at 8 mm/s; (**d**) cladding at 10 mm/s.

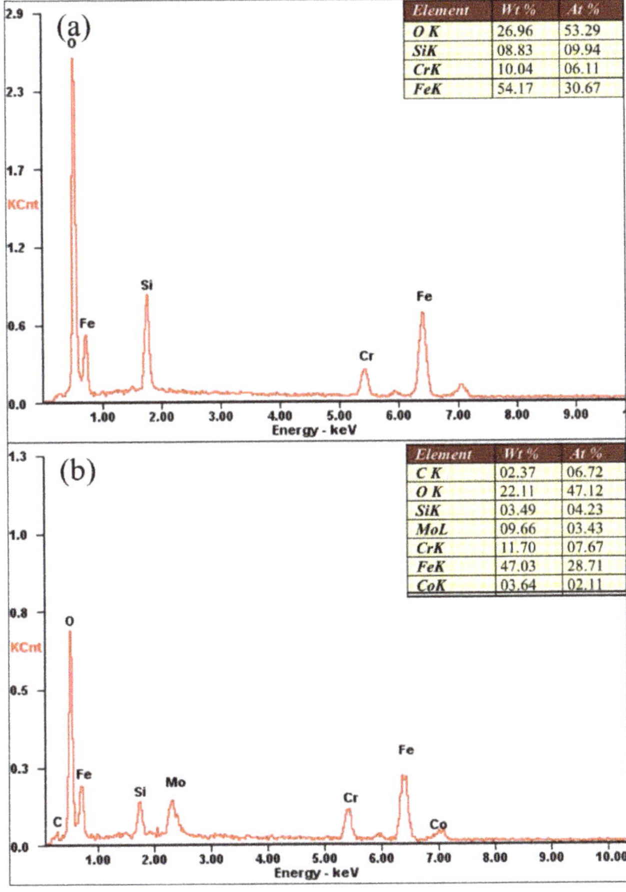

Figure 11. EDS results of different zones in Figure 10: (**a**) Zone 1; (**b**) Zone 2.

4. Conclusions

An Fe-based amorphous coating was successfully prepared by laser cladding of powder with a chemical composition of $Fe_{46.8}Mo_{22.7}Cr_{13.6}Co_{7.6}C_{4.8}B_{2.3}Y_{1.2}Si_{1.0}$ (wt.%) on a 3Cr13 stainless steel substrate. The influences of laser scanning speed on the microstructure and properties of the coatings were investigated. The main results can be summarized as follows:

(1) The laser cladding layers exhibited three distinct microstructures at various scanning speeds. At the scanning speed of 6 mm/s, the cladding layer was a mixture of amorphous and crystalline regions. For a scanning speed of 8 mm/s, the cladding layer was mainly composed of a block grain structure. For a scanning speed of 10 mm/s, the cladding layer was composed entirely of dendrites. The dilution ratios were more dominant than the scanning speed in determining the resultant microstructures.

(2) Compared with the hardness of the substrate of 200 $HV_{0.2}$, the cladding layers had an extremely high hardness of about 1300 $HV_{0.2}$ at scanning speeds of 6 mm/s and 8 mm/s, and a lower high hardness of 700 $HV_{0.2}$ at the scanning speed of 10mm/s.

(3) The wear resistance of the cladding layers was much higher than that of the substrate, and the wear resistance of the cladding layers could be improved by using a lower scanning speed due to the formation of high hardness wear resistant phases.

Author Contributions: Conceptualization, D.D. and B.C.; Formal analysis, X.H.; Investigation, X.H.; Methodology, D.D. and B.C.; Software, N.M.; Supervision, B.C.; Writing–Original Draft, X.H. and B.C.; Writing–Review & Editing, D.D. and N.M.

Funding: This research was funded by the National Natural Science Foundation of China, grant number 51675303, and the Tribology Science Fund of the State Key Laboratory of Tribology, grant number SKLT2018B05.

Conflicts of Interest: The authors declare no conflict of interest. The funders had no role in the design of the study; in the collection, analyses, or interpretation of data; in the writing of the manuscript, or in the decision to publish the results.

References

1. Amiya, K.; Nishiyama, N.; Inoue, A.; Masumoto, T. Mechanical strength and thermal stability of Ti-based amorphous alloys with large glass-forming ability. *Mater. Sci. Eng. A* **1994**, *179–180*, 692–696. [CrossRef]
2. Schuh, C.A.; Hufnagel, T.C.; Ramamurty, U. Mechanical behavior of amorphous alloys. *Acta Mater.* **2007**, *55*, 4067–4109. [CrossRef]
3. Liu, C.T.; Heatherly, L.; Horton, J.A.; Easton, D.S.; Carmichael, C.A.; Wright, J.L.; Schneibel, J.H.; Yoo, M.H.; Chen, C.H.; Inoue, A. Test environments and mechanical properties of Zr-base bulk amorphous alloys. *Metall. Mater. Trans. A* **1998**, *29*, 1811–1820. [CrossRef]
4. Guo, R.Q.; Zhang, C.; Chen, Q.; Yang, Y.; Li, N.; Liu, L. Study of structure and corrosion resistance of Fe-based amorphous coatings prepared by HVAF and HVOF. *Corros. Sci.* **2011**, *53*, 2351–2356. [CrossRef]
5. Kishitake, K.; Era, H.; Otsubo, F. Thermal-sprayed Fe-10Cr-13P-7C amorphous coatings possessing excellent corrosion resistance. *J. Therm. Spray Technol.* **1996**, *5*, 476–482. [CrossRef]
6. Cadney, S.; Brochu, M. Formation of amorphous Zr41.2Ti13.8Ni10Cu12.5Be22.5 coatings via the ElectroSpark Deposition process. *Intermetallics* **2008**, *16*, 518–523. [CrossRef]
7. Bergmann, H.W.; Mordike, B.L. Laser and electron-beam melted amorphous layers. *J. Mater. Sci.* **1981**, *16*, 863–869. [CrossRef]
8. Liang, G.Y.; Su, J.Y. The microstructure and tribological characteristics of laser-clad Ni-Cr-Al coatings on aluminium alloy. *Mater. Sci. Eng. A* **2000**, *290*, 207–212. [CrossRef]
9. Zhang, L.; Wang, C.; Han, L.; Dong, C. Influence of laser power on microstructure and properties of laser clad Co-based amorphous composite coatings. *Surf. Interfaces* **2017**, *6*, 18–23. [CrossRef]
10. Sahasrabudhe, H.; Bandyopadhyay, A. Laser processing of Fe based bulk amorphous alloy coating on zirconium. *Surf. Coat. Technol.* **2014**, *240*, 286–292. [CrossRef]
11. Yoshioka, H.; Asami, K.; Kawashima, A.; Hashimoto, K. Laser-processed corrosion-resistant amorphous Ni Cr P B surface alloys on a mild steel. *Corros. Sci.* **1987**, *27*, 981–995. [CrossRef]
12. Yue, T.M.; Su, Y.P.; Yang, H.O. Laser cladding of $Zr_{65}Al_{7.5}Ni_{10}Cu_{17.5}$ amorphous alloy on magnesium. *Mater. Lett.* **2007**, *61*, 209–212. [CrossRef]
13. Wu, X.; Xu, B.; Hong, Y. Synthesis of thick $Ni_{66}Cr_5Mo_4Zr_6P_{15}B_4$ amorphous alloy coating and large glass-forming ability by laser cladding. *Mater. Lett.* **2002**, *56*, 838–841. [CrossRef]
14. Liu, H.; Wang, C.; Gao, Y. Laser Cladding Amorphous Composite Coating of Cu-Zr-Al on Magnesium Alloy Surface. *Chin. J. Lasers* **2006**, *33*, 709–713.
15. Tan, C.; Zhu, H.; Kuang, T.; Shi, J.; Liu, H.; Liu, Z. Laser cladding Al-based amorphous-nanocrystalline composite coatings on AZ80 magnesium alloy under water cooling condition. *J. Alloy. Compd.* **2017**, *690*, 108–115. [CrossRef]
16. Shu, F.; Tian, Z.; Zhao, H.; He, W.; Sui, S.; Liu, B. Synthesis of amorphous coating by laser cladding multi-layer Co-based self-fluxed alloy powder. *Mater. Lett.* **2016**, *176*, 306–309. [CrossRef]
17. Zhu, Q.; Qu, S.; Wang, X.; Zou, Z. Synthesis of Fe-based amorphous composite coatings with low purity materials by laser cladding. *Appl. Surf. Sci.* **2007**, *253*, 7060–7064. [CrossRef]
18. Zhu, Y.; Li, Z.; Huang, J.; Li, M.; Li, R.; Wu, Y. Amorphous structure evolution of high power diode laser cladded Fe-Co-B-Si-Nb coatings. *Appl. Surf. Sci.* **2012**, *261*, 896–901. [CrossRef]
19. Wu, X.; Hong, Y. Fe-based thick amorphous-alloy coating by laser cladding. *Surf. Coat. Technol.* **2001**, *141*, 141–144. [CrossRef]
20. Li, R.; Li, Z.; Huang, J.; Zhu, Y. Dilution effect on the formation of amorphous phase in the laser cladded Ni-Fe-B-Si-Nb coatings after laser remelting process. *Appl. Surf. Sci.* **2012**, *258*, 7956–7961. [CrossRef]

21. Fang, L.; Yao, J.H.; Hu, X.X.; Chai, G.Z. Effect of laser power on the cladding temperature field and the heat affected zone. *J. Iron Steel Res. Int.* **2011**, *18*, 73–78.
22. Yang, S.; Du, D.; Chang, B. Studies of the Influence of Beam Profile and Cooling Conditions on the Laser Deposition of a Directionally-Solidified Superalloy. *Materials* **2018**, *11*, 240. [CrossRef]
23. Zeisig, J.; Schädlich, N.; Giebeler, L.; Sander, J.; Eckert, J.; Kühn, U.; Hufenbach, J. Microstructure and abrasive wear behavior of a novel FeCrMoVC laser cladding alloy for high-performance tool steels. *Wear* **2017**, *382–383*, 107–112. [CrossRef]
24. Basu, A.; Samant, A.N.; Harimkar, S.P.; Majumdar, J.D.; Manna, I.; Dahotre, N.B. Laser surface coating of Fe-Cr-Mo-Y-B-C bulk metallic glass composition on AISI 4140 steel. *Surf. Coat. Technol.* **2008**, *202*, 2623–2631. [CrossRef]
25. Wang, Y.; Li, G.; Wang, C.; Xia, Y.; Sandip, B.; Dong, C. Microstructure and properties of laser clad Zr-based alloy coatings on Ti substrates. *Surf. Coat. Technol.* **2004**, *176*, 284–289. [CrossRef]
26. Zheng, B.; Zhou, Y.; Smugeresky, J.E.; Lavernia, E.J. Processing and behavior of fe-based metallic glass components via laser-engineered net shaping. *Int. J. Powder Metall.* **2009**, *45*, 27–39. [CrossRef]
27. Steen, W.M.; Mazumder, J. Basic Laser Optics. In *Laser Material Processing*; Springer: London, UK, 2010; pp. 79–130.
28. Ueda, Y.; Murakawa, H.; Ma, N. Mechanical Simulation of Welding. In *Welding Deformation & Residual Stress Prevention*; Elsevier: Amsterdam, The Netherlands, 2012; pp. 55–98.
29. Ma, N.; Li, L.; Huang, H.; Chang, S.; Murakawa, H. Residual stresses in laser-arc hybrid welded butt-joint with different energy ratios. *J. Mater. Process. Technol.* **2015**, *220*, 36–45. [CrossRef]
30. Jin, Y.J.; Li, R.F.; Zheng, Q.C.; Li, H.; Wu, M.F. Structure and properties of laser-cladded Ni-based amorphous composite coatings. *Mater. Sci. Technol.* **2016**, *32*, 1206–1211. [CrossRef]
31. Casellas, D.; Caro, J.; Molas, S.; Prado, J.M.; Valls, I. Fracture toughness of carbides in tool steels evaluated by nanoindentation. *Acta Mater.* **2007**, *55*, 4277–4286. [CrossRef]

MDPI
St. Alban-Anlage 66
4052 Basel
Switzerland
Tel. +41 61 683 77 34
Fax +41 61 302 89 18
www.mdpi.com

Materials Editorial Office
E-mail: materials@mdpi.com
www.mdpi.com/journal/materials

www.ingramcontent.com/pod-product-compliance
Lightning Source LLC
LaVergne TN
LVHW071251170726
843515LV00006BA/1380

* 9 7 8 3 0 3 9 4 3 6 6 3 7 *